Frege's Theorem

Frege's Theorem

Richard G. Heck, Jr

CLARENDON PRESS • OXFORD

OXFORD

UNIVERSITY PRESS

Great Clarendon Street, Oxford, OX2 6DP,
United Kingdom

Oxford University Press is a department of the University of Oxford.
It furthers the University's objective of excellence in research, scholarship,
and education by publishing worldwide. Oxford is a registered trade mark of
Oxford University Press in the UK and in certain other countries

First published 2011
First published in paperback 2014

Published in the United States of America by Oxford University Press
198 Madison Avenue, New York, NY 10016, United States of America

British Library Cataloguing in Publication Data
Data available

Library of Congress Cataloging in Publication Data
Data available

ISBN 978–0–19–969564–5 (Hbk)
ISBN 978–0–19–870898–8 (Pbk)

Contents

Preface

This book has been a long time coming. The papers collected here represent the fruits of a two decade investigation of philosophical, historical, and technical questions related to Frege's Theorem. While I continue to think about these questions from time to time, my work on Frege has moved in other directions, and I feel as if I have said most of what I have to say about Frege's philosophy of arithmetic. That makes it a good time to collect these papers. Nonetheless, the process of revising them for publication has renewed my passion for problems I'd almost forgotten losing sleep over. I had a surprsing amount of pure fun writing the overview that appears as Chapter 1, as well as the postscripts that recount changes of mind or respond to published criticisms of my work. Thanks to those who have paid attention to it.

When I re-read these papers at the start of this project, I was surprised by how well they fit together. The earliest of them was drafted before I had my first job; the latest was finished only quite recently. So there are plenty of changes of direction, emphasis, and view. But there are, at the same time, dominant themes that re-appear in different forms in different places, and Chapter 1 is intended to give the reader a sense for what some of these are. The whole, I believe, is certainly greater than the sum of the parts, and that is what makes it worth reprinting the papers together. I hope that even those who have already read them separately will find that reading them together reveals something new.

While working on these topics, I have been privileged to have the support and encouragement of many friends, students, and colleagues, and I have benefitted greatly from their comments and criticisms. So a big thanks to Lawrence Abrams, Andrew Boucher, Sylvain Bromberger, Tyler Burge, John Burgess, Emily Carson, Dick Cartwright, Karin Case, Peter Clark, Roy Cook, Tony Corsentino, Annette Demby, Bill Demopoulos, Mic Detlefsen, Burt Dreben, Philip Ebert, Kate Elgin, Delia Graff Fara, Fernando Ferreira, Janet Folina, Michael Glanzberg, Warren Goldfarb, Alice Graham-Brown, Dave Grishaw-Jones, Steven Gross, Bob Hale, Jim Higginbotham, Chris Hill, David Hunter, Darryl Jung, Jinho Kang, Kathrin Koslicki, Michael Kremer, Thomas Kuhn, Øystein Linnebo, Mary Luti, Josep Macià-Fabrega, Lisa Marino, Mathieu Marion, Robert May, Ute Molitor, Charles Parsons, Paul Pietroski, Carl Posy, Ian Proops, Hilary Putnam, Agustín Rayo, Ofra Rechter, Michael Rescorla, Thomas Ricketts, Gideon Rosen, Marcus Rossberg, Tim Scanlon, Josh Schechter, Richard Schwartz, Sally Sedgwick, Lisa Sereno, Brett Sherman, Stewart Shapiro, Ori Simchen, Alison Simmons, Dan Smith, Gisela Striker, Bob Stalnaker, Zoltan Gendler Szabó, Bill Tait, Jamie Tappenden, Gabriel

Uzquiano, Albert Visser, Cathy Wearing, Kai Wehmeier, Carolyn Wilkins, and everyone on the FOM mailing list. (My apologies if I have forgotten anyone who ought to have been included.) Thanks also to Peter Momtchiloff, my editor at Oxford University Press, for his encouragement and help; to my production editor, Daniel Bourner, for all his assistance (and patience); and to my proofreader, Joy Mellor, for hers.

There are four people whose influence is felt on almost every page of this book, and so who deserve special mention.

Sir Michael Dummett was my B. Phil. supervisor at Oxford, and the influence of his teaching and writing is felt not just here but throughout my work. His ability to stitch serious scholarship together with the deepest philosophy and make it all into a seamless whole was my original inspiration for becoming a philosopher, and it is something I have tried very hard to emulate.

Jason Stanley is that person to whom I can talk about anything, including some stupid idea I had in the shower that morning. His insight has been invaluable, but his enthusiasm for my work has been even more important. Jason has often carried me through those tough stretches when it has seemed as if everything I thought I knew was false.

Without the contributions of Crispin Wright, this book would never have been conceived, let alone written. His name appears more than any other in this book, except Frege's. I am honored now to call Crispin a friend. It will be clear that we have not always agreed, but he has been a reliable supporter of me and my work since we first met by email in 1988. I am quite certain that my career, and therefore my life, would have been far different if not for his presence.

George Boolos was the person who got me working on this stuff in the first place, as is recounted in the overview. Most of the papers collected here were born in conversation with George, and even those that were written after his death were often influenced by imaginary conversations I would have with him. But as well as being my Ph.D. supervisor, my mentor, and my dear friend, George was a father-figure whose concern for me and my work knew no bound. Working on this book has not only reminded me how much I miss him but also how much he is still with me, and that experience has been precious.

Finally, let me thank my cats, Joe, Lily, Snarfy, Junebug, and Grace, as well as Bob and Cosmo, since departed, for their companionship, as well as for their attempts to help with the typing. Thanks to my daughter, Isobel, both for her love and for her curiosity about my work, which is on full display in Chapter 7. Thanks most of all, however, to my wife, Nancy Weil, for her unending support, including appropriate prodding from time to time, and for the life and love that we share. This collection is dedicated to her.

Editorial Notes

A few words about notation and terminology.

This book is concerned with philosophical, historical, and technical issues concerning a principle I call "HP": the number of Fs is the same as the number of Gs if, and only if, the Fs and Gs can be correlated one-to-one. The principle has gone by a number of names over the years. In *Frege's Conception of Numbers as Objects*, Crispin Wright calls it "$N^=$", for "numerical equality", or something of the sort, but the name did not stick. In "The Consistency of Frege's *Foundations of Arithmetic*", George Boolos dubbed it "Hume's principle", on the ground that Frege quotes from Hume when introducing it. But several people have objected that this designation is inappropriate and that something like "Cantor's principle" would have been better, since it was Cantor—not Hume, not Frege—who first realized that equinumerosity could be taken as a standard for the equality of numbers quite generally, and not just for finite sets, as Hume would have supposed. For what it's worth, I find this an unnecessary fuss. Though I never asked him, I very much doubt that Boolos thought Hume had scooped Cantor. The term "Hume's principle" was intended as a proper name, not a definite description, and has no historical implications. Perhaps that would have been clearer had Boolos written "Hume's Principle", as I generally did. But Boolos himself neatly solved this problem. Alluding to Chomsky's use of "LF" (which is not "logical form", in anything like the philosopher's sense), we may simply call the principle "HP".

We shall also have many occasions to discuss what Frege calls *Werthverläufe*. The term is translated several different ways in the existing literature: courses of values, ranges of values, and so forth. I prefer the compact "value-ranges" and so shall use it throughout, silently altering the various translations (and my own previous usage) so as to make the terminology uniform.

Frege uses the term "Begriffsschrift", throughout his career, to refer both to the formal system in which he works and to the formal language in which that system is itself formulated. In particular, the term is used not just in *Begriffsschrift*, the book, but also in *Grundgesetze*. It too is translated in several different ways. Following Thomas Ricketts, I prefer simply to render it "begriffsschrift"—it was already a neologism in German—and I use the definite article, "the begriffsschrift", when referring to the formal system, but drop it when referring to the language. I will adapt the standard translations to this terminology, too.

This book was written and edited using LyX, which is a document processor designed for academic and technical writing. Using LyX made my work on the book far more productive and enjoyable than it would have been with a traditional 'word processor'. Thanks to all my colleagues on the LyX development team, both past and present, for such a great tool, and thanks to the Free and Open Source Software movement for giving amateur programmers like me the opportunity to be involved with such things.

Here's how powerful LyX is: This book was typeset by me, directly from LyX, using a LaTeX style file I wrote to specifications provided by OUP. The real work was done by pdfLaTeX, a computer typsetting engine that is one of the available 'backends' for LyX. Thanks to the TeX and LaTeX communities for such fabulous products.

If you'd like to try LyX, you can download it, completely free of charge, from http://www.lyx.org/ and then do as you wish with it. It's not just 'free as in beer' but 'free as in speech'.

Origin of the Chapters

All of the papers in this collection have been previously published or, at least, are scheduled for publication. In but a few cases, I have made only minor changes to the published versions of the papers, correcting typos and grammatical errors, unifying notation and terminology, and clarifying a few phrases. In rare cases, I have silently corrected what I now believe to be minor errors. I have made significant changes only where I felt the earlier version was clearly in need of some help, and all such changes are noted below. Significant changes of mind are noted in postscripts to the papers themselves or in Chapter 1.

Chapter 2, "The Development of Arithmetic in Frege's *Grundgesetze der Arithmetic*", is what launched this entire investigation, as I recount in Chapter 1. It was originally published in the *Journal of Symbolic Logic* 58 (1993), pp. 579–601. Thanks to the Association for Symbolic Logic, which holds the copyright, for permission to reprint the paper. Section 2.5 is substantially different from, and hopefully better than, what was originally published.

Chapter 3, "*Die Grundlagen der Arithmetik* §§82–83", was written together with George Boolos and was originally published in M. Schirn, ed., *Philosophy of Mathematics Today* (Oxford: Oxford University Press, 1998), pp. 407–28. It also appears in George's collected papers, *Logic, Logic, and Logic* (Cambridge MA: Harvard University Press, 1998), pp. 315–38. Thanks to Sally Sedgwick for permission to reprint it. I have omitted two of the appendices. The second was entirely George's work; the third is somewhat redundant in the context of this collection.

Chapter 4, "Frege's Principle", appeared in a volume edited by Jaakko Hintikka, *From Dedekind to Gödel: Essays on the Development of the Foundations of Mathematics* (Dordrecht: Kluwer, 1995), pp. 119–42. My thanks to Springer Science+Business Media, who hold the copyright, for permission to reprint it. I have made substantial changes to the end of Section 4.4, as the original discussion made claims about the structure of Frege's proof of Theorem 49 that are just false. I have made further changes in the later sections, in an effort to highlight what I would now regard as the main point of the paper.

Chapter 5, "Julius Caesar and Basic Law V", was originally published in *Dialectica* 59 (2005), pp. 161–78, and is reprinted by kind permission of Wiley-Blackwell. It was written in the summer or fall of 1993. My views were evolving at a rapid pace at that time, and I became dissatisfied with the paper and abandoned it. Bits and pieces found their way into other papers, such as "Frege and Semantics" (Heck, 2010). I stumbled upon it while cleaning my hard drive in 2001 and thought there might yet be something of value in it, so I updated it a bit and put it on my website, otherwise not intending to do anything with it. Josep Macià-Fabrega suggested, however, that it might be appropriate to include it in a volume he was editing, and for that I owe him a great debt. It's actually pretty good.

Chapter 6, "The Julius Caesar Objection", originally appeared in a volume I edited: *Language, Thought, and Logic: Essays in Honour of Michael Dummett* (Oxford: Clarendon Press, 1997), pp. 273–308. It has undergone fairly extensive revision. For reasons discussed in the Postscript to Chapter 4, I have replaced two paragraphs in the original with a single, much improved paragraph on page 139. I have also substantially re-written the later parts of Section 6.2, beginning on page 141. The interpretation of §§55–6 of *Die Grundlagen* in the original paper now strikes me as not just wrong but, much worse, forced.

Chapter 7, "Cardinality, Counting, and Equinumerosity", originally appeared in the *Notre Dame Journal of Formal Logic* 41 (2000), pp. 187–209. Copyright is held by the University of Notre Dame, and the paper is reprinted by permission of the publisher, Duke University Press. The paper represents my attempt to come to terms with ideas I first heard expressed by Charles Parsons at a conference at Amherst College in 1991 and which ultimately appeared as "Intuition and Number". It took me nearly a decade fully to digest Charles's remarks and to formulate some sort of reply. The paper is dedicated to Charles, whose writings on the philosophy of mathematics have been a source of much pleasure, insight, and challenge.

Chapter 8, "Syntactic Reductionism", was published in *Philosophia Mathematica* 8 (2000), pp. 124–49, and is reprinted by the kind permission of Oxford University Press. An ancestor was the second of the three papers comprising my Ph. D. dissertation. I have removed most of the final section, as it overlaps substantially with parts of Chapter 9.

Chapter 9, "The Existence (and Non-existence) of Abstract Objects", will appear in a volume edited by Philip Ebert and Marcus Rossberg that is to be published by Oxford University Press. The central ideas first appeared in the last of the three papers comprising my Ph. D. dissertation, but it took a very long time before they reached a form with which I was reasonably happy. I have compressed a few things in the first section, as these are already contained in Chapter 8.

Chapter 10, "On the Consistency of Second-order Contextual Definitions", originally appeared in *Noûs* 26 (1992), pp. 491–4. I thank Wiley-Blackwell, who hold the copyright, for permission to reprint it here. It was my first publication. The paper, despite its brevity, has since had significant influence on discussion of the 'bad company' objection, of which it presents a particularly virulent strain. The Postscript recounts some of that history and offers a new proposal. Let me give special thanks to those who saw the value of this paper. It was rejected twice.

Chapter 11, "Finitude and Hume's Principle", emerged entirely from the ongoing conversations Boolos and I had about Frege's Theorem from 1991 through his death in 1996, and the paper is dedicated to his memory. It was originally published in the *Journal of Philosophical Logic* 26 (1997), pp. 589–617, and is reprinted by kind permission of Springer Science+Business Media, who hold the copyright.

Chapter 12, "A Logic for Frege's Theorem", is to appear in a *Festschrift* for Crispin Wright that is being edited by Alex Miller and will be published by Oxford University Press. Many of the ideas in this paper first surfaced in a lecture given at a special session of the European meeting of the Association of Symbolic Logic in 2000, held in Paris. I have omitted one of the appendices.

1
Frege's Theorem: An Overview

Frege was incredibly single-minded. Almost all of his intellectual work is devoted, in one way or another, to the attempt to establish logicism: the view that arithmetical truths are ultimately logical truths. As is now widely appreciated, thanks largely to Paul Benacerraf (1995), this project had both philosophical and mathematical aspects. The philosophical part involved analyzing the basic arithmetical concepts, so as to arrive at definitions of them that could be used in the mathematical part, which consisted of an attempt to prove the central theorems of arithmetic from those definitions, using nothing but logical means of inference.

The project was first announced in *Begriffsschrift*, and the innovations for which that book is celebrated are all in service of the logicist project. As Frege writes:

My initial step was to attempt to reduce the concept of ordering in a sequence to that of *logical* consequence,[1] so as to proceed from there to the concept of number. To prevent anything intuitive from penetrating here unnoticed, I had to bend every effort to keep the chain of inferences free of gaps. In attempting to comply with this requirement in the strictest possible way, I found the inadequacy of language to be an obstacle; no matter how unwieldy the expressions I was ready to accept, I was less and less able, as the relations became more and more complex, to attain the precision that my purpose required. This deficiency led me to the idea of the present begriffsschrift. (Bg, pp. 5–6, emphasis in original)

The required logic is put in place in the first two parts of the book. Frege devotes the third to developing the mentioned account of "ordering in a sequence", which is what we now call the theory of the ancestral.

Things must then have moved quickly. Just three years later, Frege says in a letter to the philosopher Anton Marty that he has "nearly completed a book in which [he] treat[s] the concept of number and demonstrates that the first principles of computation...can be proved from definitions by means of logical laws alone..." (PMC, pp. 99–100). Carl Stumpf, who was a colleague of Marty's, writes back (either on his behalf, or else in response to a similar letter) and urges Frege first to publish a prose version, rather than to publish another book filled with strange symbols (PMC, p. 172). Two years later, *Die Grundlagen der Arithmetik*

---

[1] It is off our main line, so I will not pursue this question, but does Frege really mean what he says here? What one would suppose he meant is: that he attempted to give a purely logical account of the notion of ordering in a sequence. But that is not at all what he says.

appears. This contains the philosophical work, in particular, Frege's analyses of the concept of cardinal[2] number, of the number zero, of the relation between a number and the one following it (known as predecession
or, conversely, succession), and of the concept of a natural (or finite) number. It also contains sketches of proofs, from definitions based on those
analyses, of several fundamental facts about the natural numbers, for
example, that the relation of predecession is one-one and, crucially, that
every natural number has a successor. This last is the most difficult and
most important since, given the rest, it implies that there are infinitely
many natural numbers, and the need to prove the infinity of the number-
series is what makes the project hard in the first place.

The formal presentation of these proofs would not appear for another
nine years, when the first volume of *Grundgesetze der Arithmetik* was
published, in 1893. I discuss the reasons for the long delay elsewhere
(Heck, 2005); they involve changes in the underlying logic itself. But,
with one notable exception, discussed in Chapter 3, the formal proofs
in *Grundgesetze* follow the sketch in *Die Grundlagen* quite closely. So
Frege might naturally have regarded the logicist project as more or less
complete at that point.

Unfortunately, just as a second volume was about to appear—it was
intended to tie up some loose ends and begin extending the project to
real analysis—Frege received a now famous letter from Bertrand Russell.
The gist was that (what we know as) Russell's Paradox can be derived in
Frege's logic from the 'Basic Laws' with which the proofs all begin. The
culprit is Basic Law V (five), which, for present purposes, may be stated
in the simplified form:

$$\hat{x}(Fx) = \hat{x}(Gx) \ \equiv \ \forall x(Fx \equiv Gx)$$

A term of the form "$\hat{x}(\phi x)$" is to denote the 'extension' of the concept ϕ.
The Law then says that two concepts have the same extension just in case
the same objects fall under them. That every concept has an extension
is treated as a logical truth, one that is a consequence of the fact that, in
Frege's logic, function symbols denote total functions.

One of the nice things about Law V is that it makes it particularly
easy to define the notion of membership:

$$a \in b \ \equiv \ \exists F[b = \hat{x}(Fx) \wedge Fa]$$

That is: a is in b just in case b is the extension of a concept under which a
falls. And so we have immediately:

$$a \in \hat{x}(Fx) \ \equiv \ Fa$$

---

[2] I shall omit this qualifier except when necessary, as cardinal numbers will be the main
focus of our discussion.

which is (a simplified form of) Theorem 1 of *Grundgesetze*. But then Russell's Paradox is almost immediate. Just take Fx to be: $x \notin x$, and a to be $\hat{x}(x \notin x)$. Then we have:

$$\hat{x}(x \notin x) \in \hat{x}(x \notin x) \equiv \hat{x}(x \notin x) \notin \hat{x}(x \notin x)$$

which is straightforwardly contradictory.

Frege saw the problem right away and attempted to respond to it in an appendix to the second volume. Having offered a solution, he notes that "... it will be necessary to check thoroughly all propositions discovered up to this point" (Gg, v. II, p. 265) to make sure that the changes to Law V do not undermine the proofs. Busy with other work, however, Frege does not seem to have made this check for a few years. But, on or about 5 August 1906 (WMR), he seems to have realized that the effort was hopeless. He would publish nothing for the next ten years. The three further papers he did eventually publish are concerned more with philosophy of language and mind than with philosophy of mathematics and logic.

Logicism, at least Frege's version of it, was thus a casualty of Russell's Paradox. The End.

And that was the end of the story until not very long ago. But the roots of a very different story were planted already in 1955. In "Class and Concept", which is generally devoted to clarifying the relation between these two notions, Peter Geach insists that, for Frege, "identifying numbers with certain extensions was both open to question and ... of altogether secondary importance". Indeed, Geach denies that numbers should be defined as classes (or extensions). More interestingly, though, for our purposes, Geach claims to be able to prove the infinity of the series of natural numbers without relying upon "any special set theory" (Geach, 1955, p. 569) and says that he hopes elsewhere "to explain how the infinite series of natural numbers is generated" (Geach, 1955, p. 570). So far as I know, however, no such paper ever appeared.[3]

Similar but much more explicit remarks are contained in Charles Parsons's paper "Frege's Theory of Number", which was published in 1965.[4] Parsons pays a good deal of attention to a principle he labels "(A)" but

---

[3]There are two other places that Geach comments on this matter. In his review of Austin's translation of *Die Grundlagen*, he remarks that "rejection of [Frege's view that numbers are classes] would ruin the symbolic structure of his *Grundgesetze*, but not shake the foundations of arithmetic laid down in the *Grundlagen*" (Geach, 1951, p. 541). But it is just not clear whether he meant Frege's proof of the infinity of the number series or, instead, the philosophical work done elsewhere in the book, since the latter is Geach's main focus. In his essay on Frege in *Three Philosophers*, Geach similarly remarks: "... Frege explicitly states that the identification of numbers with certain extensions is only a secondary and doubtful point, and in stating his theory of numbers I shall ignore extensions altogether" (Anscombe and Geach, 1961, p. 158). But he says almost nothing there about Frege's proofs.

[4]Parsons cites both "Class and Concept" and *Three Philosophers*, and he has remarked recently that he was much influenced by Geach's discussions.

which we shall be calling HP:[5]

$$\mathrm{N}x : Fx = \mathrm{N}x : Gx \equiv \mathrm{Eq}_x(Fx, Gx)$$

Here "$\mathrm{N}x : Fx$" abbreviates: the number of Fs; "$\mathrm{Eq}_x(Fx, Gx)$" abbreviates: the Fs and the Gs are 'equinumerous', that is, can be put into one-one correspondence. So HP says that the number of Fs is the same as the number of Gs if, and only if, the Fs and Gs are equinumerous. Parsons, even more so than Geach, emphasizes how central HP is to Frege's philosophy of arithmetic: It is the core of his analysis of number. And Parsons is the first, so far as I know, explicitly to note that, although extensions of concepts are needed for Frege's definition of numbers, they are not needed for the derivation of axioms for arithmetic from HP. Parsons remarks, almost in passing, that "...the argument could be carried out by taking [HP] as an axiom" (Parsons, 1995a, p. 198).

Though the name itself is due to George Boolos (1998k, p. 268), Parsons was thus the first explicitly to state what we now know as Frege's Theorem: Axioms for arithmetic can be derived, in second-order logic, from HP and Frege's definitions of zero, predecession, and natural number. Parsons does not, however, give a proof of Frege's Theorem. I thought until recently that Parsons simply deferred to Frege at this point, claiming that Frege's own proofs in *Die Grundlagen* have that structure already: Frege himself dervies HP from his explicit definition of numbers in terms of extensions, and then derives the arithmetical axioms from HP without making any further use of extensions. But Parsons makes no such claim; I had misremembered. And, for reasons we shall explore later, one certainly could not just have gestured at the formal arguments in *Grundgesetze*. Indeed, Goran Sundholm flatly dismisses Frege's "pious hope in [*Die Grundlagen*] to avoid the use of" extensions as "unrealistic" (Sundholm, 2001, p. 61), on the ground that at least some of his proofs—in particular, that of Theorem 263—require reference to extensions. Sundholm is wrong about this—see the postscript to Chapter 2—but showing that he is wrong requires a good deal of work.

The first published proof of Frege's Theorem would not appear until 1983, in Crispin Wright's book *Frege's Conception of Numbers as Objects* (Wright, 1983). This point alone bears some emphasis. Although initiates can nowadays rehearse the proof from memory, and we understand very well how it works, it has taken us some years to reach this stage. Indeed, Wright's own struggles illustrate this point. The idea that we might derive the axioms from HP had occurred to him, independently, by 1969.[6] But, as Wright wrote to me recently, when he attempted to write up the proof, he ran into trouble. It is not so easy to formalize the proof-sketches in *Die Grundlagen*. There is at least an apparent reference to

[5]See the editorial notes for some remarks on the terminology.
[6]*Frege's Conception* developed from Wright's B. Phil. thesis, which was submitted that year.

extensions that must be excised, though that is not hard to do. What is more problematic is the fact that Frege's proof of the crucial theorem that every number has a successor is flawed, and, as is argued in Chapter 3, there is no faithful adaptation of Frege's argument that will prove what he wanted to prove. A mere gesture towards *Die Grundlagen* therefore would not have been sufficient, even if Parsons had made it, and any claim that so-and-so had a proof at such-and-such a time has to be regarded as provisional, absent evidence that the *i*s really had been dotted and the *t*s really had been crossed.

In fact, the proof in *Frege's Conception* is not quite complete, and Wright comes harrowingly close to making the same mistake that Frege makes in *Die Grundlagen*. Let "Nx" mean: x is a natural number; and "Pab" mean: a immediately precedes b in the natural series of numbers (i.e., in the finite case, $b = a + 1$). Then, adapting his notation to ours, Wright takes induction to be:

$$F0 \land \forall x \forall y (Fx \land Pxy \to Fy) \to \forall x (Nx \to Fx)$$

rather than:

$$F0 \land \forall x \forall y (Nx \land Fx \land Pxy \to Fy) \to \forall x (Nx \to Fx)$$

and the former is, in principle, weaker.[7] However, due to an otherwise minor difference between their proofs, this oversight does not infect Wright's proof of the existence of successors, as it does infect Frege's.[8] What is worth noting, however, is that, once one has appreciated this difference, one can no longer say, as Wright does (and as I occasionally have, too), that Frege's "account of the ancestral has made it possible... to *define* the natural numbers as entities for which induction holds..." (Wright, 1983, p. 161, emphasis in original). What Frege's definition yields is just the weaker principle, whereas induction is the stronger one, and the proof of the stronger principle, though not technically difficult, requires significant logical resources. In particular, it requires impredicative comprehension (see Chapter 12).

Who first proved Frege's Theorem is not, however, a terribly interesting question, except in so far as we are asking whether Frege himself

---

[7]Saying in precisely what sense it is weaker is not so easy. But here is an example of what I have in mind. In "Ramified Frege Arithmetic" (Heck, 2011), I show that, in more or less Frege's way, we can prove the existence of successors in ramified predicative second-order logic plus HP. We have induction in the first form quite trivially, due to how N is defined. We cannot prove induction in the second form, since that would give us full PA, which we certainly cannot have. Formally, the crucial point is that interpreting induction means interpreting it with all quantifiers relativized to "N".

[8]Instead of trying to show that $Ny : P^{*=}yx$ is always the successor of x, as Frege does, Wright shows that $Ny : (Ny \land P^{*=}yx)$ is. (See his Lemma 5.) But then, in the proof of Lemma 512, we can assume that $Ny : (Ny \land P^{*=}yx)$ is non-zero, since otherwise we are in Lemma 511. And if it is non-zero, then, for some z, $Nz \land P^{*=}zx$, and then Nx, by transitivity. This very move is made, not quite explicitly, at the top of p. 163. That therefore gives us yet another way to patch Frege's proof, though it is, again, clearly not what Frege had in mind.

proved it: a question to which we shall turn shortly. But, whoever first proved it, Wright is undoubtedly responsible for the activity that has surrounded Frege's Theorem for the last quarter century or so. If Russell's Paradox, or some other contradiction, is forthcoming from HP itself, then who cares if we can replace Law V with HP? Parsons does not raise the question whether HP is consistent. Wright, however, argues in some detail that nothing like Russell's Paradox can be derived from HP[9] and then proceeds to conjecture that HP is, in fact, consistent (Wright, 1983, pp. 154–8). Once the conjecture had been made, it was quickly proved, independently, by several different people.[10] Indeed, Geach (1976, pp. 446–7) had observed a few years earlier that it has a simple model: Let the domain comprise the natural numbers plus $\aleph_0$, which is the number of natural numbers. This is a countably infinite set, and it is well-known that every subset of a countable set is countable and that countable sets are equinumerous if and only if they have the same cardinality. We can thus take the extension of the cardinality operator $Nx : \phi x$ to be the set of pairs $<S, n>$, where S is a subset of the domain and n is its cardinality. That verifies HP. The argument can easily be formalized in, say, Zermelo set-theory, so the consistency of HP is thereby assured, if we accept the axioms of Zermelo set-theory.

That is enough to establish Frege's Theorem as a legitimate piece of mathematics, but the mathematical result was never Wright's goal. Rather, Frege's Theorem was to be the basis on which Frege's logicism might be resurrected, and it was Wright's insistence on the philosophical significance of Frege's Theorem that caused all the excitement. The project had already been announced in Wright's B. Phil. thesis:

I wish. . . to vindicate Arithmetical logicism in the following form—not Frege's conception, it should be noted, but yielded by points implicit in Frege's work which in my view deserve acceptance: starting out with logical notions only, an explanation of the concept of Natural Number can be achieved (following a familiar pattern) of such a kind that foundational arithmetical truths (Peano's axioms) can be seen to be logical consequences of the explanation we adopt. It may be that this is sufficient to satisfy Frege's contention that the truths of arithmetic are analytic, even if they are not definitional equivalents of logical consequences of logical truths, as logicism is usually understood. (Wright, 1969, p. 92)

This is essentially the view labeled Number-theoretic Logicism [III], in *Frege's Conception* (Wright, 1983, p. 153). It is nowadays known as Neo-Logicism.[11]

---

[9]A similar observation, though concerning a slightly different paradox, is made by Geach (1951, p. 542).

[10]These included Harold Hodes (1981, p. 138), John P. Burgess (1984), and Allen Hazen (1985). Burgess gives the model I am about to describe; Hazen claims, but does not prove, that HP is interpretable in second-order arithmetic; the first published proof of that fact was by Boolos (1998b); Hodes claims the consistency (and even truth) of HP, but without proof.

[11]Or, sometimes, Scottish Neo-Logicism, since there are other forms. The modifier "Scot-

So the idea was roughly this. We may regard HP as "explaining" the concept of number. Such explanations are sufficiently akin to definitions that HP may be regarded as analytic of the concept of number and so as analytic in roughly the sense in which definitional truths are. But then Frege's Theorem shows that the axioms of arithmetic are logical consequences of an analytic truth and so are analytic themselves. Surely that is enough to vindicate the epistemological core of Frege's project, if not to vindicate logicism strictly so called.

Every aspect of this program has proved controversial.

Wright is not explicit, in *Frege's Conception*, about the background logic he is using in his proof of Frege's Theorem. The logic is obviously second-order, in that we are quantifying over relations and the like. But Wright does not specify a particular version of second-order logic as his background logic, preferring instead to identify the results we need our "logic of relations" to deliver (see, e.g., p. 163). It quickly became common practice, however, to assume that the proof was being given in full second-order logic, perhaps because that was the logic with which Frege himself worked. But it is famously controversial whether second-order 'logic' is properly so-called, and that means that it is controversial whether the proof of Frege's Theorem shows that the arithmetical axioms really are *logical* consequences of HP. If not, then we have no reason to regard the axioms as analytic. Some of the work collected in the present volume is concerned with this issue, especially Chapter 12, and to a lesser extent Chapter 7. We'll return to the matter below.

What has attracted the most attention, however, is Wright's claim that HP itself may be regarded as 'analytic'. It has never been entirely clear what notion of analyticity is supposed to be in play here. But what has been clear, or so I have generally supposed, is that the notion is supposed to be broadly epistemological in character, rather than metaphysical. In particular, the notion of analyticity in play is not that familiar from positivism and from Quine's criticisms thereof: a notion of 'truth in virtue of meaning'. Whereas the variety of logicism one finds in positivism is in part motivated by ontological doubts about mathematical objects, Wright's version, like Frege's own, was intended to reveal those doubts as unfounded. We are to imagine a subject, conventionally known as "Hero",[12] who is capable of second-order reasoning, but who is otherwise innocent of arithmetical concepts. Now imagine that Hero stumbles upon HP in a dream—how he discovers it is of no epistemic significance— and, impressed by its beauty, decides henceforth to use expressions of the

---

tish" reflects the fact that both Wright and Bob Hale, who soon became the second most vigorous defender of the view, were at the University of St Andrews when things were heating up. No other form of logicism will be under discussion here, though, so I shall speak simply of Neo-Logicism, meaning no disrespect to other forms.

[12]This sort of story first appears in *Frege's Conception* (Wright, 1983, pp. 141ff) and then re-appears in "On the Harmless Impredicativity of $N^=$ (Hume's Principle)" (Wright, 2001c). The name "Hero" is borrowed from a similar tale told by Gareth Evans (1985b).

form "N$x : \phi x$" as subject only to it. Then the claim is that Hero thereby acquires a concept of number. And, crucially, Wright claims that Hero can commit himself to HP, and to using names of numbers in accord with it, without epistemological presupposition. In particular, Hero does not need antecedently to make sure that there are such objects as numbers for such terms as "N$x : \phi x$" to denote. Indeed, as Wright emphasizes repeatedly, the demand is incoherent: Prior to his committing himself to HP, Hero has no concept of number and so cannot make sense of the question whether there are such objects.[13] Very roughly speaking, then, the thought is that HP is self-justifying: We need no justification to believe it.[14]

Most of *Frege's Conception*, and much of Wright's work on these issues since, is devoted to defending this conception of reference to abstract entities. I have been deeply sympathetic to it since I first encountered it in 1986, when Sir Michael Dummett, who was then my B. Phil. supervisor, suggested we read *Frege's Conception* in tutorial. By the time I arrived at MIT in 1987, I was actively working on the defense of the Fregean account of reference to abstracta, and two of the three papers that comprised my Ph. D. dissertation were directed at such issues.[15] And by luck, or manifest destiny, George Boolos just happened to be at MIT. Boolos had taken a substantial interest in both the mathematical and the philosophical aspects of Frege's Theorem before I arrived,[16] and that of course fueled my own interest. Eventually, he became my primary Ph. D. supervisor.

Boolos, in those days, was particularly concerned about what has come to be called the 'bad company' objection.[17] At that time, my own discussions of the Fregean approach to abstracta, like Wright's, was focused on what are now called 'first-order' abstraction principles, the most familiar of these being the abstraction for directions that Frege discusses in §§64–8 of *Die Grundlagen*:

$$\mathrm{dir}(a) = \mathrm{dir}(b) \equiv a \parallel b$$

First-order abstractions are, at least potentially, ontologically innocent: Nothing in the abstraction principle itself prevents one from identifying the direction of the line a with some representative line that has that

---

[13]To put it differently: It may be an intelligible question whether we, or Hero, should make use of a concept of number, but we cannot understand that question as: Are there such things as numbers? If we are not going to make use of that concept, then we cannot intelligibly ask that question. This sort of point of course goes back to Carnap (1950).

[14]Or, to put the point in the sort of language Wright might now prefer: We need to do nothing at all to earn an epistemic entititlement to believe HP (Wright, 2004).

[15]Versions of those papers appear here as Chapter 8 and Chapter 9.

[16]Indeed, I first met Boolos in the spring of 1987, while I was still in Oxford. He presented "The Consistency of Frege's *Foundations of Arithmetic*" as a lecture in 10 Merton Street.

[17]Early in his career, Boolos had been a logicist sympathizer (Boolos, 1972). Moreover, he was by then well-known for his vigorous defense of second-order logic's claim to that title (Boolos, 1998g, l). So Boolos was profoundly intrigued by Wright's position, even though he was not at all sympathetic to it.

direction, perhaps the line parallel to a that passes through some arbitrarily chosen origin. The possibility of such an identification guarantees the consistency of first-order abstractions. Second-order abstraction principles, however, can be ontologically inflationary, HP being a case in point: If, before committing himself to HP, Hero was capable of thinking of only finitely many objects, then there is no way that he can identify all the objects to which his new cognitive resources give him access with ones he could previously apprehend. What is worse, second-order abstractions can be inconsistent, Frege's own Basic Law V being but one of many salient examples.

The bad company objection, then, is that the general defense of the Fregean apporach to abstracta, if intended to include second-order abstraction, must prove too much. Surely Hero cannot simply commit himself to Basic Law V and thereby acquire a warranted belief in its truth. And, Boolos asked me one day, does Hero even have a way to tell if the principle to which he is committing himself is consistent? My very first publication, which appears here as Chapter 10, was my answer to that question. It develops a virulent form of the bad company objection.

But it was a different question Boolos asked me that has most shaped my work on Frege's Theorem.

1.1   Frege on Frege's Theorem

Dummett's long-awaited *Frege: Philosophy of Mathematics* was published in the spring of 1991. I remember seeing it in a bookstore in Amherst, Massachusetts, when I was there for a conference, and buying it excitedly. I set to reading it as soon as I got home. Many of my conversations with Boolos were soon focused on *FPM*. One day, Boolos had a particularly pressing question concerning the following passage:

Crispin Wright devotes a whole section of his book... to demonstrating that, if we were to take [HP] as an implicit or contextual definition of the cardinality operator, we could still derive all the same theorems as Frege does. He could have achieved the same result with less trouble by observing that Frege himself gives just such a derivation of those theorems. He derives them from [HP], with no further appeal to his explicit definition. (Dummett, 1991b, p. 123)

Dummett seems to be claiming that Frege himself had given a proof of Frege's Theorem. But where is that proof? Boolos had shown that the proof in *Die Grundlagen* can be reconstructed as a proof of Frege's Theorem (Boolos, 1998b), but the reconstruction is not trivial.[18] As I have already said, then, dismissing Wright's contribution with but a gesture at *Die Grundlagen* would have been worse than uncharitable. In any event, the context of Dummett's remark makes it plausible that the "derivation"

---

[18] We did not know then just how non-trivial: See Chapter 3 for that story.

of which he speaks is not that in *Die Grundlagen* but the one in *Grundge-setze*. So it looks as if Dummett is claiming that Frege proves Frege's Theorem in *Grundgesetze*. Boolos's question was: Is that true?

My dissertation had just been finished, so I had some time on my hands and quickly set to reading *Grundgesetze*. I soon discovered that, if Dummett's claim could be defended at all, it was going to take work. It was easy enough to verify that, after proving HP, Frege makes, as Dummett had said, "no further appeal to his explicit definition", but that simply does not show that Frege proved Frege's Theorem. The crucial question is whether Frege makes no further appeal *to Basic Law V*, and, strictly speaking, he most certainly does.

In *Grundgesetze*, Frege speaks not of extensions of concepts but, more generally, of the value-ranges of functions. The value-range of a function can be compared to its graph, in the set-theoretic sense: a set of ordered pairs of arguments and values. Frege does not, however, define value-ranges in terms of ordered pairs.[19] Rather, he regards the notion of a value-range as primitive, and he characterizes value-ranges simply by saying that two functions will have the same value-range just in case they have the same values for the same arguments. Where "$\grave{\epsilon}F\epsilon$" means: the value-range of the function $F\xi$, then, what Frege says is that "$\grave{\epsilon}F\epsilon = \grave{\epsilon}G\epsilon$" will be true just in case "$\forall x(Fx = Gx)$" is true (Gg, v. I, §3). Clearly, this leads quickly to Basic Law V:

$$(\grave{\epsilon}F\epsilon = \grave{\epsilon}G\epsilon) \ = \ \forall x(Fx = Gx)$$

It is the identity-sign that occurs in the middle here, rather than "$\equiv$", because, for Frege, the truth-values are objects like any others. Consequently, since concepts are functions from objects to truth-values, concepts too have value-ranges, and it is easy to see that two concepts will have the same value-range just in case they have the same extension, that is, have the value Truth for all the same arguments.

Terms denoting value-ranges appear throughout Frege's proofs. Almost every theorem in *Grundgesetze* depends upon the previously mentioned Theorem 1, which, as said earlier, leads directly to Russell's Paradox.

Nevertheless, it was clear from the outset that many of the uses Frege makes of value-ranges can easily be eliminated. Frege almost never quantifies over concepts, for example, preferring instead to quantify over their extensions.[20] So we find things like:

$$\forall f(\ldots a \in f \ldots)$$

---

[19] It's an interesting question whether, perhaps, he once did do so. As Sundholm (2001, pp. 60ff) notes, pairs seem to have played an important role in early versions of *Grundgesetze*.

[20] Is it a mere curiosity that Frege quantifies over concepts themselves in the definition of the ancestral? Is it just more convenient, technically speaking, to do so? Or is there more to be said on this point?

rather than things like:

$$\forall F(\dots Fa \dots)$$

But this is easily remedied, and it seemed likely that other uses of value-ranges would prove to no be more problematic. I therefore set out—in part because reading Frege's notation was then still a chore—to produce a complete translation of Frege's proofs into modern notation, including these sorts of mechanical 'corrections'. When I was done, I could answer Boolos's question: Frege had proved Frege's Theorem in *Grundgesetze*, *modulo* this sort of mechanical elimination of the use of extensions.

This reconstruction of Frege's proofs would appear in "The Development of Arithmetic in Frege's *Grundgesetze der Arithmetik*", which is reprinted here as Chapter 2. In particular, Section 2.1 contains a detailed discussion of how Frege uses value-ranges and Law V in his proofs of the arithmetical axioms; the central parts of the paper contain a commentary on those proofs. But the mere fact that Frege's proofs can be mechanically translated into ones in 'Frege Arithmetic'—second-order logic plus HP—does not, of course, show that Frege *knew* that arithmetic could be derived from HP. I mention some reasons to think he did in Section 2.7, but a more sustained discussion of the question was always necessary, and it appears here in Section 4.3. In brief, the philosophical significance Frege attributes to his proof of the axioms requires that those proofs depend only upon HP and not also upon Law V. It was important to Frege, for philosophical reasons, that his proof of the arithmetical axioms did not depend essentially either upon his explicit definition of numbers in terms of extensions or upon any other sort of appeal to extensions, either.

If Frege did know that arithmetic can be derived from HP, however, then the next question is: Why didn't Frege himself consider retreating from Law V to HP when confronted with Russell's Paradox? In fact, it turns out that he did consider this option. In one of his letters to Russell, Frege writes:

We can also try the following expedient, and I hinted at this in my *Foundations of Arithmetic*. If we have a relation $\Phi(\xi, \eta)$ for which the following propositions hold: (1) from $\Phi(a, b)$ we can infer $\Phi(b, a)$, and (2) from $\Phi(a, b)$ and $\Phi(b, c)$ we can infer $\Phi(a, c)$; then this relation can be transformed into an equality (identity), and $\Phi(a, b)$ can be replaced by writing, e.g., "§a = §b". If the relation is, e.g., that of geometrical similarity, then "a is similar to b" can be replaced by saying "the shape of a is the same as the shape of b". This is perhaps what you call "definition by abstraction".

What Frege is discussing here is precisely the possibility of replacing the explicit definition of numbers as extensions with HP. He concludes:

But the difficulties here are []²¹ the same as in transforming the generality of an identity into an identity of value-ranges. (PMC, p. 141)

---

[21]At this point, the translation inexplicably contains the word "not", which is not found in Frege's original letter.

Note how this language echoes the language earlier in the passage, where Frege speaks of the relation Φ's being "transfomed into an equality". There is thus an explicit comparision between "transforming the generality of an identity into an identity of value-ranges"—that is, moving from "$\forall x(Fx = Gx)$" to "$\grave{\epsilon}F\epsilon = \grave{\epsilon}G\epsilon$"—and the transition from "$\mathrm{Eq}_x(Fx, Gx)$" to "$Nx : Fx = Nx : Gx$". The same "difficulties" affect both moves.

Frege certainly is not saying here that HP too is inconsistent.[22] So these "difficulties" are independent of the contradiction, and they apply as much to HP and numbers as to Law V and value-ranges. What are they? Frege mentions them right before the two passages quoted in the last paragraph:

I myself was long reluctant to recognize the existence of value-ranges and hence classes; but I saw no other possibility of placing arithmetic on a logical foundation. But the question is, How do we apprehend logical objects? And I have found no other answer to it than this, We apprehend them as extensions of concepts, or more generally, as value-ranges of functions. I have always been aware that there were difficulties with this, and your discovery of the contradiction has added to them; but what other way is there? (PMC, pp. 140–1)

The question Frege raises here, how we apprehend logical objects, is the same question that opens §62 of *Die Grundlagen*: "How, then, are numbers to be given to us, if we cannot have any ideas or intuitions of them?" And the answer to which he here commits himself, that "[w]e apprehend [logical objects] as extensions of concepts", is the answer he had reached in §68, where numbers had been defined as extensions of concepts. He had settled on that definition because a very different proposal, made in §63, had been found wanting.

That proposal was that we apprehend numbers as 'abstracts', that is, as the referents of expressions introduced by abstraction, in this case, by HP. And what was the fatal problem with that proposal? It was the "third doubt" introduced in §66. It is known, because of the example used when it is first raised, in §56, as the "Caesar problem":

In the proposition

"the direction of a is identical with the direction of b"

the direction of a plays the part of an object, and our definition affords us a means of recognizing this object as the same again, in case it should happen to crop up in some other guise, say as the direction of b. But this means does not provide for all cases. It will not, for instance, decide for us whether England is the direction of the Earth's axis—if I may be forgiven an example which looks non-sensical. Naturally no one is going to confuse England with the direction of the Earth's axis; but that is no thanks to our definition of direction. (Gl, §66)

---

[22] Frege is talking about abstraction principles quite generally, and the examples he gives are first-order and provably consistent. The direction abstraction, in particular, is equivalent to the introduction of points at infinity into Euclidean geometry. That construction, which was well-known to Frege (Tappenden, 1995; Wilson, 1995), amounts to a consistency proof.

As said, it is Frege's inability to resolve this problem that forces him to abandon the view that numbers are abstracts in favor of the view that numbers are extensions.

At the time he wrote *Die Grundlagen*, Frege held open the possibility that there might be some other way to respond to the Caesar problem, so that reference to extensions would not be necessary (Gl, §107). In the end, however, as he wrote to Russell, he could find no other response, and so he was, by his own lights, forced to acknowledge the existence of extensions and to define numbers in terms of them.

To understand why Frege did not himself abandon logicism for Neologicism, we must thus understand the Caesar problem. Chapters 4–6 are devoted to that task. In the next section, I will try to synthesize their conclusions.

Before I continue, however, I want to emphasize again the epistemological character of Frege's discussion. What forces Frege to abandon the account of numbers as abstracts for the explicit definition of numbers as extensions is, in the first instance, an *epistemological* problem: Frege thinks that the question how we might have a kind of cognitive access to numbers that is neither sensible nor intuitive cannot be answered in terms of abstraction principles. We shall return to the importance of this point.

1.2   The Caesar Problem

1.2.1   Caesar and Value-Ranges

As we saw above, Frege explicitly compares the move from "$\forall x (Fx = Gx)$" to "$\grave{\epsilon}F\epsilon = \grave{\epsilon}G\epsilon$" to the move from "$\mathrm{Eq}_x(Fx, Gx)$" to "$\mathrm{N}x : Fx = \mathrm{N}x : Gx$", saying that the same "difficulties" affect both moves. The central difficulty with the latter, we have seen, is the Caesar problem. In what sense is it supposed to impinge upon the former?

It is easy to get confused about this matter, and I fear that I have succumbed to that confusion from time to time. At the very least, I have not been as clear as I should have been about how the Caesar problem bears upon the question how we apprehend value-ranges as objects. Let me begin by trying to do better.

The most important thing to remember about the Caesar problem is that it arises as an objection to a particular view about how we apprehend logical objects: the view introduced in §62 of *Die Grundlagen*. This view, which I shall call *abstractionism*, is a view about what is required for singular thought about certain sorts of abstract objects.[23] The objects in question are what I have already called "abstracts", but they are more

---

[23]Compare Hale and Wright's remarks about why it is important to regard "the number of Fs" as a referring expression rather than a definite description (Hale and Wright, 2009a, §1).

commonly known as *types*. Types here are supposed essentially to be 'of' objects of another sort: the tokens of that type. This relationship is reflected in how we most fundamentally refer to a type: as the type instantiated by a given token. So, for example, directions are essentially the directions *of* lines, and the fundamental way to refer to a direction is as the direction of a given line. Abstractionism then comprises two claims: First, that the capacity for singular thought about objects of a given type T derives from and is constituted by an appreciation of the truth-conditions of identity judgments about Ts, where these identity judgments involve the fundamental way of referring to Ts; Second, that the truth-conditions of such identity judgments may be given in terms of an equivalence relation on the tokens, that is, in terms of an *abstraction principle*, a statement of the form: $\phi(a) = \phi(b) \equiv Rab$.

Let me pause to insist that this is the correct way to formulate the view. The more usual formulation is that our ability to refer to (think about, apprehend) types rests upon our acceptance of an abstraction principle for them. But then the question must immediately arise what is so special about abstraction principles, which are just sentences of a particular syntactic form. Why shouldn't our capacity to refer to numbers derive from our acceptance not of HP but of the Dedekind-Peano axioms (Boolos, 1998d, p. 311)? The best answer[24] to this question is that abstraction principles are *not* special. What underlies our capacity for thought about numbers is not our acceptance of HP but our appreciation of the truth-conditions of numerical identities involving the most fundamental means of reference to numbers. What we must 'accept' is not HP itself but a specification of the truth-conditions of such identities that looks like a meta-linguistic version of HP: "$Nx : Fx = Nx : Gx$" is true iff "$Eq_x(Fx, Gx)$" is true. Given that, of course, HP's truth is immediate.[25] But acceptance of HP is not what underlies our capacity for thought about numbers, and there is no similar story to be told about the Dedekind-Peano axioms.

Now, it is clear that abstractionism could as well offer an answer to the question how we apprehend value-ranges. Indeed, Frege almost seems to endorse this view. Here is how he introduces value-ranges in *Grundgesteze*:

I use the words "the function $\Phi(\xi)$ has the same *value-range* as the function $\Psi(\xi)$" generally to denote the same as the words "the functions $\Phi(\xi)$ and $\Psi(\xi)$ always have the same value for the same argument". (Gg, v. I, §3)

It looks very much as if Frege is here explaining the truth-conditions of identities between value-ranges in terms of the co-extensiveness of the corresponding functions, and taking that to be sufficient to make refer-

---

[24]Hale and Wright (2009a) discuss this worry in a reply to a paper by MacFarlane (2009), but they do not make the present point.

[25]I do not think this is quite right, actually—see Section 1.2.3—but I am here just outlining the simplest form of abstractionism.

ence to them possible. That would be an abstractionist view about value-ranges. But Frege explicitly disowns that view later in *Grundgesetze*:[26]

...[W]e said: If a (first-level) function (of one argument) and another function are such as always to have the same value for the same argument, then we may say instead that the value-range of the first is the same as that of the second. We are then recognizing something common to the two functions, and we call this the value-range of the first function and also the value-range of the second function. We must regard it as a fundamental law of logic that we are justified in this recognizing something common to both, and that accordingly we may transform an equality holding generally into an equation (identity). (Gg, v. II, §146)

Frege is here insisting that the stipulation from §3 is inadequate, on its own, to ground reference to value-ranges. To move from "$\forall x(Fx = Gx)$" to "$\grave{\epsilon}F\epsilon = \grave{\epsilon}G\epsilon$", we must be "justified in... recognizing something common to the two functions", but Frege is here denying that we can earn a right to this recognition simply by changing the shapes of ink marks.[27] If we could, no more would need to be said in defense of the claim that we may transform "$\forall x(Fx = Gx)$" into "$\grave{\epsilon}F\epsilon = \grave{\epsilon}G\epsilon$". But, as a quick look at the broader context will show, Frege desperately wishes he could say more. It is because he can't that he can only insist on the "fundamental law of logic" he needs:[28]

If there are logical objects at all—and the objects of arithmetic are such objects—then there must also be a means of apprehending, or recognizing, them. This service is performed for us by the fundamental law of logic that permits the transformation of an equality holding generally into an equation. Without such a means a scientific foundation for arithmetic would be impossible. (Gg, v. II, §147)

And it is important to appreciate that the "fundamental law of logic" for whose acceptance Frege is arguing here is not Law V itself. It is, rather, something that is a law of logic in a quite different sense and that serves to *justify* Law V. This law is what justifies our "recognizing something common", so that "accordingly we may transform an equality holding generally into an equation" (Gg, v. II, §146).

There is a sense, then, in which the Caesar problem is still in play in *Grundgesetze*, and there is a sense in which it is not. The Caesar problem *is* in play in the sense that it continues to frustrate abstractionism, so that it prohibits Frege from adopting an abstractionist account of reference to value-ranges, just as it forced him to abandon the abstractionist account of reference to numbers. But, in a different sense, the Caesar problem is *not* in play: Frege does not hold any view to which the Caesar

---

[26]This passage is from volume II, which was published a decade later. But there is no serious doubt, it seems to me, that it reflects Frege's view in 1893, as well.

[27]It is sometimes claimed that Frege regards "$\forall x(Fx = Gx)$" and "$\grave{\epsilon}F\epsilon = \grave{\epsilon}G\epsilon$" as synonymous, the two being mere stylistic variations of one another, like active and passive. The passage we are discussing is not consistent with such an interpretation.

[28]Note the similarities to the letter to Russell from which I quoted earlier.

problem is an objection, so he does not need to solve it. What he does need is an answer to the question how we apprehend value-ranges as objects, and the sad truth, which he himself recognizes, is that he simply does not have one.

If so, then it does seem fair, in the end, to regard the explicit definition of numbers as extensions as hopeless, quite independently of the inconsistency of Law V. Why, then, did Frege insist upon painting himself into this corner? The answer at which I arrive in Chapter 5 is that, in Frege's day, at least, there were significant dialectical differences between the two positions we have been discussing. The dominant logical tradition against which Frege had himself struggled, the Boolean tradition, regarded extensions of concepts as *the* fundamental materials of logic.[29] It would thus have been entirely reasonable for Frege to expect agreement with his "fundamental law of logic" permitting the move from "$\forall x(Fx = Gx)$" to "$\dot{\epsilon}F\epsilon = \dot{\epsilon}G\epsilon$". For him simply to insist on a law permitting the move from "$\text{Eq}_x(Fx, Gx)$" to "$\text{N}x : Fx = \text{N}x : Gx$", however, would have been for him to beg the very question logicism was meant to answer.

One might object, however, that the Caesar problem must play more of a role in *Grundgesetze* than I am allowing. After all, a form of it seems to arise in §10, where Frege discusses the question whether the truth-values are value-ranges and, if so, which ones they are. But the correct conclusion, for which I argue in Section 5.2, is that the problem under discussion in §10 is not the Caesar problem, not if the 'Caesar problem' is the problem discussed in the central sections of *Die Grundlagen*.

So what is the Caesar problem?

1.2.2   The Caesar Problem as Epistemological

I wish I could say, after umpteen years of thinking about it, that I now understand the Caesar problem and why Frege thought it a fatal objection to abstractionism. Unfortunately, I don't. What I've got are a handful of thoughts about how the Caesar problem bears upon abstractionism, and how it does not. I suspect that Frege himself saw many aspects of the problem only obscurely.

Let me begin by emphasizing again that the Caesar problem arises as an objection to a certain view. Frege famously held, at least in his mature period, that every well-formed sentence must have a determinate truth-value, lest (at least) one part of that sentence not have a reference. The Caesar problem can sound as if it is just an application of this more general view, but, for reasons discussed in Section 6.1, that is wrong. This 'principle of complete determination' does indeed imply that we need to fix a sense for identities like "$0 = \text{Caesar}$", but not any way of doing so counts as resolving the Caesar problem. Frege might have

---

[29] Might Frege have been led to his acceptance of extensions in part through his reading of Boole?

stipulated truth-values for such statements, much as he stipulates in §10 of *Grundgesetze* that Truth and Falsity are to be their own unit classes, but no such stipulation will save abstractionism. Nor will rejecting the principle of complete determination.

Second, it is important to appreciate that the Caesar problem is less about deciding the truth-values of certain statements than it is about guaranteeing that those statements have a sense. On Frege's view, again, if a statement does have a sense, then that sense is a mode of presentation of a truth-value, and the statement must therefore have a truth-value as well. But the worry is not that we do not know whether Caesar is 0. It is that we do not so much as understand the question whether he is. Or, rather, the problem is that we *do* understand that question, and we *do* know the answer. But the abstractionist account of reference to numbers utterly fails to explain how we might even make sense of it. It explains identity statements of a very particular form, "$Nx : Fx = Nx : Gx$", and no others. Whence our understanding of statements of the form "$t = Nx : Gx$" when t itself is not of the form "$Nx : Fx$"?

This is one of the few things of which I am sure here: that one main lesson of the Caesar problem was supposed to be that there is more to our apprehension of numbers as objects than abstractionism can explain. I do not claim to be the only, or even the first, philosopher to appreciate this fact. The way Hale and Wright approach the Caesar problem, at least in their more recent discussions (Hale and Wright, 2001c, 2008), seems to involve at least an implicit recognition of it. But I do not think the point has been appreciated as widely as it should be. What makes a proper appreciation of it so difficult is that it is hard to keep the epistemological dimensions of the problem firmly in view. It is easy to slip back into thinking that the problem is to fix the truth-values of certain sorts of identities, using whatever resources might be available. But if the question is one about *our* apprehension of numbers as objects and abstractionism's failure to capture it, then the resources used in that account have to be ones of which ordinary folk have some sort of grasp.[30] (If one is inclined to deny that the question is about ordinary folks' apprehension of numbers as objects, see Section 1.3.)

I have been careful here *not* to say that what the Caesar problem requires is that we explain the sense of such claims as "Caesar is zero". It is, after all, a common reaction to the Caesar problem to insist that even

---

[30]Sider (2007, pp. 220ff) emphasizes a similar point, but then misplays it, or so it seems to me. Sider seems to think that Neo-logicism is a quasi-Cartesian enterprise whose goal is to "dispel doubts about mathematics" and so make it "epistemically secure". If that were the project, then, indeed, everything that went into the argument that abstract entities exist, including the "underlying metaontology", would be part of the justificatory basis for our knowledge of arithmetic, and Sider's complaint that this doesn't look epistemically secure would be justified. But that is just not the project, not as I understand it, anyway, though Sider's charge might well apply if the project were understood the way MacBride (2000, p. 158) suggests it should be. We'll discuss this issue below.

to ask whether Caesar is zero is to commit some kind of category mistake. So it is a possible view that the question whether Caesar is zero is actually unintelligible, and that our denial that he is zero involves something other than an assertion. Saying that Caesar is not a number is thus comparable, in relevant respects, to saying that the concept *horse* is not a concept. Saying that Caesar is not a number is a confused attempt to express as a substantive claim what must ultimately be understood in terms of the differences between number-words and people-words, be these logical, syntactic, or semantic.

Many people probably still think, as I once did, that this view simply isn't available to Neo-logicists. It is widely believed that Frege's proofs of the axioms of arithmetic rest essentially upon his view not just that numbers are objects, but that numbers are objects of the same logical category as concrete objects such as people. Very roughly, the thought is that, when we speak of "the number of Fs", it must be permissible for F to be true both of people and of numbers. Of people (and the like), because our initial understanding of what Frege calls "ascriptions of number" cannot presuppose a familiarity with numbers themselves. The basic instances of HP must concern things like "the number of Roman emperors". But we must also be able to speak of the number of numbers having some property, because the idea behind the proof of the existence of successors is to consider the sequence: $Nx : x \neq x$, $Nx : x = 0$, $Nx : (x = 0 \vee x = 1)$, ... If so, however, then it looks as if the numbers and the people need to be in the same domain, and that will force the Neo-logicist to regard the question whether Caesar is zero as at least intelligible.

This line of thought, perhaps bound up with various sorts of formal considerations,[31] may have played a role in leading Frege to the Caesar problem. But the lesson of Chapter 6, as I now see it, is that there is no extra problem here. Contrary to the common wisdom, Frege's proofs of the axioms of arithmetic—in particular, of the existence of successors— do *not* require there to be a single, universal domain over which all first-order variables range. The proofs do not even require that numbers be objects. If so, then the view that the question whether Caesar is zero commits a category mistake *is* available to Neo-logicists.

Contrary to what Hale and Wright (2001c, p. 346) seem to think, however, I never meant to suggest that simply making this view available somehow solved the Caesar problem in all its manifestations.[32] On the contrary, Chapter 6 opens by identifying the epistemological aspect of the Caesar problem we have been discussing and a semantical aspect we shall discuss shortly, and then suggesting that, *besides* these general

---

[31]Tappenden (2005) discusses the mathematical background of the Caesar problem.

[32]In footnote 6 of Chapter 6, I say: "Adopting the view discussed below, that numbers are simply a different *sort* from people, will not relieve one of the obligation to explain the origins of speakers' knowledge of this fact." In footnote 25, I insist that the semantic form of the Caesar problem still arises when HP is treated predicatively. These notes were contained in the original paper (Heck, 1997).

problems, which Frege raises against both the direction abstraction and HP, the Caesar problem might "raise quite specific problems in the case of numbers" (this volume, page 130). The point of Chapter 6 is to argue that those more specific problems can be resolved, and so to make room for the kind of response to the epistemological aspect of the Caesar problem just outlined. Resolving the "quite specific problems" was never intended to relieve abstractionists of the need to address the more general ones.

So I agree with Hale and Wright (2001c, p. 350) that "... we cannot avoid confronting something like the question whether numbers are or aren't of the same Sort as people". My goal, as I have been saying, was simply to put that question on the table and thereby to make a particular kind of answer to the Caesar problem available to Neo-logicists: that numbers and people are just radically different sorts of objects. The point of making this view available is that a defense of it, or so it seemed to me, would be able to appeal to quite different resources than views that assume that people and numbers must occupy a single domain. I never claimed to have provided a defense of that view, however, and I certainly was never under any illusion that it somehow didn't need one. But the view cannot just be dismissed with the suggestion that, if numbers are not of the same sort as people, then we might just as well regard "Caesar is zero" as intelligible but false (Hale and Wright, 2001c, p. 351). As Frege well knew, the problem is in no way limited to mixed identity statements: If "Caesar" and "zero" are intersubstitutable *salva significatione*, then "0 + Caesar" must also have a sense and, unless one thinks that "+" fails to refer, a reference. Which, pray tell? And if one is inclined to reply that "+" is only partially defined, then I ask: Why can't "=" be partially defined, too? And what's the difference between saying that these signs are only partially defined and saying that numbers and people are of different sorts?

1.2.3   The Caesar Problem as Semantical

Consider the formulae "$Nx : Fx = Nx : Gx$" and "$\text{Eq}_x(Fx, Gx)$". According to abstractionism, they are supposed to have the same truth-condition. Yet the former is supposed to involve reference to numbers in a way that the latter does not: It is, in particular, supposed to be by our coming to understand the former as equivalent to the latter that we acquire a capacity for singular thought about numbers. But if we really do understand "$Nx : Fx = Nx : Gx$" as an identity statment containing expressions that refer to objects, then that understanding must comprise (or at least make available) an understanding of the complex predicate "$\xi = Nx : Gx$". However, treating "$Nx : Fx = Nx : Gx$" as having the same truth-condition as "$\text{Eq}_x(Fx, Gx)$" does not, by itself, give us the ability to understand the question whether the predicate "$\xi = Nx : Gx$" is true or false of a given object. Of course, if this object is given to us as the number of ϕs, for some ϕ, then all is well. But if it is not given to us in that form, then we are completely lost, and the example of Caesar simply illustrates this point.[33] Indeed, even if the object we are given *is* a number, we cannot understand the question whether "$\xi = \mathrm{N}x : Gx$" is true or false of it unless it is given to us as the number of ϕs. But then the lesson of the Caesar problem appears to be this: If our understanding of names of numbers were adequately explained by abstractionism, then we could not understand the question whether "$\xi = \mathrm{N}x : Gx$" is true or false of *objects* at all, which means that we could not understand the predicate itself, which means that we could not understand "$\mathrm{N}x : Fx = \mathrm{N}x : Gx$" as an identity statement. So, again, abstractionism would fail fully to capture our understanding of names of numbers.

The question whether abstractionism can provide a sense for complex predicates of the form "$\xi = \mathrm{N}x : \phi x$" is particularly pressing for Neologicists, since such predicates play a critical role in Frege's proof of the existence of successors. Frege's idea is to consider the series $\mathrm{N}x : x \neq x$, $\mathrm{N}x : x = 0, \dots$. But the latter is really:

$$\mathrm{N}x : (x = \mathrm{N}y : y \neq y)$$

and the argument of the outer occurrence of "N" is thus:

$$\xi = \mathrm{N}y : y \neq y$$

If this predicate has not been provided with a sense, Frege's proof of the existence of successors consists of a series of uninterpreted squiggles.

I have introduced this issue in broadly epistemological terms since, as I have emphasized, abstractionism is fundamentally an epistemological doctrine. But the present issue is not epistemological but semantic. What this form of the Caesar problem threatens to show is that we cannot simultaneously both regard "$\mathrm{N}x : Fx = \mathrm{N}x : Gx$" as having the same truth-condition as "$\mathrm{Eq}_x(Fx, Gx)$" and treat "$\mathrm{N}x : Fx = \mathrm{N}x : Gx$" as having the 'logical form' it overtly appears to have.[34]

In the case of first-order abstractions, we can make some progress, if we accept the view, discussed in the previous section, that types are of a different logical category from their tokens. If so, then we will have a special style of variable that ranges over the types, and these variables can be eliminated in favor of variables ranging over tokens. Consider, for example, lines and their directions, and write variables ranging over directions in boldface. Then, quite generally, we can transform "$\dots \mathbf{d} \dots$" into "$\dots \mathrm{dir}(d) \dots$", trading (both free and bound) occurrences of "$\mathbf{v}$" for those of a corresponding variable "v". Repeat as necessary. The end result will then be something from which all occurrences of "$\mathrm{dir}(\xi)$" can be

---

[33] For defense of that interpretive claim, see p. 129.

[34] Sider (2007, p. 204) voices a similar concern, though he does not connect it to the Caesar problem.

eliminated *via* the abstraction principle (and related principles concerning predicates of directions).

The procedure can be extended to 'predicative' second-order abstractions, where we do not permit embeddings like: $Nx:(x = Ny:\phi y)$. That there be no such embeddings is a form of the requirement that the types introduced by the abstraction be of a different logical sort from their tokens. In this case, however, the tokens—what numbers are 'of'—are concepts, so the requirement becomes: the objects introduced by abstraction must be of a different sort from those of which these concepts are true or false. In this case, then, we can transform "...n..." into "...$Nx:Fx$..." and eliminate the cardinality operator via HP.[35,36]

This approach can easily feel like the most blatant cheating, but there is actually something very natural about it. The basic idea is just that, since directions are fundamentally and essentially *of* lines, specifying the domain over which direction-variables range shouldn't be any harder than pointing at the lines and saying: the directions of those.[37] To the extent that this explanation of variables ranging over types works, however, it works too well. Traditionally, the thought would have been that, by showing us how statements that quantify over directions can be translated into ones that only quantify over lines, the procedure shows us how to *eliminate* quantification over directions. For the reasons given in Section 8.3, I do not think that is the right way to put the point. We have to allow that abstractionism can explain our understanding of a class of abstract *terms*, by which I mean: terms that more or less look like they ought to refer to abstract objects, to types. But what the availability of the translation does show is that abstractionism cannot secure the idea that these terms refer *to abstract objects*. What the availability of the translation shows, that is to say, is that, for all we have said so far, abstractionism is compatible with the view that 'names of types' refer not to types but to representative tokens of the type. For example, a 'name of a direction' might refer to a representative line that has that direction.[38]

So here is another thing I take myself to know: So long as we

---

[35]Matters are a bit more delicate than I have indicated, since we will need to deal with problems similar to those discussed in Section 8.3. But this can be done in much the way done there.

[36]The restriction to predicative abstractions is essential. If the abstraction is regarded as impredicative, then nothing like this elimination can work. Such a reduction is precisely what Frege attempts in §§29–31 of *Grundgesetze*. Success would have amounted to a consistency proof for Basic Law V (Heck, 1998a). The problem, ultimately, is that we have no way to eliminate occurrences of abstract terms inside second-order variables, like: $G(Nx:Fx)$. In the present case, we can increase the order of "G", so that "$G(Nx:Fx)$" becomes a third-order statement. Clearly, this will not work if we can have something like: $F(Nx:Fx)$, since we will then have no way to keep the two Fs tied together.

[37]This might seem to limit the domain to the types of tokens that actually exist. This is what Hale and Wright (2001a, pp. 422–3) call the "Problem of Plenitude". My own view is that it ceases to be a problem once the strict form of abstractionism we are discussing has been abandoned in favor of a more sophisticated one. We will get to that shortly.

[38]As discussed in Section 9.2, the best version of this view is probably supervaluational.

insist that the truth-condition of "dir(a) = dir(b)" really is the same as that of "$a \parallel b$", we will be unable to argue that names of directions refer to abstract entities or, to put it in the material mode, that directions are abstract.[39] If so, then abstractionism, in this form, is inadequate as an account of our capacity for singular thought about abstract entities. Thought of the sort abstraction makes possible might just as well be about the concrete. Indeed, this sort of abstractionism looks suspiciously like Berkeley's, and he was no friend of the abstract (see Section 9.1, pages 204–206).

There is another route to much the same conclusion, one that goes through the bad company objection, mentioned earlier. The thought is very simple: If abstraction can make the truth-condition of "$Nx : Fx = Nx : Gx$" the same as that of "$\mathrm{Eq}_x(Fx, Gx)$", then why can't it also make the truth-condition of "$\hat{x}(Fx) = \hat{x}(Gx)$" the same as that of "$\forall x(Fx \equiv Gx)$"? Wright has replied, reasonably enough, that it was never any part of the view that such stipulations must always succeed. In the case of extensions, we might just regard the attempt as a failure (Wright, 2001d, p. 281). The problem then becomes to distinguish the good cases from the bad ones. Anyone familiar with this problem will know, however, that the sorts of conditions that have been proposed are highly complex and, in general, are not ones whose satisfaction is easily determined.[40] That means that we can very easily find ourselves in a position where we do not, and perhaps even cannot, know whether a particular abstraction satisfies the condition proffered and so has succeeded in making it possible for us to refer to objects of whatever kind is at issue.

There are, however, two different ways to understand what failed abstractions fail to do. Wright's various discussions of this issue are most naturally read, it seems to me, as suggesting that failed abstractions do not even specify a *sense* for the expressions they characterize.[41] So, for example, an attempt to introduce extensions by abstraction would, on this view, fail even to give a sense to "$\hat{x}(Fx)$" and so would fail to introduce even the *concept* of an extension. This kind of view is not without precedent. One might compare it to Gareth Evans's views about empty demonstratives. On Evans's view, if one is hallucinating a little green man and attempts to venture the thought *that man is laughing at me*, one fails thereby to think any thought at all (Evans, 1985c). But, for the sorts of reasons given by Gabriel Segal (2000), that view strikes me as indefensible, and similar considerations apply here.

---

[39]The problem is even worse in the second-order case, since then the representative tokens are concepts. So we cannot even argue that names of numbers refer to objects, that is, that numbers are objects.

[40]See the postscript to Chapter 10 for some of the history.

[41]For example, in one such discussion, Wright (2001d, pp. 281–2) characterizes abstraction as an attempt "to fix a new concept" and says that bad cases "misfire" or "abort". That suggests that no new concept is fixed in those cases.

Change history slightly.[42] Suppose that, instead of abandoning abstractionism, Frege had embraced it for both numbers and extensions but, for reasons of simplicity and convenience, decided to define numbers as extensions anyway. *Grundgesetze* then ends up being pretty much as it is. In particular, the formal arguments are completely unchanged. Question: Are we really to suppose that these arguments would have expressed no thoughts whatsoever? that, when Frege thinks to himself, "So $a \in \hat{x}(Fx) \equiv Fa$", he isn't actually thinking *anything*? Much the same question can of course be raised about the actual Frege and the actual *Grundgesetze*.[43] But the question has special force for the view I have attributed to Wright: the view that abstraction is a legitimate form of concept formation that just happens to misfire in some cases and so fails to introduce any concept at all. The sort of intellectual and cognitive activity abstraction makes possible does not seem to differ between the good cases and the bad ones, at least so long as we do not know that we are in a bad case.[44]

A quite different view would allow that, in both the good and the bad cases, abstraction can fix a *sense* for the introduced names, but that only in the good cases does it manage to provide those names with *reference*. If that is right, however, then it follows immediately that abstraction cannot, in fact, make the truth-condition of "$\hat{x}(Fx) = \hat{x}(Gx)$" the same as that of "$\forall x(Fx \equiv Gx)$" and, by parallel reasoning, cannot make the truth-condition of "$\mathrm{N}x : Fx = \mathrm{N}x : Gx$" the same as that of "$\mathrm{Eq}_x(Fx, Gx)$".

Isn't that conclusion just incompatible with abstractionism? With its original letter, yes, but not necessarily with its spirit. It depends upon how the extra content of "$\hat{x}(Fx) = \hat{x}(Gx)$" is to be understood. The obvious thought is that what distinguishes "$\hat{x}(Fx) = \hat{x}(Gx)$" from "$\forall x(Fx \equiv Gx)$" is that the former is committed to the existence of an extension: the common extension of $F\xi$ and $G\xi$. One way to implement this idea is to regard abstraction as conditioned on the existence of the relevant objects. So, roughly speaking, "$\hat{x}(Fx) = \hat{x}(Gx)$" will mean something like: If $\hat{x}(Fx)$ exists, then $\forall x(Fx \equiv Gx)$. Hartry Field makes a proposal of this sort, and Wright (2001a) objects that it is just incoherent: If the point of the

---

[42]Another argument here could be based upon the fact that even Law V is not inconsistent unless we accept sufficiently strong comprehension axioms (Heck, 1996). One might respond that the sort of comprehension needed in that case (Π_1^1 comprehension) is also needed for the proof of Frege's Theorem. But it seems plausible that other sorts of examples would require even stronger comprehension axioms (or even choice principles), and nothing stronger than Π_1^1 comprehension is needed for the proof of Frege's Theorem.

[43]Exactly what one might want to say about the actual case depends upon how one understands Frege's introduction of the smooth breathing. There is room, I think, for the view that Frege really does fail to explain it, since he barely gestures at what it is supposed to mean. In so far as we do understand it, perhaps that is because we fall back on the abstractionist explanation Frege himself disowns. If so, that looks like additional evidence in favor of the conception for which I am arguing.

[44]Then, I take it, it's like tic-tac-toe. You know how to play the game, but you can't really play it any more.

abstraction is to introduce (or characterize) the concept of extension, we cannot use that concept in stating the truth-condition of "$\hat{x}(Fx) = \hat{x}(Gx)$". Fair enough. But it would surely be enough to evade this objection if it could be argued that a conception of what it is for extensions to exist was implicit in the abstraction principle itself. And it would be enough, too, if such a conception could be extracted from additional materials similar in character to an abstraction principle.

Let me not try to explain here how that might go. Chapter 9 contains what I have to say about the matter. For present purposes, what matters is just the general idea: that we should seek a view that gives substantial content to the idea that directions, or shapes, or numbers exist as abstract objects (not just as representative tokens), so that the existence of such entities will *not* simply be a consequence of the fact that there are lines, or figures, or concepts. We have to do so, I have argued, if we are to have any hope at all of making sense of how an abstraction principle might fail to be true.

Doing so would also put us in a position to respond to the semantic form of the Caesar problem we were discussing earlier. We saw there that there is a promising strategy by means of which the abstraction-ist might explain quantification over directions, but that this strategy is equally available to someone like Berkeley, who wants to regard 'thought about directions' not as thought about something abstract but as abstract thought about something concrete. If "dir(a) = dir(b)" does not have the same truth-condition as "$a \parallel b$", however, then "$\exists d(d = \mathrm{dir}(b))$" does not have the same truth-condition as "$\exists l(l \parallel b)$". Unlike the latter, the former quantifies over, and therefore is committed to the existence of, directions. To say so would be completely unhelpful if we did not understand what such a commitment might involve, and abstractionism in its original form makes such an understanding unavailable in principle: The original view was that one need understand no more than the abstraction principle in order to have a concept of direction, that is, to understand what directions are and so to understand what it means to say that some (or most or all) directions are thus and so. But Chapter 9 is an attempt to provide precisely this missing piece: An account of what else we need to know to understand what it is for directions to exist as abstract objects.

Some will object that, no matter how successful this attempt, it can do nothing to secure the existence of abstracta. Indeed, it might be thought that the broadly epistemological approach I have been developing flatly misunderstands the nature of the problem, which is ontological or meta-physical: No account of our apparent capacity for singular thought about abstract entities, however successful, can do anything to guarantee that we actually succeed in thinking about anything abstract, because no such account can guarantee that the requisite objects exist. But that charge, I counter, misconstrues both the dialectical situation and what we should regard as the legitimate aspirations of philosophy.

First, it is admitted on all sides that ordinary thought is replete with what looks *prima facie* like singular reference to the abstract, and, if we are not to take it at face value, we need to be given reason not to do so. I know of no reasonably convincing route to skepticism about the abstract other than the epistemological one (see Section 8.1). I am not going to insist that the burden of proof here is on the nominalist; I hate burden of proof arguments. Nonetheless, the first order of business must surely be to answer the question how we should conceive of singular thought about entities with which we have no causal interaction. If we can come up with a plausible answer, then it is unclear what reasons might remain for doubt about the abstract.

Second, there have been several attempts in the last few years to articulate a conception of 'meta-ontology' that might serve the goals of abstractionism and so Neo-logicism. The two most prominent are due to Ted Sider and Matti Eklund. Sider (2007) proposes that Neo-logicists should adopt the view that the meanings of quantifiers are not fixed but can vary and be extended as new sorts of 'objects' are introduced. Eklund (2006) proposes instead that Neo-logicists should be ontological 'maximalists', who believe that everything that can exist does exist. But neither alternative looks plausible as an interpretation of extant Neo-logicists, or so Hale and Wright (2009b) themselves have claimed.

From the very outset, Wright's version of Platonism has been based upon a form of quietism—or, better, small-n naturalism—according to which philosophy simply ought not to be in the business of questioning the results of the sciences. Consider, for example, this remark:

If. . . certain expressions in a branch of our language function syntactically as singular terms, and descriptive and identity contexts containing them are true *by ordinary criteria*, there is no room for any ulterior failure of 'fit' between those contexts and the structure of the states of affairs which make them true. So there can be no philosophical science of ontology, no well-founded attempt to see past our categories of expression and glimpse the way the world is truly furnished. (Wright, 1983, p. 52, my emphasis)

The emphasized phrase, "by ordinary criteria", is the key to Wright's thinking here. Whether arithmetical claims really are true is simply no part of what he thinks is at issue philosophically. Here's Wright's account of what is:

For Frege, the question is. . . how we get into cognitive relations with the states of affairs which make number-theoretic statements true: a question which he rightly saw as calling for a systematic account of the content of those statements, and to which his logicism was offered as an answer. (Wright, 1983, p. 52)

The philosophical issue, then, is whether arithmetical thoughts really do involve singular thought about numbers, as they appear to do, and, if so, how such singular thought should be understood.

So, if the question is supposed to be how we know that there are any directions, or word- and sentence-types, or numbers, the answer is: Not by doing philosophy. Whereas Wright supposes, however, that the existence of such objects is essentially trivial—a consequence of the reflexivity of the equivalence relation that features in the abstraction principle—my view, which is developed in Chapter 9, is that the question whether such objects exist is ultimately a question about the extent to which substantive theorizing about them is possible.

1.2.4   The Existence of Successors

Where, then, does that leave us as regards the success of Neo-logicism?

As was mentioned earlier, the abstractionist strategy for explaining quantification over types simply does not apply when the abstraction principle is understood impredicatively. That means that the strategy does not apply, in particular, to HP, as it must be understood for the proof of Frege's Theorem: We have to be able to "count numbers" for the proof of the existence of successors to go through. The direct approach therefore does not work.

But an indirect approach might yet. As is argued in Section 6.2, the proofs of the other axioms do go through, even if we formulate HP predicatively. So, if the form of abstractionism to which I have committed myself is in fact defensible, then Neo-logicism can claim significant if limited success: We can explain how singular thought about numbers is possible and how some of the most basic laws of arithmetic are implicit in the very nature of such thought. And, as is argued in Section 6.3, construing HP predicatively does not prevent us from counting numbers. Whereas the original form of HP will apply only to concepts true or false of 'basic objects'—e.g., people, or whatever we can think about prior to becoming familiar with numbers—a new form of HP can be formulated that will apply to concepts true or false of numbers. The question is: How should we understand the relation between the number of basic objects that are F and the number of numbers that are ϕ? If we are going to continue thinking of everything predicatively, then these have to be understood as numbers of different kinds, so that the the question whether they are the same is just unintelligible. Clearly, however, that is not how we think, and it is completely obvious when those numbers will be identical: They will be the same just in case there is a one-one correspondence between the Fs and the ϕs.

It is the status of this last move that is critical. With what right, the question must be, do we here loosen the restriction on the predicativity of abstraction? It should be obvious that such loosening will not always be a good idea: A parallel move involving extensions would lead to inconsistency. So it would not be unreasonable to wonder at this point if any real progress has been made, if the goal was a logicism that would encompass the infinity of the numbers. But that need not be the goal, and I would insist, myself, that the epistemological interest of Frege's

Theorem is quite independent of the fortunes of Neo-logicism. As Frege himself makes wonderfully clear, the point of "looking for the fundamental principles or axioms upon which the whole of mathematics rests" is that, once this question has been answered, "it can be hoped to trace successfully the springs of knoweldge on which this science thrives" (PCN, op. 362). Logicism was where Frege hoped this project would lead us, but it is not intrinsic to the project itself. If HP is in some sense the fundamental principle on which our knowledge of arithmetic rests, then that fact is, it seems to me, of tremendous epistemological significance, whatever the epistemological status of HP itself (or of the more sophisticated articulation of it we have just been considering).

Nor, however, do I think that we should despair of a more encompassing logicism. Perhaps the sort of identification that is needed here, between the numbers of basic objects and the numbers of numbers, could be regarded as embodying some sort of claim that, though substantial, needs no independent warrant, but is somehow a conceptual hostage to fortune. I do not know how to develop that vague suggestion, but I think it is worth a look.

1.3 What Does HP Have To Do With Arithmetic?

"If HP is in some sense the fundamental principle on which our knowledge of arithmetic rests...". What does that mean? In what sense might HP ground arithmetical knowledge?

I do not mean to imply, by the way, that Frege's Theorem can be of no *philosophical* significance unless it is of *epistemological* significance. At the end of Chapter 7, I myself suggest, borrowing some suggestions from William Demopoulos (1998, 2000), that Frege's Theorem might feature in an explanation of why the finite cardinal numbers satisfy the Dedekind-Peano axioms. Even in this case, however, it seems clear that the success of such an explanation depends upon HP's being, in some appropriate sense, more fundamental than the Dedekind-Peano axioms and, indeed, upon HP's being *the* fundamental fact about cardinal numbers. Whether it depends, as Demopoulos (1998, p. 483) writes, upon "the thesis that [HP] expresses the preanalytic meaning of assertions of numerical identity" is not so clear.[45]

I doubt, however, that Frege would have been satisfied with only this much. His focus is clearly on an epistemological issue. This is evident at the very beginning of *Die Grundlagen*, where Frege describes his project in terms of epistemological categories that he borrows from the Kantian tradition (Gl, §3), but one might dismiss or re-interpret this allusion.[46]

---

[45]A similar claim is made by Benacerraf (1995, p. 46), though his is made on behalf of Hempel.

[46]Benacerraf (1995, pp. 53–7) is undoubtedly correct that Frege's own conception of these

One cannot, however, ignore the centrality of the question how numbers are given to us and how Frege uses that question to motivate HP (Gl, §§62ff). My own thinking about the philosophical significance of Frege's Theorem has thus always focused on the question what, if any, epistemological significance it has. And for *that* purpose, it has always seemed obvious to me that it *is* essential that HP should capture the ordinary meaning of claims of numerical identity: If the question is how numbers *are* given to us,[47] and if the answer is supposed to be that our capacity for singular thought about them is to be explained in terms of our taking the truth-condition of "$Nx : Fx = Nx : Gx$" to be (almost) the same as that of "$Eq_x(Fx, Gx)$", then the view just is that HP (or something very like it)[48] is not just implicit in our ordinary thought about numbers but is partially constitutive of our capacity for such thought.

Even if that could be established, nothing yet follows about the epistemological status of our arithmetical knowledge. Forget about analyticity and focus instead on the question whether our arithmetical knowledge is *a priori*. Even if HP is a conceptual truth in the sense that it is implicit in singular thought about numbers, it does not follow from the fact that the Dedekind-Peano axioms can be derived from HP that anyone knows those axioms *a priori*. To use a now familiar distinction, what would follow, at most, is that anyone capable of singular thought about numbers *has justification* for the Dedekind-Peano axioms,[49] not that anyone *is justified* in believing them, let alone justified *a priori*. For that to follow, HP would have to play the right sort of justificatory role with respect to arithmetical knowledge.[50] What sort of justificatory role is not at all clear to me, in part because I am no expert on epistemology. But the question does need to be asked.

Moreover, there are questions to be asked about Frege's definitions of zero, predecession, and finitude. The first of these seems pretty rea-

---

categories is importantly different from that of his predecessors. But Benacerraf sometimes gives the impression that Frege is ultimately uninterested in epistemology—see especially the discussion of "metaphysical" dependence on pages 55-6—and with that I do not agree, for the reasons about to be given.

[47] Frege's question is *not*: How *might* numbers be given to us in such a way that we *might* have *a priori* knowledge of them?

[48] This caveat, and the previous one, of course reflect my view, discussed in Section 1.2.3, that we do not really take "$Nx : Fx = Nx : Gx$" and "$Eq_x(Fx, Gx)$" to have the very same truth-condition. The qualification is not critical here, however, and so I will ignore it henceforth, in order to simplify the exposition.

[49] It is a substantial issue whether even this much is true, because it is a substantial issue whether 'having justification' is closed under logical consequence. It seems to me, however, that, if we are discussing mathematical and logical knowledge itself, then this kind of closure principle will make the notion useless. So I think there is no defensible thesis in this vicinity.

[50] Linnebo (2004, p. 168) makes an even stronger claim: "For Frege's Theorem to [establish that arithmetical knowledge is *a priori*], ... its proof must have at least a reasonable claim to being just an explication of our ordinary arithmetical reasoning". That is probably too strong.

sonable. The other two, however, are disputable. The defintion of finite, or natural, number has of course been actively disputed since not long after Frege gave it. The most famous criticisms were due originally to Poincaré, and similar concerns have been voiced ever since by those with predicativist leanings.[51] What is perhaps less obvious is that there are questions to be asked about Frege's definition of predecession, questions that focus on its logical complexity. The definition is

$$Pab \equiv \exists F \exists y[b = \mathbf{N}x : Fx \wedge Fy \wedge a = \mathbf{N}x : (Fx \wedge x \neq y)]$$

and the worry, first expressed by Linnebo (2004, pp. 172–3), is that the presence of the existential second-order quantifier here (that is, the fact that P, so defined, is Σ_1^1) makes the content of this definition depend too much on what comprehension axioms we have available: a may precede b if we accept certain comprehension axioms, but not if we do not.

Finally, there are questions to be asked about the logic used in the proof of Frege's Theorem. Traditionally, the question has been asked in the form: Is second-order 'logic' worthy of the name? But this is not the best form in which to raise the question. Rather, the question should be: Does the sort of reasoning employed in the proof of Frege's Theorem preserve whatever nice epistemological property one thinks HP has, in virtue of its being implicit in singular thought about numbers? What must be shown here depends upon what one thinks that nice property is. But it is not unreasonable to suppose that, if the 'logic' needed for the proof of Frege's Theorem does deserve the name, then the nice property will indeed be preserved, and I tend myself to discuss the question in those terms.

The questions just mentioned about Frege's logic and definitions are addressed in Chapter 12. It turns out that much less than full second-order logic is needed for the proof of Frege's Theorem. The power of second-order logic derives from the 'comprehension axioms', which are of the form:

$$\exists F \forall x[Fx \equiv \phi]$$
$$\exists F \forall x \forall y[Fxy \equiv \phi]$$

and so forth, where ϕ is some formula not containing F free.[52] Each of these axioms asserts that a given formula defines a 'concept' or 'relation': something in the range of the second-order variables. Sub-systems of second-order logic arise from restrictions on comprehension, that is, on what sort of formula ϕ may be. If, for example, we require ϕ not to contain bound second-order quantifiers, we have predicative second-order logic.

---

[51]The issue is raised, for example, by Parsons in "Frege's Theory of Number" (Parsons, 1995a, §VIII) and by Hazen in his review of *Frege's Conception* (Hazen, 1985, pp. 252–3).

[52]It generally will contain the indicated first-order variables free, and it may also contain additional free variables as parameters.

If we require ϕ to be of the form $\forall F \ldots \forall G \phi$, where ϕ contains no second-order quantifiers, then we have Π_1^1 comprehension. And so forth.

The natural question to ask is then: What is the weakest natural logic in which Frege's Theorem can be proven? The answer turns out to be that we need no more than Π_1^1 comprehension. By itself, that does not help very much, but there is a very different sort of logic that, since it has the same proof-theoretic strength, is also adequate for the proof of the axioms. This logic, which I call "Arché logic", has a stronger claim to count as 'logical' than full second-order logic does, because the standard challenges to second-order logic's right to the name do not apply to it. Moreover, Frege's definitions both of predecession and of the ancestral can be stated in such a system in an extremely natural way. The apparent complexity of the definition of predecession is then revealed as illusory.[53] The definition of the ancestral has an elegance that the usual definition lacks and that goes a long way towards making it plausible that this definition really does capture something fundamental about the notion of finitude.

I want to focus here, however, on whether there really is any sense in which an acceptance of HP is implicit in arithmetical thought. The issue seems to me to be critical, but it is poorly understood. Before we turn to that question, however, we need first to address a different one, namely, whether a Neo-logicist really does need to claim that HP is implicit in arithmetical thought. Wright has strenuously resisted this claim. Consider, for example, these remarks:

Grant that a recognition of the truth of [HP] cannot be based purely on analytical reflection upon the concepts and principles employed in *finite*[54] arithmetic. The question, however, surely concerned the reverse direction of things: it was whether access *to* those concepts and validation *of* those principles could be achieved via [HP], and whether [HP] might in its own right enjoy a kind of conceptual status that would make that result interesting. (Wright, 2001b, p. 321, emphasis original)

Wright goes on to list "four ingredient claims" he says constitute Neo-logicism, the last of which reads: "[HP] may be laid down without significant epistemological obligation: ... it may simply be stipulated as an explanation of the meaning of statements of numerical identity..." (Wright, 2001b, p. 321).

---

[53] A similar treatment can be applied to addition and multiplication, by the way. In the case of addition, for example, the definition we want is:

$$\neg \exists x (Fx \wedge Gx) \rightarrow Nx : Fx + Nx : Gx = Nx : (Fx \vee Gx)$$

Just as in the case of predecession, it is only to formulate this for the general case of "$a + b$" that we need existential second-order quantifiers.

[54] The finitude of the arithmetic does not seem to play any significant role in this particular objection, so far as I can see, but there is another line of objection in which it might figure. See p. 36, below.

The emphasis on stipulation, which one finds in many of Wright's discussions (with and without Hale), can, to some extent, be interpreted charitably. In a way, it is but a familiar sort of idealization, one that abstracts from psychological contingencies deemed philosophically irrelevant.[55] But Wright's appeal to our conceptual freedom does more substantial work. In the quote above, Wright is insisting that the important question is not whether HP *does in fact* play some central role in ordinary arithmetical thought, but whether someone *might* acquire the sort of arithmetical knowledge we have by "stipulat[ing HP] as an explanation of the meaning of statements of numerical identity" and then proving the Dedekind-Peano axioms. The interesting questions are therefore supposed to be: Could one gain access to arithmetical concepts by committing oneself to using expressions of the form "$Nx : Fx$" as HP instructs and then, by rehearsing the proof of Frege's Theorem, validate the Dedekind-Peano axioms? If so, we are told, then HP "might. . . enjoy a kind of conceptual status that would make that result interesting". But my problem is that I do not see what sort of interest *any* result of that sort might have.

Grant that someone previously innocent of numerical concepts might stumble upon HP and decide to regard it as explanatory of the concept of cardinal number. Grant that such a person might then discover the proof of Frege's Theorem and so arrive at *a priori* knowledge of the Dedekind-Peano axioms. So what? As Kripke (1980, pp. 34–5) famously pointed out, "*a priori*" is an epistemic adverb. Truths are not what are *a priori*. It is this or that person's belief that is or is not justified *a priori*. At best, then, Wright's approach leads to the conclusion that it is possible for someone to have *a priori* knowledge of the basic laws of arithmetic. But did no one know those laws *a priori* before 1983? Do more than a handful of people now? So if, as MacBride (2000, p. 158) insists, the Neo-logicist "project was never to uncover *a priori* truth in what we ordinarily think, but to demonstrate how *a priori* truth could flow from a logical reconstruction of arithmetical practice", I find myself wondering why we should care. The point here, which is essentially Quine's (1969), emerges from skepticism about the very idea of 'rational reconstruction'.[56]

And it just isn't clear that HP *can* be "stipulated as an explanation of the meaning of statements of numerical identity" (Wright, 2001b, p. 321). Statements of numerical identity already have perfectly good meanings. If one wants to stipulate HP in an effort to fix the meanings of statements of 'gnumerical' identity, then that is a different matter, but the knowledge one might then develop from HP will not be numerical knowledge but gnumerical knowledge; it will not, that is, include any knowledge even of

---

[55] In particular, it allows us easily to divorce the 'context of justification' from the 'context of discovery'.

[56] I have elsewhere voiced similar concerns about the hypothetical character of radical interpretation, if its point is understood epistemologically (Heck, 2007, §2).

numerical identities. The point, of course, is that, if we are interested in what we know and how we know it, we must individuate the objects of knowledge finely: These are questions at the level of sense, not of reference (though we shall see that there are problems even at that level).

Wright is not, of course, unaware of this point:

...[I]t is one thing to define expressions which...behave as though they express [arithmetical] notions, another to define those notions themselves. And it is the latter point, of course, that is wanted if [HP] is to be recognized as sufficient for a theory which not merely allows of pure arithmetical interpretation but to all intents and purposes *is* pure arithmetic. (Wright, 2001b, p. 322, his emphasis)

But Wright clearly thinks that little needs to be done to defend the thesis that "Nx: ...x..." really does mean: the number of ...s, or something close enough, and that none of the necessary work involves anything like conceptual analysis. Wright claims that "any doubt on the point has to concern whether the definition of the arithmetical primitives which Frege offers...[is] adequate to the ordinary *applications* of arithmetic". And to dispose of that doubt, Wright says, it will suffice to establish all instances of the following schema, which Hale once called "Nq":

$$n_f = \mathrm{N}x : Fx \; \equiv \; \exists_n x(Fx)$$

Here, "n" is a schematic variable for a numeral; $\exists_n$ is the numerically definite quantifier, "There are exactly n"; "n_f" abbreviates Frege's definition of the number that n denotes (so, e.g., "0_f" abbreviates: N$x : x \neq x$). After observing that all instances of Nq can indeed be proven—the proof is by induction on n and is not difficult—Wright then remarks: "That seems to me sufficient to ensure that [HP] itself enforces the interpretation of Fregean arithmetic as genuine arithmetic, and not merely a theory which can be interpreted as such" (Wright, 2001b, p. 322).

But this cannot possibly be sufficient. No such argument can establish any more than that the Frege-ese version of "The number of Fs is n" is *provably equivalent* to the ordinary version, and that is far weaker than showing that they have anything close to the same meaning,[57] or even that "N$x : Fx$" actually denotes a cardinal number. Just what would count as 'getting the meaning close enough to right' is a difficult question, to which we shall return, but the examples I am about to give do not split hairs.

Fix two geometrical points A and B. Now consider the following recursive definition:[58]

---

[57]To be clear, however, I am *not* assuming that there must be strict identity of sense between analysans and analysandum. Something weaker is obviously meant to be sufficient.

[58]Of course, this will only work for the finite case, but Wright too is only talking about the finite case. One might actually wonder why he thinks he can restrict attention to that case, since HP concerns infinite cases, too. This is less of a problem for me, since my own view is that what is implicit in ordinary arithmetical thought is HP restricted to the finite case. But Wright has not exactly been sympathetic to that view.

1.   If $\neg\exists x(Fx)$, then $Nx : Fx = A$

2.   If $\neg Fa$, then $Nx : (Fx \lor x = a) =$ the point bisecting the line between $Nx : Fx$ and B

Relying of course upon geometrical axioms, we can then prove HP (restricted to finite concepts),[59] define "0", "1", and the like exactly in Frege's way, and prove all instances of Nq. Does that show that "$Nx : \ldots x \ldots$", so defined, means: the number of . . . s, or something close enough?

Similarly, as W. W. Tait (1996) reminded us, the finite cardinals can be defined in terms of the finite ordinals, which is how Dedekind in fact proceeded. Indeed, the point is utterly general. So long as we are assured of the existence of (Dedekind) infinitely many objects, second-order logic will allow us to interpret the Dedekind-Peano axioms and then define the finite numbers in such a way that we can prove Nq. To give just one other example:[60]

1.   If $\neg\exists x(Fx)$, then $Nx : Fx = $ '|'

2.   If $\neg Fa$, then $Nx : (Fx \lor x = a) = Nx : Fx \frown $ '|'

Now the finite numbers are strings of strokes, as they were for Hilbert, and the proofs of HP and Nq will depend upon the laws of syntax.

It is really quite plausible that our grasp of the notion of infinity, and of a recursive process, emerges somehow from our competence with language. But surely the question whether cardinal numbers can be defined as strings of strokes is not answered by establishing Nq. Similarly, it is a serious question whether the ordinals or cardinals are more fundamental, and it is no small virtue of Dedekind's approach that, as Cantor showed, it can be extended into the transfinite. Dummett (1991b, p. 293) insists, on roughly these grounds, that, ". . . if Frege had paid more attention to Cantor's work, he would have understood. . . that the notion of an ordinal number is more fundamental than that of a cardinal number". My own view is different: I doubt that either is more fundamental. And I am not about to let Dummett off the hook by conceding that the provability of all instances of Nq shows that the definition of cardinals in terms of ordinals gets the meanings of cardinal identities close enough to right.

The short version, then, is that there are just too many ways to define "$Nx : Fx$" that make Nq provable, and very few of them have any claim at all to count as defining a 'genuinely arithmetical' notion.

So it looks as if we are left with the question what conditions a definition of "$Nx : Fx$" must satisfy if it is to count as introducing a genuinely

---

[59]This will rely upon a specification of what a finite concept is. One natural specification would use the (third-order) ancestral to define finitude recursively: Empty concepts are finite, and so are the results of adjoining an object to a finite concept. Then the proof of HP just uses the form of induction this definition makes available.

[60]Here, the frown, $\frown$, denotes concatenation.

arithmetical notion. I want to urge, however, that this way of putting the question, and the entire emphasis on our freedom to stipulate abstraction principles, is at best extremely misleading. It makes it seem as if the crucial questions are hypothetical: how reference to numbers, and *a priori* knowledge of their properties, *might* be possible. But if these things are possible, then that is because they are *actual*. We really do refer to numbers in our ordinary use of arithmetical language, and many of us we really do have *a priori* knowledge of some of their properties. The really interesting question is not how that might work but how it *does* work.

Abstractionism, as I understand it, offers an answer to part of this question. It is the view that our capacity for singular thought about types rests upon our appreciation of the truth-conditions of identity statements concerning them, where those truth-conditions can (almost) be given in the form of an abstraction principle. Maybe no form of abstractionism is true. But if, as I think, and as Wright has spent a lot of time arguing, some form of abstractionism is true, then our *actual* capacity for singular thought about numbers rests upon our *actual* appreciation of the (near) truth of some abstraction principle concerning numbers, and then the question becomes: Which one?

This question is, broadly speaking, empirical and psychological, but that makes it no less philosophical, and much of the work collected here is directed at it, in one way or another. I argue in Chapter 7 that, if we think of HP itself as restricted to the finite case, an appreciation of the connection between cardinality and equinumerosity that it reports really is fundamental to thought about cardinality. Unfortunately, that positive contribution, of which I am personally quite fond, has generated much less attention than a negative argument given in Chapter 11, where I claim to show that HP, if *not* understood as restricted to the finite case, cannot be what underlies arithmetical knowledge.

The worry, which Wright (2001b, p. 317) calls the "concern about surplus content", was first articulated by Boolos. After mentioning that HP is logically stronger than the Dedekind-Peano axioms, Boolos (1998d, p. 304) asks, "Faced with [such] results, how can we really want to call HP analytic?" It is not entirely clear why there should be any special problem for the Neo-logicist here. I suggest in Section 11.2, however, that what was really bothering Boolos rests upon a vague but nonetheless intelligible thought to the effect that the 'foundation' for a given discipline should not outstrip the discipline itself. For example, if someone said that our knowledge of elementary arithmetic rested upon our knowledge of Zermelo-Frankel set theory (ZF), one might reasonably reply that ZF is just way stronger than is needed, and so, while arithmetic might be founded on *part* of ZF, it surely isn't founded upon all of it.

In this particular case, we know how to proceed. ZF has lots of axioms, so we can look at which ones are used in interpreting arithmetic and which are not. It turns out that we can isolate a natural fragment of ZF, known as hereditarily finite set theory, and show that it is equi-interpretable with PA. That makes the view that arithmetic is founded on hereditarily finite set theory much more reasonable than the view that it is founded on ZF. The situation with HP is more difficult. HP is a single axiom, so it is less clear how we might weaken the foundation in this case. As it happens, there is a way forward: We can restrict HP's claims about identity of cardinality to the case of finite concepts, and then show that this new principle is equi-interpretable with something very close to (second-order) PA.[61]

Wright (2001b, pp. 317–18), however, expresses doubts about this entire line of argument. But I, in turn, have trouble understanding his doubts. The thought was just that, if HP really is at work in ordinary arithmetical thought, then there ought to be evidence of *its* being at work there, rather than some weaker principle that can do the same job.[62] Here is the way I put the point in Chapter 11:

> If Frege's Theorem is to have the kind of interest Wright suggests, it must be possible to recognize the truth of HP by reflecting on fundamental features of arithmetical reasoning—by which I mean reasoning about, and with, *finite* numbers, since the epistemological status of arithmetic is what is at issue. For what the logicist must establish is something like this: That there is, implicit in the most basic features of arithmetical thought, a commitment to certain principles, the (tacit) recognition of whose truth is a necessary precondition of arithmetical reasoning, and from which all axioms of arithmetic follow. (below, p. 245)

I then claim, on the basis of Boolos's results, that "...no amount of reflection on the nature of arithmetical thought could ever convince one of HP, nor even of the coherence of the concept of cardinality of which it is purportedly analytic" (below, p. 246).

Part of the argument here involves my insisting, in the second half of the long quote, on a concern with our *actual* arithmetical knowledge. Wright, as we saw, is uninterested in our actual knowledge, but we have already discussed that issue. The insistence on the restriction to *finite* arithmetic can be questioned, however. But before we get to that, let me clarify and correct my view.

As Wright is interpreting it, the argument from surplus content is precisely that: It is based upon a prohibition of surplus content.[63] I think

---

[61]I suspect that the formal arguments given for this claim could be greatly simplified if the definition of finitude mentioned in footnote 59 were used instead of the one I actually use in Chapter 11.

[62]Compare the sort of argument Evans (1982, §2.4) gives against relativizing reference: If reference were relativized to a world, say, then that would make certain readings of sentences possible that we never in fact get, so it would be a mystery why not.

[63]So "no surplus content" is supposed to be a necessary condition for successful identification of the foundation of a discipline. Wright also discusses the question whether lack of surplus content is *sufficient* for some sort of analyticity. I am not sure why that question is raised, however. I do not, so far as I can see, commit myself to such a sufficiency claim, and it does not seem very plausible.

it is an interesting question whether such a principle can be sustained, and we shall return to this question at the end of this section. But the argument was never meant to be so easy. I do not draw the conclusion that HP cannot ground our arithmetical knowledge simply from the logical facts established by Boolos. That "no amount of reflection on the nature of arithmetical thought could ever convince one of HP" is meant to follow from the fact of surplus content. But the crucial point was supposed to concern a closely related *conceptual* gap between finite arithmetic and the theory of infinite cardinality first introduced by Cantor. The reason HP has surplus content is that it answers the question when *infinite* concepts have the same cardinality. It is the fact that HP answers this question that is the problem, not the fact of surplus content to which it gives rise: Because HP answers that question the way it does, the concept of cardinality it characterizes is the Cantorian one, and what I argue in Chapter 11 is that the concept of cardinality that Cantor introduced cannot be what underlies our knowledge of finite arithmetic, because plenty of people have the latter who do not have the former.

This part of the argument can be challenged. MacBride (2000) suggests, in particular, that I overlook evidence that the Cantonian concept of cardinality is already implicit in ordinary arithmetical thought. For the reasons given in the Postscript to Chapter 11, however, I disagree, though I was no doubt too quick to draw the conclusion I did. I also think that any suggestion I might have made that the implicit commitments of ordinary arithmetical thought must be extracted by purely *philosophical* reflection should be rejected, since psychological phenomena of the sort discussed in Chapter 7 are also important.

A different way to question the argument, and one that seems implicit in some of Wright's reflections, is to ask whether finite arithmetic is really the right focus of investigation,[64] the alternative being something like the "general theory of cardinality". The idea, I take it, would be that *finite* arithmetic is not, so to speak, a natural epistemological kind; the natural kind is cardinal arithmetic generally. So what we would need to know to resolve this kind of dispute is how we should circumscribe what Frege would have called a "branch of knowledge" in such a way that its foundations might sensibly be investigated separately from those of other "branches of knowledge". Why, for example, is it at least plausible to isolate arithmetic from geometry (or syntax) and offer different accounts of our knowledge of each? Why should arithmetic, in the now familiar sense, be regarded as separable from real (or even complex) analysis? I suspect that good answers to these questions would have to draw upon psychology as well as philosophy, and I do not myself have any clear idea what such answers might be like. Nonetheless, I am reasonably confident that finite arithmetic does constitute an isolable body of knowledge in

---

[64]This is the other line of argument mentioned in footnote 54.

this sense, and the history of foundational studies, both before and after Cantor, seems to provide a good deal of evidence for this claim.

That said, I think we can now see why a prohibition on surplus content might seem reasonable. If the 'basic laws' that allegedly underlie a given, *properly circumscribed* body of knowledge were sufficient to provide for a sort of knowledge that we simply do not find exhibited, that would be reason to doubt that those really were the basic laws of that discipline. This would not be conclusive reason. We are dealing, in effect, with the distinction between competence and performance, and we might have reason to think that the competence, though there, was for some reason not manifested in performance. But one would have to tell a story about why not.[65] There is not, then, a prohibition on surplus content, but its presence should serve as a warning that something is amiss. In the present case, it is just such a warning, and there is something amiss.

1.4   Logicism and Neo-Logicism

Much of the literature on Neo-logicism has been concerned with the question whether its treatment of arithmetic can be mimicked in other areas, such as real analysis and set theory.[66] The question is not uninteresting, but my sense has long been that it has been accorded much too much significance. Casual presentations of logicism often present it as the view that "mathematics is logic". But Frege would have disagreed, since he regarded geometry—the field in which he was actually trained as a mathematician—as synthetic. By the same token, one's logicism need not encompass set theory or even analysis. Even if all of the various attempts to identify abstraction principles that are sufficient for these theories were to fail, that would in no way undermine the claim that arithmetic itself is analytic. What it would show is that other branches of mathematics depend upon "sources of knowledge" different from the ones upon which arithmetical thought draws. But maybe arithmetic just is special in that sense.

Yet another possibility is that not even all of arithmetic is analytic—that is, known on the basis of a principle constitutive of arithmetical thought—though some of it is. Perhaps no more than Robinson arithmetic is analytic. Maybe only the theory of successor is analytic. In "Ramified Frege Arithmetic" (Heck, 2011), I show that the basic axioms concerning successor can all be proven from HP in ramified predicative

---

[65]The beginnings of a story can certainly be told in the case of HP. One might want to say, much as MacBride (2000, p. 155) does, that HP was the principle guiding use of finite cardinals, but that it conflicted with an intuition regarding parts and wholes that was wrongly extrapolated from finite to infinite arithmetic. But I think there are independent problems with this line of argument. See the Postscript to Chapter 11.

[66]Several papers on these sorts of issues have been reprinted together in *The Arché Papers on the Mathematics of Abstraction* (Cook, 2007).

second-order logic, and in the context of predicative logic, there is no problem of bad company, since even Basic Law V is consistent in a predicative setting (Heck, 1996).[67] The argument has its limitations, however. Not only do we lose induction, for the sorts of reasons mentioned above, but there is also no clear way to define addition and multiplication so that the existence of sums and products can be proven. (Uniqueness is easy.)

However, it can be shown that, even in simple predicative second-order logic, we can, using standard definitions of cardinal addition and multiplication,[68] straightforwardly interpret a purely relational version of the theory known as R. The usual form of R has as axioms all true instances of the following formulae:

$$n + m = k$$
$$n \times m = k$$
$$m < k$$
$$m \neq k$$

where m, n, and k are schematic variables for numerals. In the relational version, we do not have function symbols S, $+$, and $\times$, but relations $P(a, b)$, $A(a, b, c)$, and $M(a, b, c)$ and have as axioms all true instances of:

$$P(n, m)$$
$$A(n, m, k)$$
$$M(m, n, k)$$

plus the assertions that these values are unique.[69] The resulting theory interprets R,[70] so it is sufficient for the numeralwise representability of all recursive functions and is therefore essentially undecidable (Tarski et al., 1953). What "Ramified Frege Arithmetic" shows, then, is that, if we ramify, then we do get the existence of successors and so can replace "$P(a, b)$" with a function symbol again. It would be nice if something similar could be done for sums and products, but, while I have not given up

---

[67]This paper is not reprinted here, as it consists mostly of intricate argumentation conducted in a barely intelligible formalism.

[68]For these definitions, see Burgess's book *Fixing Frege* (Burgess, 2005).

[69]Of course, the numerals cannot now be defined in the usual way—0, $S0$, $SS0$, etc.—since we do not have S. But we can define them via Russellian descriptions:

$$1 \stackrel{df}{=} \iota x(P(0, x))$$
$$2 \stackrel{df}{=} \iota x(\exists y(P(0, y) \wedge P(y, x))$$

and so forth. Since we can prove both existence and uniqueness, the descriptions are proper. The issue is less pressing in the present context, since the numerals will be defined in Frege's way.

[70]Thanks for this information to Albert Visser, whose wonderful paper on R (Visser, 2009) got me thinking about this matter.

hope, and do not have a proof that it cannot be done, my various experiments have left me skeptical.[71]

What we get in the predicative case is therefore non-trivial—essential undecidability is as good a test for non-triviality as I can imagine—but it is not very much. Still, it would be wrong to dismiss these results on that ground. Even these weak results are capable of grounding significant philosophical conclusions, it seems to me, for the simple reason that it is arithmetic's commitment to an infinity of numbers that has always seemed to set it apart from the logical. If reason itself can provide us only with access to an infinity of numbers, while we must draw upon resources from elsewhere to establish much knowledge about them, then, well, that is how things are, and reason will still have proven capable of rather more than empiricists have generally supposed.

The lesson with which I should like to close, then, is one I have already announced but shall now re-iterate: If HP, or something like it, is indeed the fundamental principle on which all arithmetical thought is founded, then that is an epistemological result of great significance, whether or not HP itself enjoys any special epistemic virtue, and whether or not our knowledge of the Dedekind-Peano axioms derives directly from our knowledge of HP.

---

[71]One *can* get the existence of sums and products by restricting the domain to numbers for which sums and products exist. There is a nice presentation of the techniques for doing so, which are originally due to Robert Solovay, in Burgess's book *Fixing Frege* (Burgess, 2005, ch. 2). The difficulty, in the present context, is that this amounts to redefining the notion of natural number. As Visser (2011) notes, there may be a coherent philosophy of arithmetic to be founded on this idea, but it is one that would need to be developed, and it is very unFregean.

2

The Development of Arithmetic in Frege's *Grundgesetze der Arithmetik*

In his *Grundlagen der Arithmetik*, Frege explicitly defines numbers as the extensions of concepts of a certain form: The number of Fs is, according to Frege, the extension of the (second-level) concept "is a concept which can be correlated one-one with the concept F" (Gl, §68). Frege then derives, from this explicit definition, what is known as HP (Gl, §73):

The number of Fs is the same as the number of Gs just in case the Fs can be correlated one-one with the Gs.

Once the proof of HP is complete, Frege sketches the proofs of a variety of facts about numbers. This sketch is intended to show that the fundamental laws of arithmetic can be formally derived from his explicit definition. In the course of sketching these proofs, however, Frege makes no further use of extensions. He appeals to his explicit definition of numbers only in the proof of HP.

It is tempting, therefore, to understand Frege's derivation of the laws of arithmetic as consisting of two parts: a derivation of the laws of arithmetic from HP and a derivation of HP from the explicit definition. If this is, indeed, the right way to interpret Frege, this is of significance for our understanding of his philosophy of mathematics. First, there is a question what the relation between the explicit definition and HP should be taken to be; the need to make a sharp separation between these two parts of his derivation of arithmetic constrains the sorts of answers we can give to this question (Dummett, 1991b, ch. 14). Second, it has been shown that the Dedekind-Peano axioms for arithmetic can indeed be derived, within second-order logic, from HP; moreover, Fregean Arithmetic—second-order logic, with HP taken as the sole 'non-logical' axiom—is consistent.[1] As is well-known, however, the formal theory in which Frege proves the axioms of arithmetic, in *Grundgesetze der Arithmetik*, is inconsistent, Russell's Paradox being derivable in it. Presumably, any axiom governing extensions to which Frege might have implicitly appealed in *Die Grundlagen* would be similarly inconsistent. Nonetheless, Frege does not appeal to extensions in his sketch of the

---

[1] See Chapter 1 for some of the history.

derivations of the axioms of arithmetic in *Die Grundlagen*, except during the proof of HP. He therefore sketches of the derivation of the axioms from HP within a consistent sub-theory of the formal theory of *Grundgesetze*. It is this that led George Boolos to suggest that "the derivability of arithmetic from HP [should be] known as *Frege's Theorem*" (Boolos, 1998k, p. 268).

One ought to be struck by a number of questions at this point. First, do Frege's formal proofs of the axioms of arithmetic, in *Grundgesetze*, also depend only upon HP, except, of course, for the proof of HP itself? Does Frege present a formal proof of Frege's Theorem in *Grundgesetze*? If so, is it just an accident that his proof can be so construed, or did Frege know that arithmetic could be derived from HP? And if he did, of what significance for our understanding of his position is this?

My purpose here is to answer the first of these questions: to show that Frege does indeed derive the axioms of arithmetic, in *Grundgesetze*, from HP, within second-order logic. The latter two questions, whether Frege himself understood the proofs in this way and what significance this fact has for our understanding of his philosophy, are discussed in Chapters 4–6 of this book.

2.1   Basic Law V in *Grundgesetze*

Our question is whether Frege derives the axioms of arithmetic from HP in *Grundgesetze*. We might naturally ask, then, whether, just as in *Die Grundlagen*, Frege makes no use of extensions after deriving HP. So formulated, the answer to our question is "No". In *Grundgesetze*, Frege has an axiom, the infamous Basic Law V, which governs what he there calls "value-ranges". Basic Law V, for our purposes, may be taken to be:[2]

$$(\grave{\epsilon}\Phi(\epsilon) = \grave{\epsilon}\Psi(\epsilon)) \equiv \forall x(\Phi(x) \equiv \Psi(x))$$

So formulated, Basic Law V governs terms which refer to the extensions of concepts (Gg, v. I, p. vii). As said above, every full second-order[3] theory containing Basic Law V is inconsistent.

Terms standing for value-ranges are used throughout *Grundgesetze*: There is hardly a page in Part II on which a term for a value-range does not occur. It is thus just not the case that Frege makes no mention of

---

[2]Strictly speaking, Basic Law V is:

$$(\grave{\epsilon}\phi(x) = \grave{\epsilon}\psi(x)) = \forall x(\phi(x) = \psi(x))$$

Since, for Frege, the truth-values are objects in the domain of the first-order variables, the Law is formulated in terms of the identity of the values of the functions $\phi\xi$ and $\psi\xi$ and of the identity of the truth-value $\forall x(\phi(x) = \psi(x))$ with the truth-value $\grave{\epsilon}\phi(x) = \grave{\epsilon}\psi(x)$.

[3]More precisely, any second-order theory that contains the Σ_1^1 comprehension axioms is inconsistent (Heck, 1996).

value-ranges once he has derived HP. Nonetheless, that does not settle the question whether Frege derives the laws of arithmetic from HP. There are different ways of making use of value-ranges, some of them eliminable, others not. What I intend to show is that, with the exception of the use in the proof of HP itself, *all* uses of value-ranges in Frege's proof of the basic laws of arithmetic can be eliminated in a uniform manner. Moreover, with that exception, Frege uses value-ranges merely for convenience.

Consider the following example, in which "$\Phi_x(\phi x)$" is a schematic variable for a second-level formula ("ϕ" being a placeholder): Suppose we have proven the formulae $\ulcorner \Phi_x(A(x)) \urcorner$ and $\ulcorner \forall x(A(x) \equiv B(x)) \urcorner$; we wish to infer $\ulcorner \Phi_x(B(x)) \urcorner$. There is, in standard axiomatic second-order logic, no uniform way to make this inference: No axiom of standard second-order logic gurantees the extensionality of second-level predicates.[4] That is not to say that $\ulcorner \Phi_x(B(x)) \urcorner$ cannot be derived from $\ulcorner \Phi_x(A(x)) \urcorner$ and $\ulcorner \forall x(A(x) \equiv B(x)) \urcorner$, whatever second-level formula $\Phi_x(\phi x)$ may be. It can be proven by induction on the complexity of the formula $\Phi_x(\phi x)$ that we shall always be able to construct such a proof. But derivations within the formal system will be complicated by the need to prove, for each case in which we wish to make inferences of this sort, specific theorems licensing them.

If we have value-ranges at our disposal, however, matters are much simplified, for Basic Law V acts as a principle of extensionality. Given a second-level concept $\Phi_x(\phi x)$, we may define a related, first-level concept as follows:

$$\Phi(z) \overset{df}{\equiv} \exists F[z = \grave{\epsilon}F(\epsilon) \wedge \Phi_x(Fx)]$$

The predicate "$\Phi(\xi)$" is thus true of an object if, and only if, that object is the value-range of a concept that falls under $\Phi_x(\phi x)$. If we then make use not of second-level concepts but of their first-order relatives, we can argue as follows:

(1) $\Phi(\grave{\epsilon}A(\epsilon))$ Premise

(2) $\forall x(A(x) \equiv B(x))$ Premise

(3) $\grave{\epsilon}A(\epsilon) = \grave{\epsilon}B(\epsilon)$ (2), Basic Law V

(4) $\Phi(\grave{\epsilon}B(\epsilon))$ (1, 3)

(5) $\Phi(\grave{\epsilon}A(\epsilon)) \wedge \forall x(A(x) \equiv B(x)) \rightarrow \Phi(\grave{\epsilon}B(\epsilon))$ (4), [1], [2]

---

[4]Charles Parsons has observed that a similar phenomenon prevents the derivation of a version of Basic Law V for second-level functions

$$\grave{\phi}(\Phi_x \phi x) = \grave{\phi}(\Psi_x \phi x) \equiv \forall F(\Phi_x Fx \equiv \Psi_x Fx)$$

from that for first-level functions. Basic Law V for first-level functions can, on the other hand, be derived from that for second-level functions, *via* type elevation.

Given that Frege is already appealing to Basic Law V, such a use of value-ranges greatly simplifies his formal system.

One might therefore want to say that this use of value-ranges allows Frege to replace second-order quantification with first-order quantification: Quantification over concepts can be replaced by quantification over their value-ranges. It is not at all clear, though, what motivation Frege might have for doing so, unless he intends somehow to eliminate second-order quantification from his system. But second-order quantification is not eliminated in this way: At best, it is merely hidden; at worst, it introduces additional (hidden) second-order quantifiers. It is entirely unclear why Frege should have any interest in hiding second-order quantification when it introduces additional second-order quantifiers. Moreover, second-order quantification occurs explicitly in Frege's definition of the ancestral (Gg, v. I, §§45, 108–9), so he has no general interest even in hiding second-order quantification.

The correct account of what is going on here is as follows: Frege is indeed using value-ranges to represent first-level functions by objects; but he is not doing so in order to replace quantification over functions by quantification over objects. Rather, he wishes to replace expressions for *second*-level functions, such as our "$\Phi_x(\phi x)$", by expressions for *first*-level functions, just as we introduced the first-level predicate "$\Phi(\xi)$" in place of "$\Phi_x(\phi x)$". Frege is explicit about this:

... [I]n further developments, instead of second-level functions, we may employ first-level functions. ... [T]his is made possible through the functions that appear as arguments of second-level functions being represented by their value-ranges.... (Gg, v. I, §34)

Why Frege would care to use first-level functions in place of second-level functions is itself a nice question: Part of the explanation, presumably, is that doing so simplifies his formal system in just the ways discussed above.[5] Such uses of value-ranges, however, are inessential to Frege's proofs and can be eliminated without difficulty.

Most of the uses Frege makes of value-ranges, except in the proof of HP, are of this sort. There is, however, one other kind of use to which he frequently puts them. To explain it, we need to introduce Frege's application-operator. We may write his definition of it as follows:[6]

$$a \frown f \overset{df}{\equiv} \exists F[f = \grave{e}F(\epsilon) \wedge Fa]$$

---

[5]Frege also appears to think that doing so will simplify the meta-theory, since he need not explain his notation for second-order parameters "in full generality" (Gg, v. I, §25). As Parsons once suggested to me, Frege is in effect using value-ranges to avoid the use of *third*-order quantification.

[6]Frege's definition covers not just concepts but functions in general. Note, too, that it is this definition which introduces additional second-order quantification into Frege's system when he uses value-ranges as discussed above.

It is not difficult, using Basic Law V, to prove Frege's Theorem 1:

$$Fa \equiv a \frown \grave{\epsilon}F\epsilon$$

Now, consider the following formula:

$$\forall x(Fx \to (Fx \lor x = c))$$

Using the application operator, Frege might write this formula thus:

$$\forall x(x \frown \grave{\epsilon}F\epsilon \to x \frown \grave{\epsilon}F\epsilon \lor x = c)$$

But he might instead write it so:

$$\forall x(x \frown \grave{\epsilon}F\epsilon \to x \frown \grave{\epsilon}(F\epsilon \lor \epsilon = c))$$

The reason is that, in the context of a given proof, our interest may be focused on the concept: ξ is an F or ξ is identical with c. Frege is emphasizing this fact by using value-ranges in the same way one might use predicate-abstraction. We might, that is, achieve the same effect thus:

$$\forall x[Fx \to \lambda y(Fy \lor y = c)(x)]$$

Such uses, either of value-ranges or of lambda-abstraction, are readily eliminable.[7]

Close examination of the proofs of the axioms of arithmetic in *Grundgesetze* will show that all uses of value-ranges within those proofs are of one of three types:[8]

1.   The ineliminable use in the proof of HP

2.   The use which allows the representation of second-level functions by first-level functions

3.   The formation of complex predicates to emphasize what one is proving

Except for those of the first sort, all these uses of value-ranges are easily, and uniformly, eliminable from Frege's proofs. Frege's proof of the axioms of arithmetic, from HP, therefore requires no essential reference to value-ranges.

---

[7]From a Fregean perspective, we may take the λ-operator to be introduced as a defined second-level relation: $\lambda_y(Fy, x) \overset{df}{\equiv} Fx$. This is the relation between an object and a concept that Frege calls "falling under".

[8]Frege does *not* use Basic Law V as a second-order comprehension principle. Comprehension is built into Frege's Rule 9 (Gg, v. I, §48), which is his Rule of Universal Instantiation (as opposed to his Axiom of Universal Instantiation, Basic Law II), which allows for the uniform replacement of a free variable, of arbitrary type, by any well-formed expression of the appropriate type, containing arbitrarily many other free variables of arbitrary types (subject, of course, to the usual restrictions, which Frege formulates precisely).

2.2   HP and Fregean Arithmetic

In *Die Grundlagen*, Frege formulates HP as follows:[9]

the Number belonging to the concept F is identical with the Number belonging to the concept G if the concept F is equinumerous with the concept G. (Gl, §72)

According to Frege, a concept F is equinumerous with a concept G just in case

there exists a relation ϕ which correlates one to one the objects falling under the concept F with the objects falling under the concept G. (Gl, §73)

And the correlation is one-one just in case (Gl, §72):

$$\forall x \forall y \forall z (\phi xy \wedge \phi xz \to y = z) \wedge \forall x \forall y \forall z (\phi xz \wedge \phi yz \to x = y)$$

Frege further explains that the relation ϕ correlates the Fs with the Gs just in case every F "stands in the relation ϕ" to some G and, conversely, for each G, there is some F which "stands in the relation ϕ" to it. Moreover, each F "stands in the relation ϕ" to some G just in case "the two propositions 'a falls under F' and 'a does not stand in the relation ϕ to any object falling under G' cannot, whatever be signified by a, both be true together. . . " (Gl, §71) Equivalently:

$$\forall x (Fx \to \exists y (Gy \wedge \phi xy)) \wedge \vee y (Gy \to \exists x (Fx \wedge \phi xy))$$

Using "$Nx : \Phi x$" as a second-level functional expression, to be read "The number of Φs", we may thus formalize HP as follows:

$$
\begin{aligned}
Nx : Fx = Nx : Gx \equiv \exists R [& \forall x \forall y \forall z (Rxy \wedge Rxz \to y = z) \wedge \\
& \forall x \forall y \forall z (Rxz \wedge Ryz \to x = y) \wedge \\
& \forall x (Fx \to \exists y (Gy \wedge Rxy)) \wedge \\
& \forall y (Gy \to \exists x (Fx \wedge Rxy))]
\end{aligned}
$$

This version of HP is the one most immediately suggested by Frege's remarks in *Die Grundlagen*.[10]

Frege's formulation of HP in *Grundgesetze* may initially strike one as rather different. But Frege does not conceive of it as a departure from his earlier informal statement. On the contrary, when giving the various definitions needed to formulate HP in the formal system of *Grundgesetze*, Frege himself quotes the passages quoted above (Gg, v. I, §38).

To state the version of HP used in *Grundgesetze*, we need to recall some of those definitions, the first of which is Frege's definition of the

---

[9]Austin's translation has "equal" for "equinumerous". The German is "gleichzahlig".

[10]It is also essentially the version used by Wright (1983, p. 105).

converse of a relation (Gg, v. I, §39):[11]

$$\mathbf{Conv}_{\alpha\epsilon}(R\alpha\epsilon)(x,y) \stackrel{df}{=} Ryx$$

The second definition is that of a relation's being *functional* (Gg, v. I, §37):[12]

$$\mathbf{Func}_{\alpha\epsilon}(R\alpha\epsilon) \stackrel{df}{=} \forall x\forall y(Rxy \to \forall z(Rxz \to y = z))$$

The third definition is that of a relation's *mapping* the objects falling under one concept into those falling under another (Gg, v. I, §38):[13]

$$\mathbf{Map}_{\alpha\epsilon\xi\eta}(R\alpha\epsilon)(F\xi, G\eta) \stackrel{df}{=} \mathbf{Func}_{\alpha\epsilon}(R\alpha\epsilon) \wedge \forall x(Fx \to \exists y(Rxy \wedge Gy))$$

In words: $R\xi\eta$ maps the Fs into the Gs just in case $R\xi\eta$ is functional and each F is related by $R\xi\eta$ to some G.[14]

We should note two important points about this definition. First, "$\mathrm{Map}(R)(F,G)$" states that $R\xi\eta$ is a functional relation which maps the Fs *into* the Gs, not, as might have seemed more natural, one which maps the Fs *onto* the Gs. That would entail that there are at least as many Fs as Gs, whence, if there is a functional relation which maps the Gs onto the Fs, there are just as many Fs as Gs (by the Schröder-Bernstein Theorem). The fact that $R\xi\eta$ maps the Fs *into* the Gs says, of itself, nothing whatsoever about the relative cardinalities of the Fs and the Gs: So long as there is at least one G, there will always be a relation which maps the Fs—whatever concept $F\xi$ may be—into the Gs, in Frege's sense.[15]

---

[11]Frege writes the converse-operator as what looks a bit like a script-U and applies it to the (double) value-range of a relation rather than to the relation itself; its value too is the (double) value-range of a relation. As said earlier, Frege uses value-ranges when the relation would be an argument of a *second*-level function. For us, "$\mathrm{Conv}_{\alpha\epsilon}(\Phi\alpha\epsilon)(\xi,\eta)$" refers to a concept of mixed level, which takes a relation and two objects as arguments. I shall usually just write: $\mathrm{Conv}(\Phi)(\xi,\eta)$, when there is no danger of confusion, or, occasionally $\mathrm{Conv}(\Phi\alpha\epsilon)(\xi,\eta)$. I shall use these same conventions throughout.

[12]Frege writes the operator as "I" and again applies it to the double value-range of a relation. Frege's word here is "eindeutig", which is often translated "many-one", but I prefer the translation "functional", especially in light of Frege's definition of mapping, to be mentioned shortly.

[13]Frege writes "Map" as "⟩". The argument of "⟩" is the double value-range of a relation; its value, in such a case, is the double value-range of a relation between value-ranges.

[14]Strictly speaking, Frege's definition is:

$$\mathrm{Map}(R)(F,G) \stackrel{df}{=} \mathrm{Func}(R) \wedge \forall x(\forall y(Rxy \to \neg Gy) \to \neg Fx)$$

Frege works exclusively with this formulation, as the mechanics of his system make it easier for him to do so. There is no rule which allows him to infer (an equivalent of) "$Fx \to \exists y(Rxy \wedge Gy)$" from "$Fx \to (Rxt \wedge Gt)$". He would contrapose to get "$\neg(Rxt \wedge Gt) \to \neg Fx$"; cite "$\forall y \neg(Rxy \wedge Gy) \to \neg(Rxt \wedge Gt)$"; infer "$\forall y\neg(Rxy \wedge Gy) \to \neg Fx$", by the transitivity of the conditional; and contrapose again (Gg, v. I, §17). It is somewhat easier for Frege to prove "$(Rxy \to \neg Gt) \to \neg Fx$" and infer "$\forall y(Rxy \to \neg Gy) \to \neg Fx$", as above.

[15]If Ga, we define: $Rxy \stackrel{df}{=} y = a$. $R\xi\eta$ then maps the whole domain into the Gs and *a fortiori* maps the Fs into the Gs. This example also illustrates the point to follow.

Secondly, "Map$(R)(F, G)$" says that $R\xi\eta$ is functional and that it relates each F to a G. It says absolutely nothing else about $R\xi\eta$: $R\xi\eta$ may, for all we know, map the entire domain into the Gs; it may relate every non-F to itself; it may not relate non-Fs to anything. In reading Frege's proofs, one must keep this fact in mind. Suppose, for example, that the relation $R\xi\eta$ maps the concept $F\xi$ into the concept $G\xi$, that c is not an F and that b is not a G; and suppose we wish to show that some relation maps $F\xi \lor \xi = c$ into $G\xi \lor \xi = b$. What one might be tempted to say is that the relation which is just like $R\xi\eta$, but which relates c to b—that is, $R\xi\eta \lor [\xi = c \land \eta = b]$—accomplishes this. But we cannot proceed so quickly: $R\xi\eta$ may already relate c to something; in particular, it may relate c to something other than b, in which case the relation so defined is not functional and so maps no concept into any other (Gg, v. I, §66). Clearly, this is not an obstacle that cannot be overcome. We shall see an example of the sort of theorem which must be proven in order to overcome it below.

It is easy to formulate HP in terms of the definitions we have just discussed:

$$\mathbf{N}x : Fx = \mathbf{N}x : Gx \equiv \exists R[\mathbf{Map}(R)(F, G) \land \mathbf{Map}(\mathrm{Conv}(R))(G, F)]$$

In words: The number of Fs is the same as the number of Gs if, and only if, there is a relation that maps the Fs into the Gs and whose converse maps the Gs into the Fs. This formulation of HP is easily seen to be equivalent to that discussed above.

What Frege has done is to group "$\forall x \forall y(Rxy \to \forall z(Rxz \to y = z))$" and "$\forall x(Fx \to \exists y(Gy \land Rxy))$" in the first conjunct and "$\forall x \forall y(Rxz \to \forall z(Ryz \to x = y))$" and "$\forall y(Gy \to \exists x(Fx \land Rxy))$" in the second. Had he instead grouped "$\forall x \forall y(Rxy \to \forall z(Rxz \to y = z))$" and "$\forall y(Gy \to \exists x(Fx \land Rxy))$", he would have had the alternative definition of "Map" which states that $R\xi\eta$ is functional and that it maps the Fs *onto* the Gs. We are more inclined nowadays to group the conjuncts "$\forall x \forall y(Rxy \to \forall z(Rxz \to y = z))$" and "$\forall x \forall y(Rxz \to \forall z(Ryz \to x = y))$" ($R\xi\eta$ is a one-one function...) and "$\forall x(Fx \to \exists y(Gy \land Rxy))$" and "$\forall y(Gy \to \exists x(Fx \land Rxy))$" (...from the Fs onto the Gs). Formulating HP in terms of either Frege's definition of Map$(R)(F, G)$, or the alternative, has technical advantages over the modern construal, however. We shall see an example shortly.

2.3   Frege's Derivation of the Axioms of Arithmetic

As in *Die Grundlagen*, Frege's first task in Part II of *Grundgesetze* is to derive HP from his explicit definition of numbers. Having done so, Frege then turns to the proofs of a number of basic truths about numbers. He says, at the beginning of *Grundgesetze*, that, in *Die Grundlagen*,

he "sought to make it plausible that arithmetic is a branch of logic" and
that "this shall now be confirmed, by the derivation of the simplest laws of
Number by logical means alone" (Gg, v. I, p. 1). One might object, rather
facetiously, that the derivation of the *simplest* laws of number hardly
confirms that arithmetic is a branch of logic: It would be more interesting were some really *complicated* laws of number derivable. Presumably,
however, Frege meant by "the simplest laws" not the simplest laws in any
syntactic or psychological sense but the most fundamental laws: those
laws from which all other laws of arithmetic follow. One might suggest,
in fact, that, to show that "arithmetic is a branch of logic", Frege must
show that *every* law of arithmetic can be derived within logic.[16] There is
surely no way to do so without isolating some (presumably, finitely many)
principles, the Basic Laws of Arithmetic, from which all other laws of
arithmetic follow, and then deriving these basic laws within logic. The
basic laws, of course, are just axioms for arithmetic. Frege's demonstration that arithmetic is a branch of logic must therefore rest upon some
axiomatization of arithmetic.

For the moment, we may take arithmetic to be axiomatized by the
Dedekind-Peano axioms. Where "$N\xi$" is a predicate to be read "ξ is a
natural number", and "$P\xi\eta$" a predicate to be read as "ξ immediately precedes η in the number-series", we may formulate these axioms as follows,
using the definitions introduced above:

1. $N0$

2. $\forall x(Nx \rightarrow \exists y(Ny \land Pxy))$

3. $\neg\exists x(Px0)$

4(a). $\text{Func}(P)$

4(b). $\text{Func}(\text{Conv}(P))$

5. $\forall F[F0 \land \forall x(Nx \land Fx \rightarrow \forall y(Pxy \rightarrow Fy)) \rightarrow \forall x(Nx \rightarrow Fx)]$

Frege proves each of these axioms in *Grundgesetze*.

To prove the axioms, we obviously need some additional definitions.
We begin with the definitions of "0" and the relation-sign "$P\xi\eta$". The definitions given in *Grundgesetze* are essentially those given in *Die Grund-*

---

[16]I say "one might suggest" because, so framed, the requirement cannot be met within
any formal theory, so long as we take the "laws" of arithmetic to be the *truths* of arithmetic:
The incompleteness theorem precludes any such demonstration. For present purposes, this
may be ignored, since the point is to argue that Frege is committed to providing an axiomatization of arithmetic. A logicist may not so characterize her project as to make it
self-fulfilling, by stipulating that the theory of arithmetic is the smallest deductively closed
set of sentences that contains the axioms of her theory. Some independently plausible characterization of arithmetic is required. I'll be claiming that Frege has one.

lagen. Frege defines zero as the number of objects which are not self-identical (Gg, v. I, §41; Gl, §74):

$$0 \overset{df}{\equiv} \mathbf{N}x : x \neq x$$

His definition of "$P\xi\eta$" is as follows (Gg, v. I, §43):

$$Pmn \overset{df}{\equiv} \exists F \exists x [Fx \wedge n = \mathbf{N}z : Fz \wedge m = \mathbf{N}z : (Fz \wedge z \neq x)]$$

That is, m precedes n if "there exists a concept F, and an object falling under it x, such that the Number which belongs to the concept F is n and the Number which belongs to the concept 'falling under F but not identical with x' is m" (Gl, §74). We shall return to the definition of "$\mathbf{N}\xi$".

Frege's proofs of the Dedekind-Peano axioms follow, for the most part, the sketch given in *Die Grundlagen*. The proof that $P\xi\eta$ is functional, Axiom 4(a), occupies section B(eta); that its converse is functional, Axiom 4(b), is proven in section Γ.[17] The proofs are straightforward, but I shall say a few things about the proof that $P\xi\eta$ is functional, primarily to illustrate some of the technical points about Frege's use of value-ranges and his definition of "Map". The proof that the converse of $P\xi\eta$ is functional is similar, both in spirit and in detail.

To show that $P\xi\eta$ is functional, we must prove that, if Pxy and Pxw, then $y = w$. Assume the antecedent. Then, by the definition of "$P\xi\eta$", there are a concept $F\xi$ and an object c such that Fc, $\mathbf{N}z : Fz = y$, and $\mathbf{N}z : (Fz \wedge z \neq c) = x$. Similarly, there is a concept $G\xi$ and an object b such that Gb, $\mathbf{N}z : Gz = w$, and $\mathbf{N}z : (Gz \wedge z \neq b) = x$. So the theorem will follow from Frege's Theorem 66:

$$Fc \wedge Gb \wedge \mathbf{N}z : (Fz \wedge z \neq c) = \mathbf{N}z : (Gz \wedge z \neq b) \rightarrow$$
$$\mathbf{N}z : Fz = \mathbf{N}z : Gz$$

For, if so, then since $\mathbf{N}z : (Gz \wedge z \neq b)$ and $\mathbf{N}z : (Fz \wedge z \neq c)$ are both x, they are identical; so, $\mathbf{N}z : Fz = \mathbf{N}z : Gz$; hence, $y = w$.

The proof of Theorem 66 requires two lemmas, the first of which is Theorem 63:

$$\neg \exists z (Qbz) \wedge Fc \wedge \mathbf{Map}(Q)(G\xi \wedge \xi \neq b, F\xi \wedge \xi \neq c) \rightarrow$$
$$\mathbf{Map}(Q\xi\eta \vee (\xi = b \wedge \eta = c))(G, F)$$

In words: If there is no object to which $Q\xi\eta$ relates b, if $Q\xi\eta$ maps the Gs other than b into the Fs other than c, and if c is an F, then the relation

---

[17] Frege's proofs show that $P\xi\eta$ is one-one, not that it is one-one if it is restricted to natural numbers. At this point in Part II of the *Grundgesetze*, the concept of a natural number has not been defined: The proof shows that zero, $\aleph_0$ and all other cardinal numbers have at most one predecessor and at most one successor, in Frege's sense. Frege's notion of succession, as applied to infinite cardinals, does not coincide with that now common in set-theory, however. On Frege's definition, the successor of $\aleph_0$ is $\aleph_0$. (It then follows, from Theorem 145, to be proven below, that $\aleph_0$ is not a finite number.)

which is just like $Q\xi\eta$, except that it also relates b to c, maps the Gs into the Fs. The second lemma is Theorem 56:[18]

$$\neg Fc \wedge \neg Gb \wedge \mathrm{N}z : Fz = \mathrm{N}z : Gz \rightarrow$$
$$\exists Q[\neg\exists z(\mathbf{Conv}(Q)(c, z)) \wedge \neg\exists z(Qbz)\wedge$$
$$\mathbf{Map}(\mathbf{Conv}(Q))(F, G) \wedge \mathbf{Map}(Q)(G, F)]$$

In words: If c is not an F and b is not a G and the number of Fs is the same as the number of Gs, then there is a relation $Q\xi\eta$, which relates b to no object and whose converse relates c to no object and which correlates the Fs one-one with the Gs.[19] Theorem 56 is an example of the sort of result that is required if we are to allow that a relation may map the Gs into the Fs, yet relate non-Gs to objects as well (and possibly to Fs).

The proof of Theorem 66 from Theorems 63 and 56 provides us with an example of the sort of purely technical advantage Frege's version of HP has: Frege is able to substitute into Theorem 63 to prove results about the *converse* of Q. Substituting "$\mathbf{Conv}(Q)$" for "Q", swapping "F" and "G", and swapping "b" and "c" in Theorem 63, we have:

$$\neg\exists z(\mathbf{Conv}(Q)(c, z)) \wedge \mathbf{Map}(\mathbf{Conv}(Q))(F\xi \wedge \xi \neq c, G\xi \wedge \xi \neq b) \wedge Gb \rightarrow$$
$$\mathbf{Map}(\mathbf{Conv}(Q)(\xi, \eta) \vee (\xi = c \wedge \eta = b))(F, G)$$

But we have also Frege's Theorem 64ι:[20]

$$\forall x \forall y \{\mathbf{Conv}(Q)(x, y) \vee (x = c \wedge y = b) \equiv$$
$$\mathbf{Conv}(Q\xi\eta \vee (\xi = b \wedge \eta = c))(x, y)\}$$

Hence, Theorem 64λ:[21]

$$\neg\exists z(\mathbf{Conv}(Q)(c, z)) \wedge Gb \wedge \mathbf{Map}(\mathbf{Conv}(Q))(F\xi \wedge \xi \neq c, G\xi \wedge \xi \neq b) \rightarrow$$
$$\mathbf{Map}(\mathbf{Conv}(Q\xi\eta \vee (\xi = b \wedge \eta = c)))(F, G)$$

Combining this with Theorem 63 and applying HP, we have Theorem 64ν:

$$Gb \wedge Fc \wedge \neg\exists z(\mathbf{Conv}(Q))(c, z)) \wedge \neg\exists z(Qbz)\wedge$$
$$\mathbf{Map}(\mathbf{Conv}(Q))(F\xi \wedge \xi \neq c, G\xi \wedge \xi \neq b)\wedge$$
$$\mathbf{Map}(Q)(G\xi \wedge \xi \neq b, F\xi \wedge \xi \neq c) \rightarrow$$
$$\mathrm{N}x : Gx = \mathrm{N}x : Fx$$

---

[18]In fact, this is a contrapositive of Theorem 56. I shall make such minor changes without special mention. Frege often works with contrapositives of the natural results for technical reasons, as noted above.

[19]When we have $\mathbf{Map}(Q)(G, F)$ and $\mathbf{Map}(\mathbf{Conv}(Q))(F, G)$, we say that $Q\xi\eta$ correlates the Gs one-one with the Fs. Note the order here.

[20]Theorem 64ι is the theorem with index "ι" that occurs *during* the proof of Theorem 64 and so appears in the text *before* Theorem 64.

[21]This is the sort of point at which Basic Law V is used to ease the transition. We should need here to prove a theorem which allows substitution of co-extensive relational expressions in the relevant argument place of "Map". It would not be difficult to prove such a theorem, but even stating it in general form requires third-order logic, unless one simply wants to prove a meta-theorem.

Finally, substituting, "$F\xi \wedge \xi \neq c$" for "$F\xi$" and "$G\xi \wedge \xi \neq b$" for "$G\xi$" in Theorem 56, we have:

$$\neg(Fc \wedge c \neq c) \wedge \neg(Gb \wedge b \neq b)\wedge$$
$$\mathbf{N}x : (Fx \wedge x \neq c) = \mathbf{N}x : (Gx \wedge x \neq b) \rightarrow$$
$$\exists Q[\neg\exists z(\mathbf{Conv}(Q)(c, z)) \wedge \neg\exists z(Qbz)\wedge$$
$$\mathbf{Map}(\mathbf{Conv}(Q))(F\xi \wedge \xi \neq c, G\xi \wedge \xi \neq b)\wedge$$
$$\mathbf{Map}(Q)(G\xi \wedge \xi \neq b, F\xi \wedge \xi \neq c)]$$

The first two conjuncts are obvious, so Theorem 66 follows immediately from this and Theorem 64ν.

We shall leave the proof that $P\xi\eta$ is functional here: The proofs of the two lemmas, Theorems 63 and 56, pose no difficulty.

2.4   Frege's Derivation of the Axioms of Arithmetic, continued

We turn now to Frege's proofs of the other axioms. Axiom 3, that zero has no predecessor, is perhaps the easiest of all to prove; Frege proves it in section E(psilon) as Theorem 108. Suppose that $Pa0$; by definition, there are a concept F and an object x falling under it, such that the number of Fs is 0 and the number of Fs other than x is a:

$$\exists F\exists x(Fx \wedge a = \mathbf{N}z : (Fz \wedge x \neq z) \wedge 0 = \mathbf{N}x : Fx)$$

A fortiori, there is a concept F whose number is 0 and under which some object falls:

$$\exists F\exists x(Fx \wedge 0 = \mathbf{N}x : Fx)$$

But that yields a contradiction, for it is easy to show that, if something is F, the number of Fs is not zero (Theorem 93):

$$\exists x(Fx) \rightarrow 0 \neq \mathbf{N}x : Fx$$

We prove the contrapositive. If 0 is the number of Fs, there is a relation R that maps the Fs into the non-self-identicals (and whose converse maps the non-self-identicals into the Fs). But then, by definition:

$$\forall x[Fx \rightarrow \exists y(y \neq y \wedge Rxy)]$$

But nothing is non-self-identical, so nothing is F.

To make any further progress, Frege must define the predicate "$\mathbb{N}\xi$", that is, define the predicate "ξ is a natural number". His definition is again the same as that given in *Die Grundlagen*. First, Frege introduces the ancestral: Given a relation $Q\xi\eta$, we say that a concept F is *hereditary in the Q-series* just in case, whenever x is F, each object to which $Q\xi\eta$ relates it is F; i.e.:

$$\forall x[Fx \rightarrow \forall y(Qxy \rightarrow Fy)]$$

We now say that an object b *follows* an object a in the Q-series just in case b falls under every concept that is hereditary in the Q-series and under which each object to which Q relates a falls. Formally, writing "Q^*ab" for "b follows a in the Q-series", Frege's definition of the *strong*[22] ancestral is (Gg, v. I, §45):[23]

$$Q^*ab \overset{df}{\equiv} \forall F[\forall x(Qax \rightarrow Fx) \wedge \forall x \forall y(Fx \wedge Qxy \rightarrow Fy) \rightarrow Fb]$$

Frege then defines the *weak* ancestral as (Gg, v. I, §46):[24]

$$Q^{*=}ab \overset{df}{\equiv} Q^*ab \vee a = b$$

Frege reads "$Q^{*=}ab$" as "b is a member of the Q-series beginning with a" or, equivalently, "a is a member of the Q-series ending with b".

The concept $N\xi$—that is, ξ is a finite (or natural) number—is then definable as $P^{*=}0\xi$: An object is a natural number just in case it belongs to the P-series beginning with 0.[25] Axiom 1, which states that zero is a natural number, follows immediately from the definition of the weak ancestral. Famously, induction too follows almost immediately from this definition. Matters are slightly more complicated than one might have thought, however. The hypothesis of induction is not that, *whenever* x is F, its successor is F; it is only that, whenever x is a natural number that is F, its successor is F. But this is not a substantial difficulty; for, as Frege shows, the following, his Theorem 152, is provable:[26]

$$Q^{*=}ab \wedge Fa \wedge \forall x \forall y[Q^{*=}ax \wedge Fx \wedge Qxy \rightarrow Fy] \rightarrow Fb$$

That is: If b is a member of the Q-series beginning with a, if a is F, and if F is hereditary (as one might put it) in the Q-series beginning with a, then b is F. Induction is an instance of Theorem 152: Take Q to be P; take a to be zero.[27]

---

[22]So-called because we need not have that Q^*aa; take Q, for example, to be the empty relation.

[23]For Frege, *, which he writes as $\backsim$, is a one-place function taking an object as argument (a double value-range in the interesting cases) and returning a double value-range as value. For us, "*" is a mixed-level predicate, taking a first-order relational expression and two objects as arguments and binding the argument places of the predicate. Hence, it would be more accurate to write it, as I did in the original version of this chapter, as something like: $\Im_{\alpha\epsilon}(Q\alpha\epsilon)(x,y)$. That, however, becomes tiresome, and the asterisk notation is well-established. See note 11 for similar remarks about other defined notions.

[24]Frege writes the weak ancestral as "$\smile$".

[25]Frege has no special symbol for our predicate "$N\xi$": He does, however, regularly read "$P^{*=}0\xi$" as "ξ is a finite number" (Gg, v. I, §108).

[26]For the proof, see Section 2.5.2. The difference I am emphasizing here does become important in the context of predicative theories (Heck, 2011).

[27]Oddly enough, Frege never actually writes this instance of Theorem 152 down, though it is, of course, applied.

2.5   An Elegant Proof that Every Number has a Successor

2.5.1   The Strategy of the Proof

To complete the discussion of Frege's proofs of the axioms of arithmetic, we have now only to discuss Axiom 2, which states that every natural number has a successor. The basic idea behind the proof is that each natural number is succeeded by the number of numbers less than or equal to it; more precisely, what we want to prove is Frege's Theorem 155:

$$P^{*=}0b \rightarrow P(b, \mathrm{N}x : (P^{*=}xb))$$

In words: If b is a natural number, then b precedes the number of members of the P-series ending with b. The proof proceeds by induction, the induction justified by Theorem 152, mentioned above. The relevant concept for the induction (that is, what we substitute for "F") is:

$$P(\xi, \mathrm{N}x : (P^{*=}x\xi))$$

So the goal is to show that zero falls under this concept and that it is hereditary in the P-series beginning with zero. We thus need to prove Theorem 154:
$$P(0, \mathrm{N}x : P^{*=}x0)$$

and Theorem 150:

$$\forall y[P^{*=}0y \wedge P(y, \mathrm{N}x : (P^{*=}xy)) \rightarrow \forall z(Pyz \rightarrow P(z, \mathrm{N}x : P^{*=}xz))]$$

We first need to collect some subsidiary results.

2.5.2   The Basic Facts about the Ancestral

The proofs of many of the results that follow draw heavily upon certain basic facts about the ancestral. To avoid undue prolixity, it is worth collecting all these facts here, so that they may later be referred to simply as 'the basic facts about the ancestral'. All of these are, in one way or another, manifestations of the ancestral's transitivity. The transitivity of the ancestral itself is not, however, among the facts to which Frege needs to appeal here: It does not appear in *Grundgesetze* until much later.

Two of the most important facts about the strong ancestral are given in Theorem 123:

$$Q^*ab \wedge \forall x \forall y(Fx \wedge Qxy \rightarrow Fy) \wedge \forall x(Qax \rightarrow Fx) \rightarrow Fb$$

and Theorem 128:

$$Q^*ab \wedge \forall x \forall y(Fx \wedge Qxy \rightarrow Fy) \wedge Fa \rightarrow Fb$$

These are forms of induction for the strong ancestral: If one knows, for example, that Q^*ab, then (128) allows one to show that Fb by establishing a basis case and an induction step: by showing that Fa and that $F\xi$ is hereditary in the Q-series.

Theorem 123 is immediate from the definition of the strong ancestral. Theorem 128 differs from it only in that, where (123) contains the condition that everything immediately following a in the Q-series must be F, (128) contains the condition that a itself must be F. But plainly, if a is F and F is hereditary in the Q-series, then everything that immediately follows a in the Q-series is F.

A very simple fact about the strong ancestral is recorded in Theorem 131:

$$Qab \to Q^*ab$$

In words: If $Q\xi\eta$ relates a to b, then b follows after a in the Q-series. To prove this, assume the antecedent. To show that Q^*ab, we prove the formula that defines it:

$$\forall F[\forall x(Qax \to Fx) \land \forall x \forall y(Fx \land Qxy \to Fy) \to Fb]$$

So let $F\xi$ be arbitrary and assume that $\forall x(Qax \to Fx)$ and $\forall x \forall y(Fx \land Qxy \to Fy)$. Then since Qab, by the first of these assumptions, certainly Fb. The latter assumption is not needed.

Perhaps the most basic manifestation of transitivity is Theorem 129:

$$Q^*ab \land Qca \to Q^*cb$$

This says that if b follows after a in the Q-series and if a follows *immediately* after c, then b follows after c in the Q-series, as well. Suppose the antecedent. We want to show that Q^*cb. So let $F\xi$ be an arbitrary concept satisfying the two conditions:

$$\forall x(Qcx \to Fx)$$
$$\forall x \forall y(Fx \land Qxy \to Fy)$$

We must show that Fb. By (128), however:

$$Q^*ab \land \forall x \forall y(Fx \land Qxy \to Fy) \land Fa \to Fb$$

So we need only show that Fa. But we have supposed that $\forall x(Qcx \to Fx)$ and that Qca, so certainly Fa, and we are done.

Theorem 132 strengthens (129) by weakening the first clause of the antecedent:

$$Q^{*=}ab \land Qca \to Q^*cb$$

Assume the antecedent. Then either Q^*ab or $a = b$. The first case is (129). And if $a = b$, then since Qca we have Qcb; but then the consequent follows by (131).

Theorems 133 and 134 are, as it were, the obverses of (129) and (132):

$$Q^*ab \land Qbc \to Q^*ac$$
$$Q^{*=}ab \land Qbc \to Q^*ac$$

Their proofs are sufficiently similar that I shall omit them. Theorem 137 is a weakening of (134):

$$Q^{*=}ab \land Qbc \to Q^{*=}ac$$

The weakening is justified by Theorem 136:

$$Q^*ab \to Q^{*=}ab$$

which obviously follows from the definition of the weak ancestral, as do Theorems 139 and 140:

$$a = b \to Q^{*=}ab$$
$$Q^{*=}aa$$

Frege does not here record the weakenings of (129), (132), and (133) corresponding to that of (134).

The final facts I shall mention here are forms of induction for the weak ancestral. The first of these is Theorem 144:

$$Q^{*=}ab \land Fa \land \forall x \forall y (Fx \land Qxy \to Fy) \to Fb$$

Theorem 144 is the same as (128), except that the first conjunct of the antecedent has been weakened from "Q^*ab" to "$Q^{*=}ab$". So suppose the antecedent. As before, if $Q^{*=}ab$, then either Q^*ab or $a = b$. The former case is (128), so we need only consider the latter. But we have supposed that Fa, so certainly Fb, by identity.

Finally, then, let us consider Theorem 152:

$$Q^{*=}ab \land Fa \land \forall x \forall y[Q^{*=}ax \land Fx \land Qxy \to Fy] \to Fb,$$

which was mentioned above. Take $F\xi$ in (144) to be $F\xi \land Q^{*=}a\xi$ and assume the antecendent of (152), that is, that $Q^{*=}ab$, that Fa, and that $F\xi$ is hereditary in the Q-series, restricted to members of the Q-series beginning with a. To establish the antecedent of (144), we must show that:

(i) $Fa \land Q^{*=}aa$

(ii) $\forall x \forall y[(Fx \land Q^{*=}ax) \land Qxy \to (Fy \land Q^{*=}ay)]$

The former follows from (140). Suppose, then, that $Fx \land Q^{*=}ax$ and that Qxy. Since $F\xi$ is hereditary in the Q-series beginning with a, Fy; moreover, we have $Q^{*=}ax$ and Qxy, so $Q^{*=}ay$, by (137). So, by (144), Fb.

So, in summary, the 'basic facts about the ancestral' include four forms of induction:

$$Q^*ab \wedge \forall x \forall y (Fx \wedge Qxy \rightarrow Fy) \wedge Fa \rightarrow Fb$$
$$Q^*ab \wedge \forall x (Qax \rightarrow Fx) \wedge \forall x \forall y (Fx \wedge Qxy \rightarrow Fy) \rightarrow Fb$$
$$Q^{*=}ab \wedge Fa \wedge \forall x \forall y (Fx \wedge Qxy \rightarrow Fy) \rightarrow Fb$$
$$Q^{*=}ab \wedge Fa \wedge \forall x \forall y [Q^{*=}ax \wedge Fx \wedge Qxy \rightarrow Fy] \rightarrow Fb$$

various manifestations of transitivity:

$$Qab \rightarrow Q^*ab$$
$$Q^*ab \wedge Qca \rightarrow Q^*cb$$
$$Q^{*=}ab \wedge Qca \rightarrow Q^*cb$$
$$Q^{*=}ab \wedge Qbc \rightarrow Q^*ac$$
$$Q^{*=}ab \wedge Qbc \rightarrow Q^*ac$$
$$Q^{*=}ab \wedge Qbc \rightarrow Q^{*=}ac$$

and some simple facts about the weak ancestral:

$$Q^*ab \rightarrow Q^{*=}ab$$
$$a = b \rightarrow Q^{*=}ab$$
$$Q^{*=}aa$$

As said earlier, these shall henceforth all be cited simply as 'basic facts'. Sometimes, they will not be cited explicitly at all.

2.5.3   Theorem 154

To prove that every natural number has a successor, we need to prove Theorem 154:

$$P(0, \mathbf{N}x : P^{*=}x0)$$

and Theorem 150:

$$\forall y [P^{*=}0y \wedge P(y, \mathbf{N}x : (P^{*=}xy)) \rightarrow \forall z (Pyz \rightarrow P(z, \mathbf{N}x : P^{*=}xz))]$$

The proof of Theorem 154 is relatively easy. It relies only upon the fact that nothing ancestrally precedes zero in the P-series, which is Theorem 126:

$$\neg P^*x0$$

This follows immediately from Theorem 124, whose proof, like those of other minor results to follow, is confined to the notes:[28]

$$Q^*ab \rightarrow \exists x (Qxb)$$

---

[28]The proof of this theorem is entirely analogous to the usual proof, in first-order arithmetic, that every natural number other than zero has a predecessor. In this case, the role of induction is played by Frege's Theorem 123. The relevant concept, for the induction, is $\exists z(Qz\xi)$. The proofs that $\forall x(Qax \rightarrow \exists z(Qzx))$ and $\forall x[\exists z(Qzx) \rightarrow \forall y(Qxy \rightarrow \exists z(Qzy))]$ are then pretty trivial.

Since nothing immediately precedes zero in the P-series, then, nothing ancestrally precedes it, either; and so, by the definition of the weak ancestral, the only member of the P-series ending with zero is zero itself (Theorem 154β):

$$P^{*=}x0 \to x = 0$$

Hence, nothing is a member of the P-series ending with zero, other than zero (Theorem 154δ):

$$\neg\exists x[P^{*=}x0 \land x \neq 0]$$

Frege's Theorem 97 tells us, however, that if nothing is F, then the number of Fs is zero:

$$\neg\exists x Fx \to \mathbf{N}x : Fx = 0$$

So we have Theorem 154ϵ:

$$\mathbf{N}x : (P^{*=}x0 \land x \neq 0) = 0$$

Now, a straightforward consequence of Frege's definition of predecession is Theorem 102:[29]

$$Gb \land m = \mathbf{N}x : (Gx \land x \neq b) \to P(m, \mathbf{N}x : Gx)$$

Taking $G\xi$ to be $P^{*=}0\xi$ and both b and m to be 0, we have:

$$P^{*=}00 \land \mathbf{N}x : (P^{*=}x0 \land x \neq 0) = 0 \to P(0, \mathbf{N}x : P^{*=}x0)$$

So we have reached Theorem 154ζ:

$$P^{*=}00 \to P(0, \mathbf{N}x : P^{*=}x0)$$

But zero is trivially a member of the P-series ending with zero, so Theorem 154 follows.

Theorem 97 follows from Frege's Theorem 96:

$$\forall x(Fx \equiv Gx) \to \mathbf{N}x : Fx = \mathbf{N}x : Gx$$

which states that, if the Fs are the Gs, the number of Fs is the same as the number of Gs. Theorem 97 follows: If nothing is F, then the Fs are the non-self-identicals.

The proof of Theorem 96 is straightforward but, in a sense, surprising. As noted earlier, there is, in standard systems of second-order logic, no principle of extensionality for second-order predicates and functions. If there were one, then the proof of (96) would be utterly trivial, for extensionality would immediately imply that the cardinality operator, being a second-level function, is extensional, and that this operator is extensional

---

[29]To prove Theorem 102, we need to show that, if the antecedent holds, then, for some F and some y, $\mathbf{N}x : Fx = \mathbf{N}x : Gx$, Fy, and $m = \mathbf{N}x : (Fx \land x \neq y)$. Just take G to be F and y to be b.

is precisely what (96) says. As we also noted, however, one of the important roles value-ranges play in Frege's system is to allow him to overcome the absence of just such a principle: Basic Law V does not guarantee the extensionality of second-level functions, but it does guarantee the extensionality of the first-level functions Frege uses in their place. What is surprising, then, is that Frege *does not* appeal to Basic Law V in the proof of (96). Instead, he shows that, if the Fs are the Gs, then identity correlates the Fs one-one with the Gs, whence the number of Fs is, by HP, the same as the number of Gs.

In *Grundgesetze*, the cardinality operator applies to value-ranges, not to concepts: Frege does not have a second-level function $\mathrm{N}x:\Phi x$ but a first-level function $\mathcal{N}(\xi)$. Using the latter, something akin to (96) can be formulated as follows:

$$(96^*)\quad \forall x(Fx \equiv Gx) \to \mathcal{N}(\grave{\epsilon}F\epsilon) = \mathcal{N}(\grave{\epsilon}G\epsilon)$$

This follows from Basic Law V immediately. As it happens, though, Theorem 96 actually reads as follows:

$$\forall a(a \frown u \equiv a \frown v) \to \mathcal{N}u = \mathcal{N}v$$

This version of (96) is more general than (96*), since u and v here may be any objects one wishes: In particular, they need not be value-ranges. But if u and v are value-ranges—say $u = \grave{\epsilon}U\epsilon$ and $v = \grave{\epsilon}V\epsilon$—then, since Theorem 1 tells us that $a \frown \grave{\epsilon}U\epsilon \equiv Ua$, the antecedent will imply that $u = v$, and the consequent will then follow by Leibniz's Law. And the case in which u and v are not value-ranges is irrelevant. Frege might therefore just as well have proven (96) using Basic Law V. That he does not do so suggests that he was self-consciously avoiding appeal to Law V in his proofs of the basic laws of arithmetic.

2.5.4 An important lemma?

We turn now to Frege's proof of Theorem 150:

$$\forall y[P^{*=}0y \wedge P(y, \mathrm{N}x:P^{*=}xy) \to \forall z(Pyz \to P(z, \mathrm{N}x:P^{*=}xz))]$$

The proof shows (150) to be a consequence of the fact that P is one-one, together with the fact that the cardinality operator is extensional, that being (96). As we have just seen, however, the only facts about numbers to which Frege appeals in his proof of (154) are (96) and the axiom stating that zero has no predecessor, so Frege's proof of (155) shows *it* to be a consequence of the extensionality of $\mathrm{N}x:\Phi x$, the fact that P is one-one, and the fact that zero has no predecessor. Let me say that again: Once P has been shown to be one-one and zero has been shown to have no predecessor, the only further appeal to HP that is needed for the proof that every number has a successor is that required to prove Theorem 96.

But as Boolos (1998c, 1998h) has emphasized, one might plausibly regard (96) as a logical truth.

The formal proof Frege gives is not as general as the proof just indicated, though remarks Frege makes in the associated text indicate that he was aware of this generalization. I shall therefore present the proof in its more general form.

Perhaps the most crucial result needed for the the proof of (150) is Theorem 145, to whose proof Frege devotes section Zeta. Theorem 145 says that no natural number follows itself in the P-series:

$$P^{*=}0b \to \neg P^* bb$$

The proof is by induction, the induction justified by (144):

$$Q^{*=}ab \wedge Fa \wedge \forall x \forall y (Fx \wedge Qxy \to Fy) \to Fb$$

To prove (145), Frege takes $F\xi$ to be $\neg P^* \xi\xi$, $Q\xi\eta$ to be $P\xi\eta$, and a to be 0. Substituting, then, we have:

$$P^{*=}0b \wedge \neg P^*00 \wedge \forall x \forall y (\neg P^* xx \wedge Pxy \to \neg P^* yy) \to \neg P^* bb$$

The second conjunct follows from (126). The third conjunct follows from:

$$P^* yy \wedge Pxy \to P^* xx$$

which is Frege's Theorem 145α. Frege derives it from the following two propositions:

$$P^* yy \wedge Pxy \to P^{*=}yx$$
$$P^{*=}yx \wedge Pxy \to P^* xx$$

The latter follows immediately from the basic facts. The former is an instance of Frege's Theorem 143:

$$P^* yz \wedge Pxz \to P^{*=}yx$$

And (143), in turn, is a consequence of the following more general result, which I shall call Theorem 143*:

$$\mathrm{Func}(\mathrm{Conv}(Q)) \wedge Q^* yz \wedge Qxz \to Q^{*=}yx$$

To derive (143) from (143*), make the obvious substitution and note that $\mathrm{Func}(\mathrm{Conv}(P))$.

Frege is perfectly aware that this more general proposition is provable. He does not give a formal derivation of it, but he does give an informal proof of it during his discussion of (143):

Evidently, the corresponding proposition $[Q^*ab \land Qcb \to Q^{*=}cb]$ would not hold, in general, in an arbitrary series. It is here essential that predecession in the number-series takes place functionally (Theorem 88). We rely upon the proposition that, if in some (Q-)series an object (b) follows after a second object (a), there there is an object which belongs to the (Q-)series beginning with the second object (a) that stands to the first object (b) in the series-forming (Q-)relation. In signs [Theorem 141]:

$$Q^*ab \to \exists x(Qxb \land Q^{*=}ax)$$

Now, if one knows that there is no more than one object which stands to the first object (b) in the (Q-)relation, then this object must also belong to the (Q-)series beginning with the second object (a). (Gg, v. I, §112)

In words: Suppose that Func(Conv(Q)), that Q^*ab, and that Qcb. By Theorem 141, there is some object, call it w, such that Qwb and $Q^{*=}aw$. But, since the converse of Q is functional and Qcb, we have $c = w$, and hence $Q^{*=}ac$. That establishes Theorem 143*.

Boolos dubbed Theorem 141 'the roll-back theorem'. It is of quite general utility. The proof is, of course, by induction, the induction justified by (123). The relevant concept is $\exists z(Qz\xi \land Q^{*=}az)$. We must thus establish that:

(i) $\forall x(Qax \to \exists z[Qzx \land Q^{*=}az])$

(ii) $\forall x[\exists z(Qzx \land Q^{*=}az) \to \forall y[Qxy \to \exists z(Qzy \land Q^{*=}az)]]$

The former is obvious. For the latter, suppose that $\exists z(Qzx \land Q^{*=}az)$ and that Qxy. The former yields $Q^{*=}ax$, by transitivity. So we have $Q^{*=}ax \land Qxy$ and so $\exists z(Qzy \land Q^{*=}az)$, as wanted.

Frege's proof of (145) thus consists essentially in its derivation from a much more general result, which I shall call Theorem 145*:[30]

$$\text{Func}(\text{Conv}(Q)) \land \neg Q^*aa \to (Q^{*=}ab \to \neg Q^*bb)$$

In words: If the converse of Q is functional and a does not follow itself in the Q-series, then no member of the Q-series beginning with a follows itself in the Q-series. For note that, if we take $F\xi$ in (144) to be $\neg Q^*\xi\xi$, then we have the following, Theorem 144*:

$$Q^{*=}ab \land \neg Q^*aa \land \forall x\{\neg Q^*xx \to \forall y(Qxy \to \neg Q^*yy)\} \to \neg Q^*bb$$

We then have the following instances of Theorems 132 and 143*, just as in the proof of Theorem 145α:

$$Q^{*=}yx \land Qxy \to Q^*xx$$
$$\text{Func}(\text{Conv}(Q)) \land Q^*yy \land Qxy \to Q^{*=}yx$$

---

[30]Substituting, we have:

$$\text{Func}(\text{Conv}(P)) \land \neg P^*00 \to [P^{*=}0b \to \neg P^*bb]$$

But the converse of P is functional; and, as mentioned earlier, 0 does not follow itself in the Pred-series. Theorem 145 then follows by *modus ponens*.

Hence:
$$\text{Func}(\text{Conv}(Q)) \wedge Q^*yy \wedge Qxy \to Q^*xx$$
Theorem 145* then follows easily from this and Theorem 144*.

2.5.5 Another important lemma

Given Theorem 145, our next goal is Theorem 149:
$$P^{*=}0a \wedge Pda \to \mathrm{N}x : P^{*=}xd = \mathrm{N}x : (P^{*=}xa \wedge x \neq a)$$

In words: If a is a natural number and d precedes a, then the number of numbers less than or equal to d is the same as the number of numbers less than or equal to a, other than a. This follows from (96) and Theorem 149α:
$$P^{*=}0a \wedge Pda \to \forall x\{P^{*=}xd \equiv (P^{*=}xa \wedge x \neq a)\}$$

Theorem 149α, in turn, follows immediately from the following general fact about the ancestral, which we may call Theorem 149*:
$$\text{Func}(\text{Conv}(Q)) \wedge \neg Q^*aa \wedge Qda \to \forall x\{Q^{*=}xd \equiv (Q^{*=}xa \wedge x \neq a)\}$$

In words: If the converse of Q is functional, if a does not follow itself in the Q-series, and if Qda, then the members of the Q-series ending with d are the members of the Q-series ending with a, other than a. Substituting, we have:
$$\text{Func}(\text{Conv}(P)) \wedge \neg P^*aa \wedge Pda \to \forall x\{P^{*=}xd \equiv (P^{*=}xa \wedge x \neq a)\}$$

Again, the converse of P is functional, and, if a is a natural number, it does not follow itself in the P-series, by (145).

Frege does not prove Theorem 149*, but, as in the case of Theorem 145*, his proof can easily be generalized to yield it. He derives Theorem 149α, from the following two results:
$$Pda \to [(P^{*=}xa \wedge x \neq a) \to P^{*=}xd]$$
$$Pda \wedge P^{*=}0a \to [P^{*=}xd \to (P^{*=}xa \wedge x \neq a)]$$

For the former: If $P^{*=}xa$ and $x \neq a$, then P^*xa, by the definition of the weak ancestral. Hence, if Pda, then $P^{*=}xd$, by (143). For the latter: If Pda and $P^{*=}xd$, then P^*xa, by the basic facts. Since P^*xa, $P^{*=}xa$, by definition; and if $x = a$, then P^*aa, contradicting (145), since a is a natural number.

The proof of (149*) is identical, except we prove the more general results:
$$\text{Func}(\text{Conv}(Q)) \wedge Qda \to \{[Q^{*=}xa \wedge x \neq a] \to Q^{*=}xd\}$$
$$Qda \wedge \neg Q^*aa \to [Q^{*=}xd \to (Q^{*=}xa \wedge x \neq a)]$$

Again, it seems reasonable to suppose Frege knew of this generalization.

2.5.6   Completion of the proof

We now complete the proof of Theorem 150. My explanation of the proof will closely follow that given by Frege himself in section 114 of *Grundgesetze*.

Recall that Theorem 150 is:

$$\forall y[P^{*=}0y \wedge P(y, \mathbf{N}x : P^{*=}xy) \to \forall z(Pyz \to P(y, \mathbf{N}x : P^{*=}xz))]$$

Much of the elegance of Frege's proof lies in the the ease with which he derives Theorem 150 from Theorem 149, which is, again:

$$P^{*=}0a \wedge Pda \to \mathbf{N}x : P^{*=}xd = \mathbf{N}x : (P^{*=}xa \wedge x \neq a)$$

The formal derivation, in *Grundgesetze*, takes just six lines.

Theorem 150 follows, by generalization, from Theorem 150ϵ:

$$P^{*=}0d \wedge P(d, \mathbf{N}x : P^{*=}xd) \wedge Pda \to P(a, \mathbf{N}x : P^{*=}xa)$$

For the proof of (150ϵ), suppose that d is a natural number, that d precedes the number of members of the P-series ending with d, and that d precedes a. We must show that a precedes the number of members of the P-series ending with a. To do so, we must find some concept F and some object x falling under F such that a is the number of Fs other than x and the number of Fs is the same as the number of members of the P-series ending in a. That is, we must show that:

$$\exists F \exists x[a = \mathbf{N}z : (Fz \wedge z \neq x) \wedge Fx \wedge \mathbf{N}z : P^{*=}za = \mathbf{N}z : Fz]$$

The concept in question is to be $P^{*=}\xi a$; the object in question is to be a itself. Hence, we must show that:

$$a = \mathbf{N}z : (P^{*=}za \wedge z \neq a) \wedge P^{*=}aa \wedge \mathbf{N}z : P^{*=}za = \mathbf{N}z : P^{*=}za$$

The last two conjuncts are trivial. The first we may derive, by the transitivity of identity, from:

$$a = \mathbf{N}x : P^{*=}xd$$
$$\mathbf{N}x : P^{*=}xd = \mathbf{N}x : (P^{*=}xa \wedge x \neq a)$$

The former follows from the functionality of P, since, by hypothesis, d precedes both a and $\mathbf{N}x : P^{*=}xd$. The latter, in turn, is the consequent of (149): And, since $P^{*=}0d$ and Pda, $P^{*=}0a$, by the basic facts. Since we are assuming that Pda, that establishes the antecedent of (149).

That, then, completes the proof of Theorem 150 and so completes Frege's proof that every number has a successor.

2.6 Frege's Axiomatization of Arithmetic

Part II of *Grundgesetze* is entitled "Proofs of the Basic Laws of Cardinal Number". The proof of the axioms of arithmetic occupies just one third of Part II. A discussion of the rest will have to wait for another occasion,[31] but one of the results proven in the remainder deserves special mention, namely, Frege's Theorem 263. Where "∞", read "Endlos", is a name for the number of natural numbers,[32] Theorem 263 is essentially:

$$\mathbf{Func}(Q) \wedge \forall x(Gx \equiv Q^{*=}ax) \wedge \neg\exists x(Q^*xx) \wedge \forall x(Gx \to \exists y(Qxy)) \to$$
$$Nx:Gx = \infty$$

In words: If Q is functional, if the Gs are the members of the Q-series beginning with a, if no object follows itself in the Q-series, and if each G is related by Q to some object, then the number of Gs is Endlos.

It is essential, if we are to understand *Grundgesetze*, that we ask ourselves, concerning such theorems, why they are here: *Grundgesetze* is not a random collection of results, and Frege did not just include whatever came to mind; each of the results surely has some purpose. Unfortunately, he rarely stops to tell us what purposes the various results have. In the case of this Theorem, its significance is not at all apparent from its statement; to understand its significance, we must look at its proof. What Frege proves is that, if Q is functional, if the Gs are the members of the Q-series beginning with a, if no member of the Q-series beginning with a follows itself in the Q-series,[33] and if each G is related, by Q, to some object, then the number of Gs is Endlos because the Gs, ordered by Q, are *isomorphic to the natural numbers*, ordered by succession. If we write "Q" as "S", "G" as "N", and "a" as "0", then what Frege proves is that the following four conditions determine a structure isomorphic to the natural numbers:[34]

1. $\mathbf{Func}(S)$

2. $\neg\exists x[Nx \wedge S^*xx]$

3. $\forall x[Nx \to \exists y(Sxy)]$

4. $\forall x[Nx \equiv S^{*=}0x]$

That is to say, Theorem 263 establishes that (1)–(4) are axioms for arithmetic, ones that are different from, though closely related to, those due

---

[31]I have continued the discussion elsewhere (Heck, 1995a, 1998b). A book-length treatment is in preparation.

[32]That is: $\infty \overset{df}{=} Nx:P^{*=}0x$ (Gg, v. I, §122).

[33]The Theorem, with the stronger condition, that no object follow itself in the Q-series, is inferred from one with this weaker condition.

[34]See the Postscript for some remarks on the relation between this result and a famous result due to Dedekind (1963).

to Dedekind and Peano. The accusation considered earlier, that Frege's argument that 'arithmetic is a branch of logic' suffers from a failure to isolate *the* basic laws, *the* axioms, of arithmetic, would thus be completely unjustified. Frege does not explicitly say so, but it is clear from his proof of Theorem 263 that he conceived of (1)–(4) as 'the simplest laws of finite cardinal numbers'. It is for this reason that Theorem 145 is as heavily emphasized by Frege as it is (Gg, v. I, §108), why Frege even devotes a separate section to its proof: Theorem 145 is one of his axioms for arithmetic.

Frege's axioms capture our intuitions about the structure of the natural numbers at least as well as the Dedekind-Peano Axioms. The natural numbers are the members of a series beginning with the number zero (Axiom 4). Each natural number is immediately followed in that series by one, and only one, natural number (Axioms 1 and 3). And, finally, each natural number is followed by a *new* natural number, one which has not previously occurred in the series; that is, no natural number follows itself in the series (Axiom 2).[35] These general principles are, it seems to me, entirely consonant with our ordinary understanding of the natural numbers, as well they should be. After all, Frege intends to derive the basic laws of *arithmetic*, not the basic laws of some formal theory which bears no obvious relation to arithmetic as we ordinarily understand it. It is therefore essential that his axioms not only characterize the right kind of structure but that they successfully capture our ordinary understanding of the natural numbers.

Not that Frege would have cared, but it should be said that his axioms are, compared with Dedekind's, more involved with second-order notions. Of course, the induction axiom, as formulated by Frege and Dedekind, is second-order as it stands: But, as is well-known, it can for many purposes be weakened and written as a first-order axiom schema. Frege's Axiom 2, on the other hand, contains a second-order universal quantifier in negative position. It is, however, not difficult to see that it follows from "$\neg \exists x(Sx0)$" and "$\text{Func}(\text{Conv}(S))$", the two missing axioms from the Dedekind-Peano axiomatization. It is just this which is established by the proof of Theorem 145.

2.7   Closing

Frege's derivation of the axioms of arithmetic in *Grundgesetze* has been unjustly neglected. Not only *can* his proofs be re-constructed in second-order logic, but the proofs he gives are themselves derivations of the ax-

---

[35]Note here that Axiom 2 is equivalent to:

$$\forall x[Nx \rightarrow \neg \exists y(Sxy \land S^{*=}yx)]$$

If $\exists y(Sxy \land S^{*=}yx)$, then S^*xx; conversely, if S^*xx, then the 'roll-forward' theorem, which is closely related to the roll-back theorem, implies that $\exists y(Sxy \land S^{*=}yx)$.

ioms for arithmetic—both his and the Dedekind-Peano axioms—from HP, within second-order logic. So Frege proved Frege's Theorem, which is as it should be.

I said earlier that this fact has no significance for our understanding of Frege's philosophy unless Frege *knew* that the axioms could be proven from HP. Surely, however, he did: He says, in *Die Grundlagen*, that he "attach[es] no decisive importance even to bringing in the extensions of concepts at all" (Gl, §107). This would be a strange remark for Frege to make if extensions were required, not just for the formulation of the explicit definition and the derivation of HP from it, but for his proofs of the axioms from HP itself.[36]

That said, there are a variety of questions that an appreciation of the role Basic Law V plays in Frege's derivation of the axioms of arithmetic might raise. The most important of these is: How can an axiom which plays such a limited *formal* role be of such fundamental importance to Frege's philosophy of mathematics? Rarely, so far as I know, is it said that Frege's abandonment of the logicist project was due to some realization that, with Basic Law V refuted and no suitable weakening forthcoming, he could no longer derive the laws of arithmetic within logic: It is said so rarely because it can seem so little worth saying; it can seem so much common sense, and that, in some sense, it must be. But there was, in a clear sense, no *formal* obstacle to the logicist program, even after Russell's discovery of the contradiction, and Frege knew it. It is surely right that Frege realized he could no longer derive the laws of arithmetic *within logic*. But the question is why Frege was prepared to accept Basic Law V, but not HP, as a fundamental truth of logic, as a 'primitive truth', one which, as he puts it in *Die Grundlagen*, "neither need[s] nor admit[s] of proof" (Gl, §3).

A particularly nice statement of the philosophical significance of Basic Law V is made in the letter to Russell mentioned a bit ago:

...[T]he question is, How do we apprehend logical objects? And I have found no other answer to it than this, We apprehend them as extensions of concepts, or more generally, as value-ranges of functions. I have always been aware that there were difficulties with this, and your discovery of the contradiction has added to them; but what other way is there? (PMC, pp. 140–1)

It is to answer this question, first raised in the famous section §62 of *Die Grundlagen*, that Frege introduces extensions of concepts in the first place. What demands their introduction, then, is the so-called Julius Caesar problem (Gl, §§56, 66–8).

The questions to which we *really* need answers are thus: What, exactly, does Frege mean by his question how we apprehend logical objects? What is the real point of the Caesar problem? In what, for Frege, does it consist that a truth is a 'primitive' truth? one which "neither need[s]

---

[36]For more on this issue, see Section 4.3.

nor admit[s] of proof"? Or, if the notion of a primitive truth is not abso-
lute, but relative to a particular systematization of logic (or arithmetic),
in what does it consist that some truth is even a *candidate* for being a
'primitive' truth? And, if Frege would not have been opposed to accepting
HP as a primitive truth, perhaps there was some obstacle, in his view, to
accepting it as a primitive *logical* truth. So, in what does it consist that
a 'primitive' truth *is* a primitive logical truth?—I could hazard a guess at
answers to such questions, but it would, at this point, only be a guess.

What I do know is that the questions lately raised need answers. Until we have such answers, we shall not understand the significance Basic
Law V had for Frege, since we shall not understand why he could not
abandon it in favor of HP; and that is to say that we shall not understand
how he conceived the logicist project. We shall thus not fully understand
Frege's philosophy until we understand the enormous significance the
question how we apprehend logical objects, and the Caesar problem, had
for him; until, that is, we understand how he could consistently be so hos-
tile to psychologism, yet hold his own philosophy of mathematics hostage
to epistemological considerations.

Postscript

The results discussed in Section 2.6, related to and including Frege's
Theorem 263, are obviously reminiscent of famous results of Richard
Dedekind's, in particular, of his proof that all structures satisfying the
Dedekind-Peano axioms for arithmetic are isomorphic. Dedekind pub-
lished this result in *Was Sind und Was Sollen die Zahlen?* six years before
Grundgesetze appeared. That makes it a nice question how Dedekind's
result is related to Frege's.

In the original version of this chapter, I remarked that "[i]t is difficult
to know whether Frege knew of Dedekind's proof before he composed his
own" (Heck, 1993b, p. 598, note 44). In fact, I was inclined to think he
didn't and, in a later reprint, the claim was strengthened: "It seems unlikely that Frege knew of Dedekind's proof before he composed his own,
but it is impossible to be certain of this" (Heck, 1995b, p. 284, note 44).
I based my suspicion on Frege's remark, in the preface to *Grundgesetze*,
that Dedekind's book had only "lately come to [his] notice" (Gg, p. vii).
Goran Sundholm, however, has reported that Frege taught Dedekind's
book in a seminar in 1889 and so suggests that it is "a virtual certainty
that Frege *did* know of Dedekind's proof when composing his own" (Sundholm, 2001, p. 61, footnote 17, emphasis in original). He may have, but I
do not think things are so clear.

The argument Sundholm gives for this claim is as follows:

[Volume I of *Grundgesetze*] reworks formally a lot of themes from Dedekind. Especially Frege's beautiful Theorem 263 is important in this context. Frege, when

attempting to formalize Dedekind's theory of chains, using the 1879 techniques of the *Begriffsschrift*, will speedily have discovered that his earlier pious hope, in [*Die Grundlagen*], to avoid the use of [extensions of concepts] turns out to be unrealistic. (Sundholm, 2001, p. 61)

Dedekind's theory of chains is just a special case of Frege's theory of the ancestral, however, so that was surely not the problem. But perhaps what Sundholm has in mind is Frege's use of ordered pairs in the proof of (263). The crucial chains needed in the proof of (263) are chains of pairs, and Frege does define pairs in *Grundgesetze* in terms of extensions, as Sundholm (2001, p. 62) notes. But, as I have shown in detail elsewhere (Heck, 1995a), Frege's proof does not *require* pairs at all. Not only are the modifications required to remove them ones that would have been well within Frege's abilities, the structure of his proofs suggests that he was aware that pairs were a convenience here, not a necessity.

In fact, I suspect that this use of pairs was a kind of leftover from an earlier draft. Sundholm emphasizes that, according to notes taken by Heinrich Scholz before Frege's *Nachlass* was lost, pairs were discussed very early in a draft of *Grundgesetze* from 1889, at which time Frege had not yet come to the view that the truth-values are objects. That implies, for reasons discussed in Section 6.1, that Frege would not have had anything quite like 'double value-ranges' in his system in 1889. I conjecture, therefore, that pairs were used to define the extensions of relations, in a now familiar way: First, we define a concept true of pairs:

$$\rho(<a, b>) \equiv Rab$$

and then take the extension of R to be that of ρ.[37] This is *exactly* how Frege uses pairs in the proof of (263). So, as said, it's a leftover.

Nonetheless, Sundholm is surely right that Frege knew Dedekind's result when composing the precise version of the proof that appears in the published version of *Grundgesetze*. That is not the question I meant to address, however. The interesting question is whether Frege had come to a similar result independently or whether he means simply to be formalizing Dedekind's. The latter would have been an achievement, and Frege might simply have wanted to make it clear that anything Dedekind can do he can do better.[38] On the other hand, however, Frege's proof is far more general than Dedekind's (Heck, 1995a), and that makes it at least somewhat plausible that the proof is Frege's own.

The truth, however, is that we just do not know how many of the theorems present in the published version of *Grundgesetze* were present in

---

[37]Note, moreover, that, if pairs were used to define the extensions of relations, then pairs could not have been defined as Frege defines them in *Grundgesetze*. He would therefore have had to treat them as an additional primitive, perhaps subject to the ordered pair axiom. That would have made the treatment of truth-values as objects even more attractive, since it allows pairs to be defined.

[38]...as Annie Oakley and Frank Butler famously insist in the song "Anything You Can Do I Can Do Better", written by Irving Berlin for the 1946 musical *Annie Get Your Gun*.

the "nearly complete" version Frege had already in 1889 but famously had to "discard" (Gg, p. ix). Maybe Frege had Theorem 263 before he taught *Was Sind?* and maybe he did not. I am not sure there is any way for us to know without Frege's *Nachlass* being rediscovered.

3

Die Grundlagen der Arithmetik §§82–83

(With George Boolos)

Reductions of arithmetic, whether to set theory or to a theory formulated in a higher-order logic, must prove the infinity of the sequence of natural numbers. In his *Was sind und was sollen die Zahlen?*, Dedekind attempted, in the notorious proof of Theorem 66 of that work, to demonstrate the existence of infinite systems by examining the contents of his own mind. The axioms of General Set Theory, a simple set theory to which arithmetic can be reduced, are those of Extensionality, Separation ("Aussonderung"), and Adjunction:

$$\forall w \forall z \exists y \forall x [x \in y \equiv x \in z \lor x = w)$$

It is Adjunction that guarantees that there are at least two, and indeed infinitely many, natural numbers. The authors of *Principia Mathematica*, after defining zero, the successor function, and the natural numbers in a way that made it easy to show that the successor of any natural number exists and is unique, were obliged to assume an axiom of infinity on those occasions on which they needed the proposition that different natural numbers have different successors.

In §§70–83 of *Die Grundlagen der Arithmetik*, Frege outlines the derivations of some familiar laws of the arithmetic of the natural numbers from principles he takes to be "primitive" truths of a general logical nature. In §§70–81, he explains how to define zero, the natural numbers, and the successor *relation*; in §78 he states that it is to be proved that this relation is one-one and adds that it does not follow that every natural number *has* a successor; thus, by the end of §78, the existence, but not the uniqueness, of the successor remains to be shown. Frege sketches, or attempts to sketch, such an existence proof in §§82–3, which would complete his proof that there are infinitely many natural numbers.

§§82–3 offer severe interpretive difficulties. Reluctantly and hesitantly, we have come to the conclusion that Frege was at least somewhat confused in these two sections and that he cannot be said to have outlined, or even to have intended, any correct proof there. We will discuss two (correct) proofs of the statement that every natural number has a successor which might be extracted from §§82–3. The first is quite similar to a proof of this proposition that Frege provides in *Grundgesetze der Arithmetik*, differing from it only in notation and other relatively minor respects. We will argue that fidelity to what Frege wrote in *Die Grundlagen* and in *Grundgesetze* requires us to reject the charitable suggestion that it was this (beautiful) proof that he had in mind when he wrote *Die Grundlagen*. The second proof we discuss conforms to the outline Frege gives in §§82–3 more closely than does the first. But if it had been the one he had in mind, the proof-sketch in these two sections would have contained a remarkably large gap that was never filled by any argument found in *Grundgesetze*. In any case, it is certain that Frege did not know of this proof.

We begin by discussing §§70–81.

In §70, Frege begins the definition of equinumerosity by explaining the notion of a relation, arguing that like (simple) concepts, relational concepts belong to the province of pure logic. In §71, he defines "the objects falling under F and G are correlated with each other by the relation ϕ". Using modern notation, but strictly following Frege's wording, we would write:

$$\forall a \neg (Fa \wedge \neg \exists b(a\phi b \wedge Gb)) \wedge \forall a \neg (Ga \wedge \neg \exists b(Fb \wedge b\phi a))$$

To put the definition slightly more transparently, the objects falling under F and G are correlated by ϕ iff

$$\forall x(Fx \rightarrow \exists y(Gy \wedge x\phi y)) \wedge \forall y(Gy \rightarrow \exists x(Fx \wedge x\phi y))$$

In §72, Frege defines what it is for the relation ϕ to be one-one ("beiderseits eindeutig", "single-valued in both directions"): It is for it, as we should say, to be a function, i.e., $\forall d \forall a \forall e(d\phi a \wedge d\phi e \rightarrow a = e)$, that is one-one, i.e., $\forall d \forall a \forall e(d\phi a \wedge b\phi a \rightarrow d = b)$. Frege then defines "equinumerous" ("gleichzahlig"): F is equinumerous with G iff there is a relation that correlates the objects falling under F one-one with those falling under G:

$$\exists \phi [\forall x(Fx \rightarrow \exists y(Gy \wedge x\phi y)) \wedge \forall y(Gy \rightarrow \exists x(Fx \wedge x\phi y)) \wedge$$
$$\forall d \forall a \forall e(d\phi a \wedge d\phi e \rightarrow a = e) \wedge \forall d \forall a \forall e(d\phi a \wedge b\phi a \rightarrow d = b)]$$

We abbreviate this formula: $F \approx G$.

At the end of §72, Frege defines the number that belongs to F as the extension of the concept "equinumerous with the concept F". He also defines "n is a (cardinal) number": there is a concept F such that n is the number that belongs to F. His next task, attempted in §73, is to prove a principle that Crispin Wright (1983) once called $N^=$ (for numerical equality), Michael Dummett (1991b) calls "the original equivalence", and we call "HP": the number belonging to F is identical with that belonging to G iff F is equinumerous with G.

The trouble with the definition of number given in §72 and the proof of HP given in §73 is that they implicitly appeal[1] to an inconsistent theory of extensions of second-level concepts. Russell of course demonstrated the inconsistency of Frege's theory, presented in *Grundgesetze der Arithmetik*, of extensions of first-level concepts; a routine jacking-up of Russell's argument shows that of the theory Frege tacitly appeals to in *Die Grundlagen*.[2] It is by now well-known, however, that Frege Arithmetic, i.e., the result of adjoining a suitable formalization of HP to axiomatic second-order logic, is consistent if second-order arithmetic is and is strong enough to imply second-order arithmetic (as of course Frege can be seen as attempting to prove in *Die Grundlagen*). Indeed, Frege Arithmetic and second-order arithmetic are equi-interpretable (Boolos and Heck, 1998, appendix 2).

Writing: $\#F$ to mean: the number belonging to the concept F, we may symbolize HP: $\#F = \#G \equiv F \approx G$.[3]

The development of arithmetic sketched in §§74–81 makes use only of Frege Arithmetic and can thus be reconstructed in a consistent theory (or one we believe to be so!). Nothing will be lost and much gained if we henceforth suppose that Frege's background theory is Frege Arithmetic.

In §74, Frege defines 0 as the number belonging to the concept "not identical with itself": $0 = \#[x : x \neq x]$. ($[x : \ldots x \ldots]$ is the concept *being an object x such that $\ldots x \ldots$*.) Frege notes that it can be shown on logical grounds that nothing falls under $[x : x \neq x]$. In §75, he states that $\forall x(\neg Fx) \to [\forall x(\neg Gx) \equiv F \approx G]$ has to be proved, from which $\forall x(\neg Fx) \equiv 0 = \#[x : Fx]$ follows. These have easy proofs. Frege outlines that of the former in detail.

§76 contains the definition of "n follows immediately after m in the 'natürliche Zalenreihe'":

$$\exists F \exists x \, (Fx \wedge \#F = n \wedge \#[y : Fy \wedge y \neq x] = m)$$

It is advisable, we think, to regard the relation so defined in this section as going from m to n, despite the order of "n" and "m" in both the definiens and the definiendum of "n immediately follows m in the natural

---

[1]The appeal is made when Frege writes "In other words:" at the end of the second paragraph of §73.

[2]Let (V) be $\forall \mathcal{F} \forall \mathcal{G}(\hat{\mathcal{F}} = \hat{\mathcal{G}} \equiv \forall X(\mathcal{F}X \equiv \mathcal{G}X))$. Then (V) is inconsistent (in third-order logic). For let $\mathcal{F}$ be $[X : \forall \mathcal{H}(\forall x(Xx \equiv x = \hat{\mathcal{H}}) \to \neg \mathcal{H}X)]$ and let X be $[x : x = \hat{\mathcal{F}}]$. Suppose $\mathcal{F}X$. Then $\forall \mathcal{H}(\forall x(Xx \equiv x = \hat{\mathcal{H}}) \to \neg \mathcal{H}X)$. So $\forall x(Xx \equiv x = \hat{\mathcal{F}}) \to \neg \mathcal{F}X$, whence $\neg \mathcal{F}X$ by the definition of X. Thus $\neg \mathcal{F}X$. So for some $\mathcal{H}$, $\forall x(Xx \equiv x = \hat{\mathcal{H}})$ and $\mathcal{H}X$, and then $X(\hat{\mathcal{F}}) \equiv \hat{\mathcal{F}} = \hat{\mathcal{H}}$. By the definition of X again, $\hat{\mathcal{F}} = \hat{\mathcal{F}} \equiv \hat{\mathcal{F}} = \hat{\mathcal{H}}$, $\hat{\mathcal{F}} = \hat{\mathcal{H}}$, and by (V), $\forall X(\mathcal{F}X \equiv \mathcal{H}X)$, contra $\neg \mathcal{F}X$ and $\mathcal{H}X$. (We use "$\hat{\ }$" to mean "the extension of" and "$[: \ldots]$" to denote concepts (of whatever level).)

[3]Our notation here is different from elsewhere in the book, where I have used the variable binding, term-forming operator $Nx : \phi x$, rather than the functional $\#$. The difference is merely notational in the presence of full second-order comprehension. When that is lacking, the difference becomes substantial (Burgess, 2005, §2.6).

series of numbers". We shall thus symbolize this relation: mPn ("*P*" for "(immediately) precedes").[4]

Call a concept *Dedekind infinite* if it is equinumerous with a proper subconcept of itself; equivalently, if it has a subconcept equinumerous with the concept *being a natural number*. With the aid of the equivalence of these definitions of Dedekind infinity, it is not difficult to see that nPn if and only if n is the number belonging to a Dedekind infinite concept. Thus the number of finite numbers, which Frege designates roughly: ∞,[5] but which we shall as usual denote: $\aleph_0$, follows itself in the "natürliche Zahlenreihe", in symbols: $\aleph_0 P \aleph_0$. Since $\aleph_0$ is not a finite, i.e., natural, number, we shall translate "in der natürliche Zahlenreihe" as "in the natural sequence of numbers".[6]

§77 contains the definition of 1, as $\#[x : x = 0]$, and a proof that $0P1$. In §78, Frege lists a number of propositions to be proved:

1. $0Pa \to a = 1$

2. $1 = \#F \to \exists x(Fx)$

3. $1 = \#F \to (Fx \wedge Fy \to x = y)$

4. $\exists x(Fx) \wedge \forall x \forall y(Fx \wedge Fy \to x = y) \to 1 = \#F$

5. P is one-one ("beiderseits eindeutig"), i.e., $mPn \wedge m'Pn' \to (m = m' \equiv n = n')$[7]

Frege observes that it has not yet been stated that every number immediately follows or is followed by another. He then states:

6.   Every number except 0 immediately follows a number in the natural sequence of numbers.

---

[4]This notation also differs slightly from that used elsewhere in the book.

[5]This is not quite the symbol Frege uses, which looks like a very open omega with a frown across the top.

[6]Timothy Smiley (1988) observed that "in the natural series of numbers" is to be preferred as a translation of "in der natürliche Zahlenreihe" to Austin's "in the series of natural numbers". We have substituted "sequence" for "series" throughout.

[7]Frege does not indicate what proof of 78.5 he might have intended. Here is an obvious one that he might have had in mind. (It is essentially the proof Frege gives in *Grundgesetze*, for which see Section 2.3.)

Suppose mPn and $m'Pn'$. Then for some F, F', x, x', Fx, $F'x'$, $\#F = n$, $\#F' = n'$, $\#[y : Fy \wedge y \neq x] = m$, and $\#[y' : F'y' \wedge y' \neq x'] = m'$.

Assume $m = m'$. Then by HP, there is a one-one correspondence ϕ between the objects y such that Fy and $y \neq x$ and the objects y' such that $F'y'$ and $y' \neq x'$. We may assume that if $y\phi y'$, then Fy, $y \neq x$, $F'y'$, and $y' \neq x'$. Let $y\psi y'$ iff $(y\phi y' \vee [y = x \wedge y' = x'])$. Then ψ is a one-one correspondence between the objects falling under F and those falling under F', and so by HP, $n = n'$.

Assume $n = n'$. By HP, let ψ be a one-one correspondence between the obejcts falling under F and those falling under F'. We may assume that if $y\phi y'$, then Fy, and $F'y'$. Let $y\phi y'$ iff $(Fy \wedge y \neq x \wedge F'y' \wedge y' \neq x' \wedge [y\psi y' \vee (y\psi x' \wedge x\psi y')])$. Then ψ is a one-one correspondence between the objects y such that Fy and $y \neq x$ and the objects y' such that $F'y'$ and $y' \neq x'$, and so by HP, $m = m'$.

It is clear from §44 of *Grundgesetze*[8] that Frege did not take (6) to imply that 0 does not immediately follow a number, that $\neg x P 0$. This proposition is proved separately in *Grundgesetze*, as Theorem 108, and will be used later on here, at a key point in the argument.

§79 contains the definition of the strong ancestral of ϕ, "x precedes y in the ϕ-sequence" or "y follows x in the ϕ-sequence":

$$\forall F \left(\forall a \left(x \phi a \to F a\right) \to \forall d \forall a \left(F d \to d \phi a \to F a\right) \to F y\right)$$

which was Definition (76) of the *Begriffsschrift*. Frege will use this definition in §81 to define "member of the natural sequence of numbers ending with n". We shall use the standard abbreviation: $x \phi^* y$ for the strong ancestral. To prove that if $x \phi^* y$, then $\dots y \dots$, it suffices, by the comprehension scheme $\exists F \forall a (F a \equiv \dots a \dots)$ of second-order logic, to show that $\forall a (x \phi a \to \dots a \dots)$ and $\forall d \forall a (\dots d \dots \to d \phi a \to \dots a \dots)$. We call this method of proof *Induction 1*. (Induction 2 and Induction 3 will be defined below.)

Here and below, we associate iterated conditionals to the right. Thus, e.g., "$A \to B \to C$" abbreviates "$(A \to (B \to C))$". This convention provides an easy way to reproduce in a linear symbolism one major notational device of both *Begriffsschrift* and *Grundgesetze*.

Frege mentions in §80 that it can be deduced from the definition of "follows" that if b follows a in the ϕ-sequence and c follows b, then c follows a; the transitivity of the strong ancestral is Proposition (98) of the *Begriffsschrift*. The proof Frege gives there can be formalized in second-order logic only with the aid of the comprehension schema (or something to the same effect); however, there is an easier proof that makes uses only of the ordinary quantifier rules, applied to the universal quantifier in the definition of ϕ^* (Boolos, 1998i, pp. 158–9). For the proof in §§82–3, Frege will also need Propsition (95) of *Begriffsschrift*: if $x \phi y$, then $x \phi^* y$, which easily follows from the definition of ϕ^*.

At the very end of §80 Frege states that only by means of the definition of following in a sequence is it possible to reduce the method of inference ("Schlussweise", which Austin mistranslates as "argument") from n to $n + 1$ to the general laws of logic. Of course, the method of inference from n to $n + 1$ is what we call mathematical induction; Frege's remarks may be taken to be a claim that mathematical induction can be proved with the aid of the definition of the ancestral of P.

In §81, Frege defines the weak ancestral: "y is a member of the ϕ-sequence beginning with x" and "x is a member of the ϕ-sequence ending with y" are to mean: $x \phi y \vee y = x$. We shall use the abbreviation: $x \phi^{*=} y$. He states at the beginning of the section that if ϕ is P, then he will use the term "natural sequence of numbers" instead of "P-sequence". We thus have five terms: "y follows x in the natural sequence of numbers", "x

---

[8]All reference to sections of *Grundgesetze* in this chapter are to Volume I.

precedes y...", "y immediately follows x...", "x is a member of the natural sequence of numbers ending with y", and "y is a member...beginning with x". We shall abbreviate these as: xP^*y, xP^*y, xPy, $xP^{*=}y$, and $xP^{*=}y$, respectively.

Induction 2 is the following method of proof, in which weak ancestrals occur as hypotheses: To prove that if $x\phi^{*=}y$, then $...y...$, it suffices to prove:

(i) $...x...$

(ii) $\forall d \forall a (...d... \rightarrow d\phi a \rightarrow ...a...)$

Induction 2 follows quickly from Induction 1: If (i) and (ii) hold, then so does $\forall a (x\phi a \rightarrow ...a...)$; thus, if $x\phi^*y$, then $...y...$, by Induction 1. But if $x = y$, then by (i), $...y...$ again. Frege proves Induction 2 as Theorem 144 of *Grundgesetze*.

A basic fact about the weak ancestral, to which we shall repeatedly appeal, is that $x\phi^*a$ and thus $x\phi^{*=}a$, provided that $x\phi^{*=}d$ and $d\phi a$, for then either $x\phi^* d\phi a$, $x\phi^* d\phi^*a$, and $x\phi^*a$, or $x = d\phi a$, $x\phi a$, and $x\phi^*a$, by (95) and (98) of *Begriffsschrift*. That $x\phi^*a$ if $x\phi^{*=}d$ and $d\phi a$ is Theorem 134 of *Grundgesetze der Arithmetik*; that $x\phi^{*=}a$ if $x\phi^*a$ is Theorem 136.

Frege has not yet defined finite, or natural, number. He will do so only at the end of §83, where "n is a finite number" is defined as "n is a member of the natural sequence of numbers beginning with 0", i.e., as: $0P^{*=}n$. By Induction 2, to prove that $...n...$ if n is finite, it suffices to prove $...0...$ and $\forall d \forall a (...d... \rightarrow dPa \rightarrow ...a...)$.

In the formalism in which we are supposing Frege to be working the existence and uniqueness of 0, defined in §74 as $\#[x : x \neq x]$, are given by the comprehension scheme for second-order logic and the standard convention of logic that function signs denote *total* functions. Thus $\#$ denotes a total function from second-order to first-order entities, and the existence of $\#[x : x = x]$, that of $\#[x : x \neq x]$, and that of $\#[x : x = \#[x : x \neq x]]$ will count as truths of logic. The propositions that 0 is a natural number and that any successor of a natural number is a natural number follow immediately from the definition of "natural number"; 78.5 says that P is functional and one-one. So apart from the easily demonstrated statement that nothing precedes zero, by the end of §81 Frege can be taken to have established the Dedekind-Peano axioms for the natural numbers, except for the statement that every natural number *has* a successor.

Using the notation we have introduced, we may condense §§82–3 as follows:

§82. It is now to be shown that—subject to a condition still to be specified— (0) $nP\#[x : xP^{*=}n]$. And in thus proving that there exists a Number k such that nPk, we shall have proved at the same time that there is no last member of the natural sequence....

...It is to be proved that (1) $dPa \wedge dP\#[x : xP^{*=}d] \rightarrow aP\#[x : xP^{*=}a]$.

It is then to be proved, secondly, that (2) $0P\#[x : xP^{*=}0]$. And finally, it is to be deduced that (0') $0P^{*=}n \to nP\#[x : xP^{*=}n]$. The method of inference ("Schluss-weise") here is an application of the definition of the expression "y follows x in the natural sequence of numbers", taking $[y : yP\#[x : xP^{*=}y]]$ for our concept F.[9]

§83. In order to prove (1), we must show that (3) $a = \#[x : xP^{*=}a \wedge x \neq a]$. And for this again it is necessary to prove that (4*) $[x : xP^{*=}a \wedge x \neq a]$ has the same extension as $[x : xP^{*=}d]$. For this we need the proposition (5') $\forall a(0P^{*=}a \to \neg aP^*a)$. And this must once again be proved by means of our definition of following in a sequence, along the lines indicated above.

We are obliged hereby to attach a condition to the proposition $nP\#[x : xP^{*=}n]$, the condition that $0P^{*=}n$. For this there is a convenient abbreviation...: n is a finite number. We can thus formulate (5') as follows: no finite number follows itself in the natural sequence of numbers.

(We have added some reference numbers; (1) is Frege's own. Primes indicate the presence of a finiteness condition in the antecedent; the asterisk in (4*) indicates (what at least appears to be) a reference to extensions.)

It might appear that Frege proposes in these two sections to prove, not (0), but (0'), as follows: First, prove (5') by an appeal to the definition of P^*. Then derive (4*) from (5') and (3) from (4*). From (3) derive (1). Prove (2). Then, finally, infer (0') from (2) and (1), by a similar appeal to the definition of P^*.

However, it will turn out that this precise strategy cannot succeed. It cannot be (4*) and (3) that Frege wishes to derive—(3), e.g., is false if $a = \aleph_0$, as we shall see—but certain conditionals (4') and (3'), whose consequents are (4*) (or rather an equivalent of it) and (3).

We do not, of course, know how Frege might have tried to fill in the details of this proof-sketch at the time of composition of *Die Grundlagen*. In particular, we do not know exactly how he would have proved (5'). (We can be reasonably certain that his proof of (2), however, would have been at least roughly like the proof we shall give below.) But, since he later proved a version of the following lemma as Theorem 141 of *Grundgesetze*, it seems plausible to us to speculate that he might have intended to appeal to something rather like it in his proof of (5'). The lemma is a logicized version of the arithmetical truth: if $i < k$, then for some j, $j + 1 = k$ and $i \leq j$.

Lemma. $xP^{*=}y \to \exists z(zPy \wedge xP^*z)$

Proof. Let $Fa \equiv \exists z(zPa \wedge xP^{*=}z)$. Then $xPa \to Fa$, for if xPa, then certainly Fa: take $z = x$. And $Fd \wedge dPa \to Fa$: Suppose Fd and dPa. Then

---

[9]This sentence seems to throw Austin. But we take its last half to mean: when one takes for the concept F what is common to the statements about d and about a, about 0 and about n, and thus that the concept in question is $\#[y : yP\#[x : xP^{*=}y]]$. Austin's translation makes it sound as if some binary relation holding between d and a and also between 0 and n were meant. However good his German and English may have been, Austin was no logician. It is time for a reliable English translation of *Die Grundlagen*.

for some z, zPd and $xP^{*=}z$. By the basic fact about the weak ancestral, $xP^{*=}d$. But since dPa, Fa. The lemma follows by Induction 1.   □

With the aid of the lemma, we can now use Induction 2 to prove (5'):

Proof. $0 = \#[x : x \neq x]$. By HP and the definition of P, $\forall z(\neg zP0)$, and therefore, by the lemma, $\neg 0P^*0$.

Now suppose dPa and aP^*a. Then by the lemma, for some z, zPa and $aP^{*=}z$, i.e., either aP^*z or $a = z$, and therefore either $zPaP^*z$ or $zPa = z$. In either case, zP^*z. Since dPa and zPa, $z = d$ by 78.5, and so dP^*d. Thus, $\neg dP^*d \wedge dPa \rightarrow \neg aP^*a$.

(5') now follows by Induction 2.   □

(5') merits a digression. The part of *Die Grundlagen der Arithmetik* entitled "Our definition completed and its worth proved" begins with §70 and ends with §83; the concluding sentence of §83 reads: "We can thus formulate the last proposition above as follows: No finite number follows itself in the natural sequence of numbers." Apart from its position in the book and the fact that Frege mentions it in both the table of contents and the recapitulation of the book's argument at the end of *Die Grundlagen*, there are a number of reasons for thinking that Frege regarded this proposition as especially significant.

First, there is, according to Frege, an interesting connection with *counting*. When we count, he points out in §108 of *Grundgesetze*, we correlate the objects falling under a concept $\Phi(\xi)$ with the number words in their normal order from "one" up to a certain one, "N"; N is then the number of objects falling under $\Phi(\xi)$. Since correlating relations between concepts are not in general unique,

the question arises whether one might arrive at a different number word "M" with a different choice of this relation. By our stipluations, M would then be the same number as N, but at the same time one of the two number words would follow after the other, e.g., "N" after "M". Then N would follow in the sequence of numbers after M, i.e., after itself. Our Proposition [(5')] excludes this for finite numbers.

We find this argument of considerable interest, but will not enter into a discussion of its correctness here.

Second, one of Frege's major philosophical aims, as is well known, was to show that reason, under the aspect of logic, could yield conclusions for which many philosophers of his day might have supposed some sort of Kantian intuition to be necessary. The proof of (5') is a paradigm illustration of how the role of intuition in delivering knowledge can be played by logic instead.

One might think that the truth of (5') could be seen by the following sort of mixture of reason and intuition: (5') says that there is no (non-null) loop of P-steps leading from a back to a whenever a is a finite number. So if a is finite but not zero and there is a loop from a to a, then within the loop, there is some number x that (immediately) precedes a, and therefore there is a loop from x (through a, back) to x. But since a is finite, there is a finite sequence of P-steps from zero to some number d preceding a; since *precedes* is one-one, $d = x$, and therefore there is a loop from d to d. Thus a loop "rolls back" from a to d, and then all the way back to zero. But there is no loop from zero to zero; otherwise, some number would precede zero, and that is impossible.

Of course, Frege's proof of Theorem 145 avoids any appeals to intuition like those found in the foregoing argument.

Finally, in the proof of Theorem 263 of *Grundgesetze*, Frege shows that any structure satisfying a certain set of four conditions is isomorphic to that of the natural numbers. We find it quite plausible to think that Frege realized that the statement that the natural numbers satisfy these conditions constitutes an axiomatization of them and regarded them as *the* basic laws of arithmetic. (See Section 2.6.) Since one of these conditions is the one (5′) shows to be satisfied, there is considerable reason to think that Frege regarded (5′) as one of the basic laws of arithmetic.

End of digression.

(4*) at least appears to mention extensions of (first-level) concepts and may well do so.[10] But (4*) is unlike the definition of cardinal number and proof of HP in that any mention of extensions it contains is readily eliminable without loss: Frege could have written to exactly the same point, "a member of the natural sequence of numbers ending with a, but not identical with a, is a member of the natural sequence of numbers ending with d, and vice versa".

It is evident that Frege cannot be proposing to derive (4*) or the equivalent

(4) $\forall x\,([xP^{*=}a \land x \neq a] \equiv xP^{*=}d)$

from (5′) since both (4*) and (4) contain free occurrences of "d". Since the supposition of §82 that dPa is still clearly in force, it might be thought that Frege wishes to derive

(4†) $dPa \rightarrow \forall x\,([xP^{*=}a \land x \neq a] \equiv xP^{*=}d)$

from (5′).

However, if $d = a = \aleph_0$, then, as we have observed, dPa; and then, since $aP^{*=}a$, (4†) has a true antecedent and false consequent. Thus it cannot be (4†) that Frege is proposing to derive from (5′).

We may note, though, that $\forall x\,([xP^{*=}a \land x \neq a] \equiv xP^{*=}d)$ can be derived from dPa and $\neg aP^*a$. So we may take it that Frege is proposing to derive

---

[10]This is not so clear in the German, which reads: "Und dazu ist wieder zu beweisen, dass dieser Begriff gleichen Umfanges mit dem Begriffe 'der mit d endenden natürlichen Zahlenreihe angehörend' ist". A more literal translation would have Frege say that we must prove that one concept "is the same in extension" as another: A natural translation of "gleichen Umfanges mit... ist" would be is "is co-extensional with".

(4') $0P^{*=}a \to dPa \to \forall x\,([xP^{*=}a \land x \neq a] \equiv xP^{*=}d)$

from (5').

Proof. Suppose $0P^{*=}a$ and dPa. Assume $xP^{*=}a \land x \neq a$. Then xP^*a. By the lemma, for some c, cPa and $xP^{*=}c$. By 78.5, $c = d$. Thus, $xP^{*=}d$. Conversely, assume $xP^{*=}d$. Since dPa, xP^*a, by the basic fact about the weak ancestral, and so $xP^{*=}a$. If $\neg aP^*a$, then also $x \neq a$. But since $0P^{*=}a$, it follows from (5') that indeed $\neg aP^*a$. Thus (4') is proved.   □

Nor could Frege be proposing to derive (3) $a = \#[x : xP^{*=}a \land x \neq a]$ from any proposition he takes himself to have demonstrated. For (3) is false if "a" has $\aleph_0$ as value. In fact, $\#[x : xP^{*=}\aleph_0 \land x \neq \aleph_0] = 0$. For if $xP\aleph_0$, then since $\aleph_0 P\aleph_0$, $x = \aleph_0$, by 78.5. Let S be the converse of P. Then if $\aleph_0 Sx$, $x = \aleph_0$. Thus if $\aleph_0 S^*x$, $x = \aleph_0$. (Let $Fa \equiv a = \aleph_0$ in the definition of S^*.) But the ancestral is the converse of the ancestral of the converse. So if $xP^*\aleph_0$, then $x = \aleph_0$. Thus $xP^{*=}\aleph_0$ iff $x = \aleph_0$, and therefore for *no* x, $xP^{*=}\aleph_0 \land x \neq \aleph_0$. By a proposition given in §75, $\#[x : xP^{*=}\aleph_0 \land x \neq \aleph_0] = 0$.

However, it is important to observe that at this point it is not only the conjunct dPa of the antecedent of (1) that is assumed to be in force; the other conjunct $dP\#[x : xP^{*=}d]$ is also assumed to hold. (It is easy to be oblivious to this further assumption since (3) does not mention d. But it is supposed at this point that a is such that dPa, and it is likely also supposed that d is such that $dP\#[x : xP^{*=}d]$.) Since (3) follows from these two conjuncts and the consequent of (4'), we may take it that Frege wishes to prove:

(3') $0P^{*=}a \to dPa \to dP\#[x : xP^{*=}d] \to a = \#[x : xP^{*=}a \land x \neq a]$

Proof. Suppose dPa and $dP\#[x : xP^{*=}d]$. Then by 78.5 (the other way), $a = \#[x : xP^{*=}d]$. Suppose further that $0P^{*=}a$. Then by (4'), $\forall x([xP^{*=}a \land x \neq a] \equiv xP^{*=}d)$. By HP, $\#[x : xP^{*=}a \land x \neq a] = \#[x : xP^{*=}d]$. Thus $a = \#[x : xP^{*=}a \land x \neq a]$.   □

We come now to the difficult question how Frege proposes to derive (1) from (3'). Frege tells us that to prove (1), we must show (3). But (3) is not unconditionally true. However, (3'), whose consequent is (3) and whose antecedent contains a conjunct stating that the value of "a" satisfies the condition of finiteness, can be proved. Thus it might seem reasonable to think that Frege may be proposing, as in the case of (4) and (3), not to derive (1) from (3), but some conditional whose antecedent expresses a finiteness condition and whose consequent is (1). Moreover, since dPa is one of the clauses of the antecedent, it we take $0P^{*=}d$ as another conjunct of the antecedent, we need not also take $0P^{*=}a$. So we have

(1') $0P^{*=}d \to dPa \to dP\#[x : xP^{*=}d] \to aP\#[x : xP^{*=}a]$

(1') readily follows from (3').

Proof. Suppose that $0P^{*=}d$, dPa, and $dP\#[x : xP^{*=}d]$. By the basic fact about the weak ancestral, $0P^{*=}a$. By (3′), $a = \#[x : xP^{*=}a \wedge x \neq a]$. Since $a = a$, $aP^{*=}a$. By the definition of P, $\#[x : xP^{*=}a \wedge x \neq a] \, P \, \#[x : xP^{*=}a]$. Thus $aP\#[x : xP^{*=}a]$.   □

It may be useful to recapitulate here our (somewhat intricate) derivation of (1′) from (5′) and the other propositions to which Frege appeals.

Proof. Suppose $0P^{*=}d$, dPa, and $dP\#[x : xP^{*=}d]$. By the basic fact about the weak ancestral, $0P^{*=}a$, and thus by (5′), $\neg aP^*a$. If $xP^{*=}a \wedge x \neq a$, then xP^*a, and so by the lemma, for some z, $xP^{*=}z$ and zPa. By one half of 78.5, $z = d$, and so $xP^{*=}d$. Conversely, if $xP^{*=}d$, then by the basic fact, xP^*a, whence $x \neq a$ (since $\neg aP^*a$) and $xP^{*=}a$. Thus $\forall x \, ([xP^{*=}a \wedge x \neq a] \equiv xP^{*=}d)$, which is (4), and so by HP $\#[x : xP^{*=}a \wedge x \neq a] = \#[x : xP^{*=}d]$, and therefore $dP\#[x : xP^{*=}a \wedge x \neq a]$. By the other half of 78.5, $a = \#[x : xP^{*=}a \wedge x \neq a]$, which is (3). Since $aP^{*=}a$ (trivially), by the definition of P, $\#[x : xP^{*=}a \wedge x \neq a] \, P \, \#[x : xP^{*=}a]$, and therefore $aP\#[x : xP^{*=}a]$.   □

(2) is proved much more easily.

(2) $0P\#[x : xP^{*=}0]$

Proof. $0 = \#[x : x \neq x]$. By HP and the definition of P, $\forall z(\neg zP0)$. By the lemma, $\forall x(\neg xP^*0)$, and so $\forall x \neg(xP^{*=}0 \wedge x \neq 0)$. By a result of §75 mentioned above, $\#[x : xP^{*=}0 \wedge x \neq 0] = 0$. But $0P^{*=}0$, whence $\#[x : xP^{*=}0 \wedge x \neq 0] \, P \, \#[x : xP^{*=}0]$, and therefore $0P\#[x : xP^{*=}0]$.   □

(0′) must now be derived from (1′) and (2). It is not possible to appeal to Induction 2 because of the presence of "$0P^{*=}d$" in the antecedent of (1′). But, it might be supposed, Frege can appeal here to *Induction 3*, which he explicitly demonstrated in *Grundgesetze* as Theorem 152: To prove that if $x\phi^{*=}y$, then $\ldots y \ldots$, it suffices to prove:

(i) $\ldots x \ldots$

(ii) $\forall d \forall a(x\phi^{*=}d \to \ldots d \ldots \to d\phi a \to \ldots a \ldots)$

Note the formula $x\phi^{*=}d$, whose presence weakens (ii) and so strengthens the method. The derivation of Induction 3 from Induction 2 is significantly more interesting than that of Induction 2 from Induction 1. It appeals to the comprehension scheme of second-order logic and uses a technique sometimes called "loading the inductive hypothesis". (At the beginning of §116 of *Grundgesetze*, Frege writes, "To prove proposition (γ) of §114, we replace the function mark '$F(\xi)$' with '$\neg(aP^{*=}x \to \neg F(\xi))$'.")

Proof of Induction 3. Suppose $x\phi^{*=}y$ and, moreover, (i) and (ii). Let $Ga \equiv \ldots a \ldots \wedge x\phi^{*=}a$ (second-order comprehension). Now, $x\phi^{*=}x$ trivially; thus by (i), Gx. We now show $\forall d \forall a(Gd \to d\phi a \to Ga)$: Suppose $d\phi a$ and Gd, i.e., $\dots d \dots$ and $x\phi^{*=}d$. By (ii), $\dots a \dots$. By the basic fact about the weak ancestral, $x\phi^{*=}a$. Thus, $\forall d \forall a(Gd \to d\phi a \to Ga)$. By Induction 2, Gy, whence $\dots y \dots$. $\qquad\qquad\qquad\qquad\qquad\qquad\qquad\qquad\qquad\qquad\qquad\square$

We believe that no one will seriously dispute that this proof of $(0')$, which features a derivation of $(1')$ from $(5')$ and an appeal to Induction 3, is Fregean in spirit, ingenious, and of a structure that fits the proof-sketch found in §§82–3 rather well. But there are a number of strong reasons for doubting that Frege had *this* proof in mind while writing these two sections. Accordingly, we shall refer to it as the *conjectural* proof.[11]

First of all, Frege twice *says* that (1) is to be proved, once in §82 and again in §83. He says, moreover, "The method of inference here is an application of the definition of the expression 'y follows x in the natural sequence of numbers', taking $[y : yP\#[x : xP^{*=}y]]$ for our concept F". It would thus seem natural to take Frege as arguing by appeal to Induction 1 or Induction 2 (with P as ϕ). Frege mentions the condition that n be finite, but does not also mention, as he might easily have done, the need to assume that d (or a) is finite as well. Thus it would seem to be overly charitable to assume that the argument he really intended proceeds via Induction 3.

Second, notice that Frege says in §83 that $(5')$, which he proves in *Grundgesetze* by appeal to Induction 2, "must likewise ('ebenfalls') be proved by means of our definition of following in a series, as indicated above". It seems plain that Frege does not intend to use Induction 3 to prove $(5')$; "ebenfalls" suggests that the induction used to prove $(0')$ would be like the one used for $(5')$.

The most telling objections to the suggestion that Frege was intending to sketch the conjectural proof in *Die Grundlagen*, however, arise from a close reading of Section H (Eta) of Part II of *Grundgesetze*. We quote and comment upon part of Section H.[12]

H. *Proof of the Proposition*

$$0P^{*=}b \to bP\#[x : xP^{*=}b]$$

§114. Analysis

We wish to prove the proposition that the Number that belongs to the concept

member of the number-series ending with b

follows after b in the number-series if b is a finite number. Herewith, the conclusion that the number-series is infinite follows at once; i.e., it follows at once that there is, for each finite number, one immediately following after it.

---

[11]Of course, what is conjectural is whether the proof is Frege's, not whether it is a (correct) proof.

[12]The present translation is based upon one due to Richard Heck and Jason Stanley. We have changed Frege's notation to ours and added some material in brackets.

We first attempt to carry out the proof with the aid of Theorem (144) $[aq^{*=}b \rightarrow \forall d(Fd \rightarrow \forall a(dqa \rightarrow Fa) \rightarrow (Fa \rightarrow Fb)]$, replacing the function-mark "$F\xi$" with "$\xi P\#[x : xP^{*=}\xi]$". For this we need the proposition "$dP\#[x : xP^{*=}d] \rightarrow dPa \rightarrow aP\#[x : xP^{*=}a]$".

That is to say, one's "first" idea might be to prove (0′) by applying Induction 2 to the concept $[y : yP\#[x : xP^{*=}y]]$, which would, among other things, require a proof of (1). (A footnote, to which we shall return, is attached to this last sentence.)

Substituting... in (102) $[\#[x : Fx \wedge x \neq b] = c \rightarrow Fb \rightarrow cP\#F]$, ... we thus obtain

$$\#[x : P^{*=}xa \wedge x \neq a] = a \wedge aP^{*=}a \rightarrow a = \#[x : xP^{*=}a]$$

from which we can remove the subcomponent "$aP^{*=}a$" by means of (140) $[aP^{*=}a]$. The question arises whether the subcomponent "$\#[x : xP^{*=}a \wedge x \neq a] = a$" can be established as a consequence of "dPa" and "$dP\#[x : P^{*=}xd]$".

Put differently, the problem reduces to that of proving

(3†) $dPa \rightarrow dP\#[x : xP^{*=}d] \rightarrow a = \#[x : xP^{*=}a \wedge x \neq a]$

which is (3′) without the finiteness condition $0P^{*=}a$, and which, together with the relevant instance of Frege's Theorem 102, implies (1).

By the functionality of progression in the number-series..., we have

$$dP\#[x : P^{*=}xd] \wedge dPa \rightarrow a = \#[x : P^{*=}xd]$$

... We thus attempt to determine whether

$$\#[x : xP^{*=}a \wedge x \neq a] = \#[x : xP^{*=}d]$$

can be shown to be a consequence of "dPa". ... For this it is necessary to establish

$$[bP^{*=}a \wedge b \neq a] \equiv bP^{*=}d$$

as a conseqence of "dPa"....

That is, (3†) will follow from

(4†) $dPa \rightarrow \forall x ([xP^{*=}a \wedge x \neq a] \equiv xP^{*=}d)$

an easy consequence of HP, and the one-one-ness of P.

For this it is necessary to establish

$$bP^{*=}a \wedge b \neq a \rightarrow bP^{*=}d$$

and

$$bP^{*=}d \rightarrow bP^{*=}a \wedge b \neq a$$

as consequences of "dPa". But it turns out that another condition must be added if "$b \neq a$" is to be shown to be a consequence of "$bP^{*=}d$" and "dPa". By (134) we have

$$bP^{*=}d \wedge dPa \rightarrow bP^{*}a$$

If b coincided with a, then the main component [consequent] would transform into "aP^*a". By (145) [our (5')], this is impossible if a is a finite number. Thus the subcompent "$0P^{*=}a$" is also added.

Admittedly, the desired application of (144) thereby becomes impossible; but, with (137) [viz., $aq^{*=}e \to eqm \to aq^{*=}m$], we can replace this subcomponent with "$0P^{*=}d$" and derive from (144) the Proposition [(152)]

$$(aq^{*=}b \to \forall d\,(Fd \to aq^{*=}d \to \forall a\,(dqa \to Fa))) \to (Fa \to Fd)$$

which takes us to our goal.

That is, to establish the first half of (4'), we need to know that $\neg aP^*a$; this will follow from (5') and the additional assumption that a is finite. However, this new assumption must then be carried along throughout the proof, transforming (4†) into (4'), (3†) into (3'), and (1) into "$0P^{*=}a \to dP\#[x : xP^{*=}d] \to dPa \to aP\#[x : xP^{*=}a]$", from which (1') easily follows. The attempt to prove (0') via Induction 2 then fails, since we simply have not proved (1), though we can still complete the proof by making use of Induction 3 instead.

It is, we think, difficult to read these passages without supposing that they reveal Frege's *second thoughts* about his idea in *Die Grundlagen* of applying Induction 2 to prove $0P^{*=}n \to nP\#[x : xP^{*=}n]$ by substituting $[y : yP\#[x : xP^{*=}y]]$ for F. The attempt won't work, he says, because we need the hypothesis that a is finite in order to derive $\neg aP^*a$, which is needed for $bP^{*=}d \to dPa \to b \neq a$, which is in turn necessary for the rest of the proof. Read side by side with §§82–3 of *Die Grundlagen*, Frege's discussion in these paragraphs strikes us as penetrating and direct criticism of his earlier work. Moreover, the criticism suggests a way in which the conjectural proof can be regarded as Frege's after all: it is the proof obtained on amending the proof-sketch of §§82–3 in the way suggested in this section of *Grundgesetze*.

It is striking that the formal proof Frege actually gives in *Grundgesetze*, though closely related to the conjectural proof, is not quite the same proof.[13] The formal proof, given in §§115, 117, and 119, does proceed by deriving (0') by means of Induction 3 (Frege's Proposition 152), from (1'), which is (150ε),[14] and (2), which is (154). And the proof of (1') does begin with a derivation of (4'), which is (149ε), from (5'), which is (145). But (1') is not derived from (4') via (3'); the argument is slightly different.

This part of the *Grundgesetze* proof, translated into English plus our notation, runs as follows. By the basic fact about the weak ancestral, it suffices to show that if $0P^{*=}a$, dPa, and $dP\#[x : xP^{*=}d]$, then $aP\#[x : xP^{*=}a]$. By (4') and (an easy consequence of) HP, we have that $\#[x : xP^{*=}a \land x \neq a] = \#[x : xP^{*=}d]$ (cf. 149). But substituting into Proposition

---

[13]For a fuller account, see Chapter 2.

[14]By proposition nx we mean the proposition labeled with Greek letter x that occurs *during*, as opposed to *after*, the proof of proposition number n.

(102) quoted above, we have

$$\#[x : xP^{*=}a \wedge x \neq a] = \#[x : xP^{*=}d] \rightarrow aP^{*=}a \rightarrow$$
$$\#[x : xP^{*=}d] \, P \, \#[x : xP^{*=}a]$$

Hence, by (140), $\#[x : xP^{*=}d] \, P \, \#[x : xP^{*=}a]$ (cf. 150β). Since dPa and $dP\#[x : xP^{*=}d]$, $a = \#[x : xP^{*=}d]$ (cf. 150γ), whence $aP\#[x : xP^{*=}a]$ (cf. 150δ), and we are done.

Comparing this argument with the relevant portion of the conjectural proof, one sees immediately how little they differ from each other; one might therefore overlook (or ignore) the fact that (3′) does not actually appear in the proof given in *Grundgesetze*. But the omission of (3′) is significant, since the "proof" discussed in §114 explicitly highlights (3†) as what must be proved if (1) is to be derived from (4†). The typical point of a section of *Grundgesetze* headed "Analysis" is to describe a formal proof found in "Construction" sections that follow it. Thus on reading §114, one would naturally expect the following proof to include, not just proofs of the results of adding a finiteness condition to (4†) and to (1), but also, as part of the derivation of the latter from the former, a proof of a proposition similarly related to (3†). As we said, however, the derivation of (1′) from (4′) in §115 does not go via (3′). That (3†) is so much as mentioned in §114 is therefore bound to seem mysterious unless one reads it as we have suggested: as criticism of Frege's *own* "first attempt" to prove (0′) in §§82–3 of *Die Grundlagen*, for (3†) or (3′) is indeed an intermediate step in *that* proof.

This observation concerning how the *Grundgesetze* proof differs from the conjectural proof also suggests a plausible explanation of the origin of the mistake of which we have accused Frege. Consider the two lists of propositions in   Table 3.1 on page 84. As we have seen, (4′) follows from (5′), (3′) from (4′), and (1′) from (3′). But notice also that (4†) follows from (5†), (3†) from (4†), and (1) from (3†), as obvious modifications of our proofs show. Frege, able to prove (5′) and desirous of proving (1), may well have lost sight of the need for a finiteness condition somewhere in the middle of his argument—perhaps he had not yet fully written out the argument in his conceptual notation—and mistakenly concluded that he could deduce (1) from (5′). If forced to guess, we would suppose that it was between (4′) and (1), i.e., at (3†) or (3′), that the finiteness condition vanished, for it is there that the *Grundgesetze* proof differs from the conjectural proof.

The first sentence of the second paragraph of §83 calls for some discussion. Frege writes there that we are obliged "hereby" ("hierdurch") to attach to the proposition that $nP\#[x : xP^{*=}n]$ the condition that $0P^{*=}n$. One might be forgiven for thinking that, in so stating, Frege is indicating that this condition is required by the presence of the finiteness condition in (5′), since it is with an indication of how (5′) is to be proven that the previous paragraph ends. But this thought cannot be right. Frege says in §82 that, once (1) and (2) are proved, "it is to be deduced that

Table 3.1: The Origin of Frege's Mistake

(1') $0P^{*=}d \to dPa \to dP\#[x : xP^{*=}d] \to aP\#[x : xP^{*=}a]$

(3') $0P^{*=}a \to dPa \to dP\#[x : xP^{*=}d] \to a = \#[x : xP^{*=}a \wedge x \neq a]$

(4') $0P^{*=}a \to dPa \to \forall x\,([xP^{*=}a \wedge x \neq a] \equiv xP^{*=}d)$

(5') $0P^{*=}a \to \neg aP^*a$

(1) $dPa \to dP\#[x : xP^{*=}d] \to aP\#[x : xP^{*=}a]$

(3†) $dPa \to dP\#[x : xP^{*=}d] \to a = \#[x : xP^{*=}a \wedge x \neq a]$

(4†) $dPa \to \forall x\,([xP^{*=}a \wedge x \neq a] \equiv xP^{*=}d)$

(5†) $\neg aP^*a$

$0P^{*=}n \to nP\#[x : xP^{*=}n]$" by means of Induction 2. Thus what subjects n in (0) to a finiteness condition is not the presence of such a condition in (5'), but the kind of proof (0') being given in the first place. "Hierdurch" refers to the use in the proof of (0') of the "definition of following in a series, on the lines indicated above", that is, as was discussed in §82.

There is one final piece of textual evidence to which we should like to draw attention. As we said earlier, a footnote is attached to (1) when it is first mentioned in §114: "This proposition is, as it seems, unprovable, but it is not here being asserted as true, since it stands in quotation marks". The natural explanation for this remark of Frege's is that he once *did* believe (1) to be provable, namely when he wrote *Die Grundlagen*, and any defender of the view that Frege was outlining the conjectural proof in §§82–3 will have the occurrence of this remark to explain away.

Apart from the light it may throw on the question whether Frege made a reparable error, the footnote is astonishing. Note that Frege says, not that (1) seems to be false, but that it "seems to be *unprovable*" [emphasis ours]. There is, moreover, reason to suppose Frege believed (1) to be not false, but *true*. For one thing, had Frege believed it to be false, he presumably would have said so. Furthermore, Frege's difficulty was probably not that he did not know how to prove (1), but rather that he did not know how to prove it *in his formal system*. There is a very simple proof of (1) that depends only upon (1'), Dedekind's claim that every infinite number is (the number of a concept that is) Dedekind infinite, and the observation, made earlier, that "d is Dedekind infinite" is equivalent to "dPd". We may take (1') to be one half of a dilemma, the other half of which is:

$$\neg 0P^{*=}d \to dPa \to dP\#[x : xP^{*=}d] \to aP\#[x : xP^{*=}a].$$

This proposition may be proved as follows. Suppose the antecedent. Since d is not finite, it is Dedekind infinite. So dPd, and since dPa, $d = a$, and now the consequent is immediate.

This proof is one Frege might well have known. It is not at all difficult and once (1') has been proved, a proof of (1) by dilemma suggests itself. Moreover, Frege was familiar with Dedekind's claim and, at least while he was working on Part II of *Grundgesetze*, believed it to be true (RevCan, op. 271).[15] As for the observation, not only is it easily proved, it is natural, in Frege's system, just to use "dPd" as a *definition* of "d is Dedekind infinite" (cf. *Grundgesetze*, Proposition 426). We conclude that Frege believed (1) to be a true but unprovable formula of Frege Arithmetic.

Frege's belief that (1) is unprovable in Frege Arithmetic is mistaken, however. A proof of (1) can be given that makes use of techniques that are different from any found in §§82–3 of *Die Grundlagen* or in relevant sections of *Grundgesetze*, but with which Frege was familiar. What we shall prove is that the hypothesis $0P^{*=}d$ of (1'), that d is finite, is dispensable. More precisely, we shall prove that if $dP\#[x : xP^{*=}d]$, then $\#[x : xP^{*=}d]$ is finite, from which it follows that d is finite, since, by Proposition (143) of *Grundgesetze* (viz, $dPb \to aP^{*}b \to aP^{*=}d$), any predecessor of a finite number is finite.

Theorem (FA). *Suppose $dP\#[x : xP^{*=}d]$. Then $\#[x : xP^{*=}d]$ is finite.*[16]

Proof. In FA, define $h : [x : 0P^{*=}x] \to [x : xP^{*=}d]$ by:

$$h(0) = d; \quad h(n+1) = \begin{cases} y & \text{if } yPh(n) \\ h(n) & \text{if } \neg\exists y(yPh(n)) \end{cases}$$

The definition is OK since P is one-one.

Since in general $yR^{*}z \equiv z(R^{\cup})^{*}y$,[17] $\forall x(xP^{*=}d \equiv d(P^{\cup})^{*=}x)$, and so h is onto. Therefore $[x : xP^{*=}d]$ is countable, i.e., either finite or countably infinite. If the latter, then $\#[x : xP^{*=}d] = \aleph_0$, and by the supposition of the theorem, $dP\aleph_0$. But as we saw just after the proof of (4'), $xP^{*=}\aleph_0 \to x = \aleph_0$. Since $dP\aleph_0$, $dP^{*=}\aleph_0$, $d = \aleph_0$, and $\#[x : xP^{*=}d] = 1$, contra $\#[x : xP^{*=}d] = d$. Therefore $\#[x : xP^{*=}d]$ is finite. □

Thus Frege could have proved (1) after all and thus appealed to Induction 2 to prove (0'). Of course, the technology borrowed from second-order

---

[15]Of course, if Frege did know of our proof and believed (1) to be unprovable, then he must have believed Dedekind's result too to be unprovable, which he (rightly) did. For further discussion, see Heck (1995a).

[16]This result is due to Heck; the present proof, to Boolos.

[17]$R^{\cup}$ is the converse of R.

arithmetic used in the proof just given, particularly the inductive definition of h, is considerably more elaborate than that needed to derive
Induction 3 from Induction 2. The conjectural proof is unquestionably to
be preferred to this new one on almost any conceivable grounds.

So, Frege erred in §§82–3 of *Die Grundlagen*, where an oversight
marred the proof he outlined of the existence of the successor. Mistakes of
that sort are hardly unusual, though, there are four or five ways the proof
can be patched up, and Frege's way of repairing it cannot be improved on.
But even if one ought not to make too much of Frege's mistake, there is
lots to be made of his belief that (1) was true but unprovable in his system. One question that must have struck Frege is: If there are truths
about numbers unprovable in the system, what becomes of the claim that
the truths of arithmetic rest solely upon definitions and general logical
laws? Another that may have occurred to him is: Can the notion of a
truth of logic be explained otherwise than via the notion of provability?

Appendix: Counterparts in *Grundgesetze* of Some Propositions of *Die Grundlagen*

Proposition of this chapter	Proposition of *Grundgesetze*
HP	32, 49
$\forall x(\neg Fx) \rightarrow 0 = \#[x : Fx]$	94, 97
781.	114
78.2	113
78.3	117
78.4	122
78.5	71, 89, 90
78.6	107
$\neg zP0$	108
Induction 1	123
The basic fact about the weak ancestral	134, 136
The Lemma	141
Induction 2	144
(5′)	145
(4′)	149α
(1′)	150
Induction 3	152
(2)	154
(0′)	155

Postscript

"Reluctantly and hesitantly, we have come to the conclusion that Frege was at least somewhat confused in these two sections and that he cannot be said to have outlined, or even to have intended, any correct proof" in §§82–3 of *Die Grundlagen*. So Boolos and I said back on page 69. That was true then, and it remains true now. Every time I re-read this paper, I am reluctant and hesitant again. By the time I am done with it, however, I am convinced again. Perhaps the story of its genesis will add a bit to the case.

This chapter began life as a paper of George's, one he presented at the conference *Philosophy of Mathematics Today*, held in Munich in the summer of 1993. Much of the first half of the chapter, up to where we begin to discuss the proof presented in §§82–3, was then as it is now. But the punch line was very different: It was that Frege's argument suffers from a very large redundancy.

As George was reading it, the proof of what we call (1′) went via a proof of

(3#) $\quad 0P^{*=}a \rightarrow aP\#[x : xP^{*=}a]$

which is our (3) plus a finiteness condition. In reading the proof this way, George was not alone. Crispin Wright had read the proof the same way in *Frege's Conception of Numbers as Objects* (Wright, 1983, §xix), as I had, in the original version of Chapter 2 (Heck, 1993b, pp. 591–2). Nor did anyone present when George presented the paper voice any objection to that reading, and, though the audience was small, it included many of the finest philosophers of mathematics in the world,[18] as a look at the proceedings (Schirn, 1998) will show.

What George observed, however, was that (3#) implies our (0′) almost immediately. Substituting into (102), we have:

$$aP^{*=}a \wedge \#[x : P^{*=}xa \wedge x \neq a)] = a \rightarrow aP\#[x : P^{*=}xa])$$

The first conjunct is trivial, and (3#) tells us that the second follows from $0P^{*=}a$. So the proofs of (1′) and (2), and the derivation of (0′) from them, are all unnecessary. Talk about redundancy! It was stunning.

I taught a course on Frege the next fall and took the opportunity to read George's paper closely. Somewhere along the way, I realized that we had all been misreading Frege's proof. I think the realization must have been due to how intensively I was then working on *Grundgesetze*. I knew the structure of Frege's argument there like I knew my way home, and my strategy, in effect, was to impose the later argument onto the unclear presentation in *Die Grundlagen*. That made it obvious that Frege was

---

[18]Many of the finest, and one guy two years out of graduate school. Thanks again to Mattias Schirn for inviting me, and the other presenters for welcoming me into their midst.

carrying along his earlier assumptions, so that (3#) was no part of the argument; only (3′) was. I sent George an email the day I had my epiphany. He responded with a note inviting me to co-author a new version of the paper.

The next version took the charitable route, reading Frege as sketching the conjectural proof in §§82–3. That was the view on which I'd settled in the fall, and it surfaces in a postscript I wrote for the reprint of Chapter 2 in Demopoulos's collection *Frege's Philosophy of Mathematics* (Heck, 1995b, pp. 288–94). But George was never very happy with this interpretation. Something about it simply didn't sit right with him. His original worries were the ones we recount, starting on page 80, as the first and second. I wasn't moved. Somewhere along the way, though, we started reading *Grundgesetze* section H closely, and it was this reading that firmly convinced both of us. What did it for me was the realization that the "Analysis" in §114 did not match the "Construction" given in §§115, 117, and 119. Not only do I know of no other place in *Grundgesetze* where there is such a mismatch, it is entirely out of the spirit of how the book is put together.

There is, in fact, even more evidence in favor of our reading. Frege says on a couple of occasions that his argument involves an "application of the definition of the" ancestral. And when he gives that definition, of course, he refers the reader to *Begriffsschrift* for the details. So one would naturally suppose that the "application" to which he is alluding would be of the sort one finds in *Begriffsschrift*. But, as we note, the conjectural proof requires an appeal to Induction 3, which is Proposition (144) of *Grundgesetze*, and nothing remotely like it is present in *Begriffsschrift*. Had Frege intended to appeal to Induction 3, therefore, it seems to me that he would have mentioned it explicitly. But he does not.

What has continued to bother me, though, is the question how Frege could have made such a mistake. He could not have had a formal proof to consult, which means that he probably had not produced a full proof in his conceptual notation. This is somewhat puzzling, though, since Frege says, in a letter to the philosopher Anton Marty, written in 1882:

I have now nearly completed a book in which I treat of the concept of number and demonstrate that the first principles of computation, which up to now have generally been regarded as unprovable axioms, can be proved from definitions by means of logical laws alone.... (PMC, p. 99)

This book presumably contained *formal* demonstrations of these "first principles of computation".

As Frege says, however, he had not finished his formal treatment in 1882, and one of the things left unfinished may well have been the proof of the existence of successor. Still, one might wonder how he could have made such a mistake. Part of the answer, it seems to me, is that *Die Grundlagen* must have been written very quickly. Marty's colleague Carl Stumpf, in a reply to Frege's letter, urges him "to explain [his] line of thought first in ordinary language and then... in conceptual notation" (PMC, p. 172, letter xl/1). It was this suggestion that seems to have given birth to *Die Grundlagen*. Stumpf's letter is dated September 9, 1882, and *Die Grundlagen* was published in 1884. Publication took time back then, and Frege must surely have done a good deal of reading in preparing to write the first three, more critical and philosophical, chapters of the book. He would have had very little time to work on his other manuscript.

But one would still like to have a somewhat better sense how, even if he had no more than the sketch given in *Die Grundlagen*, a logician of Frege's stature might have made such a mistake. In re-thinking this material yet again, though, I had an idea about this, one that I think has real plausibility.

As we mention in our discussion of "hierdurch" (see page 83), there is a sense in which the presence of the finiteness condition in $(0')$ is over-determined. Any proof that goes *via* $(1')$ will require it for *two* reasons. On the one hand, the finiteness condition must be there in $(0')$ in order to license its presence in the antecedent of $(1')$; on the other, it has to be there anyway, since the the proof is by induction. It's not all that hard to see why you might think that, once the finiteness condition has been inserted into $(0')$, that will somehow guarantee of itself that all the numbers we are discussing in the course of the proof are finite, in particular, that a is finite, which is what we need to know in order to apply $(5')$. And, indeed, the finiteness condition in $(0')$ *does* license the later claim that a is finite, but it does not do so in any immediate way, but only *via* a very complicated argument: the one recounted at the end of the paper.

So what I think happened is something like this. In 1884, Frege says to himself, "Since we're only talking about finite numbers here, we can assume that a is finite". A little while later he asks, "Why exactly was that?" And then it turns out that the assumption made "without loss of generality" is one it takes real work to justify. It's hardly unheard of.

How much work does it take to justify that assumption? Not that much, in a sense. But it has become clear to me in the last few years that Induction 3 is a *much* stronger method of proof than Induction 2. In particular, the derivation of Induction 3 from Induction 2 requires *impredicative* comprehension: We need to be able to instantiate the initial universal quantifier in Induction 2 with something like $aQ^{*=}\xi \wedge F\xi$, and the definition of the ancestral involves a universal second-order quantifier. So $aQ^{*=}\xi \wedge F\xi$ is Π_1^1, and the proof of Induction 3 needs Π_1^1 comprehension. The proof of Induction 3 therefore cannot be carried out in predicative second-order logic, and that is what makes impredicative comprehension essential to Frege's proofs of the axioms of arithmetic.

See Chapter 12 for more on this issue, and my paper "Ramified Frege Arithmetic" (Heck, 2011).

4

Frege's Principle

In my *Grundlagen der Arithmetik*, I sought to make it plausible that arithmetic is a branch of logic and need not borrow any ground of proof whatever from either experience or intuition. In the present book this shall now be confirmed, by the derivation of the simplest laws of Numbers by logical means alone. (Gg, v. I, p. 1)

In his *Grundgesetze der Arithmetik*, Frege does indeed prove the "simplest laws of Numbers", the axioms of arithmetic being among these laws. However, as is well known, Frege does not do so "by logical means alone", since his proofs appeal to an axiom that is not only not a logical truth but is a logical falsehood. The axiom in question is Frege's Basic Law V, which governs terms of the form "$\grave{\epsilon}\Phi(\epsilon)$", terms that purport to refer to what Frege calls "value-ranges". For present purposes, Basic Law V may be written:[1]

$$(\grave{\epsilon}F\epsilon = \grave{\epsilon}G\epsilon) \equiv \forall x(Fx \equiv Gx)$$

The formal theory of *Grundgesetze*, like any (full)[2] second-order theory containing this sentence, is thus inconsistent, since Russell's Paradox is derivable from Basic Law V in (full) second-order logic.

In *Die Grundlagen*, Frege does not present a formal proof of the axioms of arithmetic. Instead, he merely sketches the proofs of a number of basic facts about numbers. The proofs of the corresponding results in *Grundgesetze* follow these sketches closely, for the most part.[3] In his proof-sketches, Frege does not refer to value-ranges, but to what he calls "extensions of concepts", of which value-ranges are a generalization; as we shall see below, value-ranges and Basic Law V are later introductions to his system. Presumably, however, any other formal principle by means of which Frege might have intended to formalize his informal use of extensions of concepts in *Die Grundlagen* would also have been inconsistent.

---

[1]Strictly speaking, Basic Law V is:

$$(\grave{\epsilon}F\epsilon = \grave{\epsilon}G\epsilon) = \forall x(Fx = Gx)$$

For Frege, truth-values are objects in the domain of the first-order variables, so the Law states that the truth-value of "$\forall x(Fx = Gx)$" is identical with that of "$\grave{\epsilon}F\epsilon = \grave{\epsilon}G\epsilon$". So formulated, it introduces, not just extensions of concepts, but the value-ranges of functions in general.

[2]This restriction is needed because Basic Law V is consistent both with simple and with ramified predicative second-order logic (Heck, 1996).

[3]As noted in the postscript to Chapter 3, my views on this have changed over the years.

Frege's informal proofs, in *Die Grundlagen*, begin with his showing how the notion of one-one correspondence can be defined in logical terms: Frege explains, as is now standard, that the *F*s can be correlated one-one with the *G*s just in case there is a one-to-one relation *R* that relates each *F* to a *G* and which is such that, for each *G*, there is some *F* that *R* relates to it (Gl, §§70–2). This done, Frege reminds the reader that, according to his usage, a concept *F* is equinumerous with a concept *G* just in case the *F*s can be correlated one-one with the *G*s. He then gives his explicit definition of names of numbers, which is:

the number belonging to the concept *F* is the extension of the concept "equinumerous with the concept *F*". (Gl, §72)

The number of *F*s is thus, as it were, the class of concepts having the same cardinality as *F*. Frege then turns immediately to the derivation, from this definition, of HP:

the number belonging to the concept *F* is identical with the number belonging to the concept *G* if [and only if] the concept *F* is equinumerous with the concept *G*. (Gl, §73)

We will discuss the role HP plays in Frege's philosophical views below.

Once he has proven HP, Frege outlines proofs that each number has at most one predecessor, that each natural number has exactly one successor, and so forth. In these proofs, Frege makes no further appeal to his explicit definition: The proofs depend only upon HP itself. Frege's method suggests that we may divide his proofs of the axioms of arithmetic, in *Die Grundlagen*, into two parts: First, a proof of HP from the explicit definition; Second, a proof of the axioms of arithmetic from HP, a proof in which the explicit definition plays no role whatsoever and in which, perhaps, the extensions of concepts too play no role.[4] Indeed, not only do Frege's informal proofs of the axioms neither appeal to the explicit definition nor make any explicit use of the notion of an extension, proofs of the axioms from HP really can be given within second-order logic (Wright, 1983). It is this beautiful and surprising result that George Boolos has urged us to call *Frege's Theorem*.

It is natural to wonder at this point whether Frege's formal arguments in *Grundgesetze* have a similar structure, that is, whether Frege derives HP from his explicit definition (now given in terms of value-ranges) and then derives the axioms of arithmetic from HP alone, no further use being made either of the explicit definition or of value-ranges. The explicit definition does not, it is true, play any further role. But the answer to our question, so formulated, is "No", because Frege makes use of value-ranges throughout Part II of *Grundgesetze* (in which he proves the axioms). But the question ought really to be: Does Frege derive the axioms of arithmetic from HP without making any *essential* use of value-ranges? Does

---

[4]The first to note that Frege's proofs can be so read was Peter Geach (1955), though Charles Parsons (1995a) is more explicit about the point.

he use value-ranges, in those proofs, merely for convenience? The answer to this question is "Yes", as we saw in Chapter 2. So the only essential appeal Frege makes to Basic Law V is in the proof of HP: All the proofs in Part II of *Grundgesetze*, except that of HP itself, are therefore carried out in a consistent sub-theory of the formal theory of *Grundgesetze* (excepting inessential uses of value-ranges), namely, in Frege Arithmetic—second-order logic plus HP. So Frege proved Frege's Theorem, which is as it should be.[5]

This fact has enormous importance for our understanding of Frege's philosophy, in particular, for our understanding of his philosophy of mathematics. What I wish to do here is to begin an investigation of its import by discussing certain historical questions that arise immediately. Before turning to that discussion, however, we must address a prior question, namely, whether Frege knew that the axioms of arithmetic could be proven in Frege Arithmetic. For, if Frege did not know that—if it were, so to speak, a happy accident that Part II of *Grundgesetze* can be read as a formal proof of Frege's Theorem—the fact that it can be so read would presumably have no significance for our understanding of Frege's thought.

There are many more questions about the role HP plays in Frege's thought than we shall even begin to answer here. My hope is that our discussion will demonstrate that a proper understanding of Frege's work requires an understanding of his attitude toward Basic Law V; of why it came to occupy so central a place in his philosophy; of why its refutation, in his own opinion, brought a large part of his life's work to ruin. It is commonly assumed that Frege abandoned his attempt to prove the axioms of arithmetic within (higher-order) logic because he believed that, with Basic Law V refuted, he could no longer do so. But the importance of Basic Law V does not lie in its formal role, for there was, in a clear sense, no *formal* obstacle to the logicist program, even after Russell's discovery of the contradiction, and Frege knew it. Indeed, we can only understand Frege's attitude toward Basic Law V once we appreciate the ridiculously meager formal role it plays.

4.1   Numbers as Extensions of Concepts

Crispin Wright has suggested that Frege's logicism may be reformulated, in the wake of Russell's Paradox, as follows:

[I]t is possible, using the concepts of higher-order logic with identity, to explain a genuinely sortal notion of cardinal number; and hence to deduce appropriate

---

[5]Actually, Frege does make one use of value-ranges that it is not trivial to eliminate, namely, in his definition of ordered pairs. But Frege does not use ordered pairs in his proofs of the axioms of arithmetic, only in his proof of their categoricity. And, in any event, Frege arguably knew his use of ordered pairs to be eliminable, too (Heck, 1995a).

statements of the fundamental truths of number-theory... in an appropriate system of higher-order logic with identity to which a statement of that explanation has been added as an axiom. (Wright, 1983, p. 153)

Wright's suggestion is that Frege's heirs may, and Frege should, just abandon the explicit definition of number and install HP as the fundamental axiom of the theory of arithmetic. Since Basic Law V is used only in the formulation of the explicit definition and the derivation of HP from it, no appeal whatsoever to Basic Law V is then required, and Russell's Paradox ceases to be an obstacle.

The first question which ought to strike one, once one realizes that Frege knew he could derive the axioms of arithmetic from HP, is: Why did he not adopt this course himself?

To answer this question, we must first understand why Frege introduces extensions of concepts in the first place, namely, to resolve the so-called 'Caesar problem'. According to Frege, a proper explanation of names of numbers must yield an explanation of the senses of identity statements containing names of numbers (Gl, §62). Frege first argues that "the sense of the proposition 'the number which belongs to the concept F is the same as that which belongs to the concept G'" is given by HP: "The number of Fs is the same as the number of Gs" is true just in case the Fs can be correlated one-one with the Gs (Gl, §63). The question of the central sections of *Die Grundlagen* is whether HP, on its own, can be taken as a complete explanation of the senses of identity statements concerning numbers.

Famously, Frege considers three objections to the claim that it can, quickly rejecting the first two (Gl, §§63–5). The third objection is that HP provides only for the resolution of questions of the form, "Is the number of Fs the number of Gs?" It does not determine the answer to such questions as, "Is the number of Fs Julius Caesar?" Hence, HP does not provide a general explanation of the senses of numerical identities: It explains only certain such statements, ones in which what are recognizably names of numbers—in the first instance, names of a quite particular form, viz., "the number of Fs"—occur on both sides of the identity-sign (Gl, §§66–7). Frege then considers several attempts to resolve this problem but finds them lacking, eventually settling for an explicit definition of names of numbers, namely, that mentioned earlier: The number of Fs is the extension of the concept "concept which can be correlated one-one with the concept F" (Gl, §68). The question whether Caesar is a number then reduces to the question whether he is an extension of a certain kind. Frege says that he "assume[s] that it is known what the extension of a concept is" (Gl, §68, note), and it is natural to interpret him as assuming it known whether Caesar is an extension and, if so, which extension he is.

Now, it would be natural to suppose that the explicit definition makes HP otiose. However, this would merely be an understandable mistake.

When Frege recapitulates the argument of *Die Grundlagen* at the end of the book, he writes:[6]

The possibility of correlating one-to-one the objects falling under a concept *F* with the objects falling under a concept *G*, we recognized as the content of a recognition-statement concerning numbers. Accordingly, our definition had to lay it down that a statement of this possibility means the same as [*als gleichbedeutend mit*] a numerical identity. (Gl, §106)

A proper definition of (names of) numbers is, in a sense to be explained, required to specify that the sense of an identity statement connecting names of numbers is given by HP.[7] It need not do so directly, and the Caesar problem obstructs any attempt to explain statements of numerical identity by means of HP alone: Nonetheless, the explanation of the senses of such statements must, in a sense to be explored below, yield HP. Frege's explicit definition yields HP in the strongest possible sense, since it provably implies it.

Frege's objection is thus not that HP fails to capture the senses of those identity statements which *do* have numerical terms (of a certain kind) on either side of the identity-sign. His objection is that it fails to explain the senses of other sorts of identity statements involving names of numbers. Most importantly, we cannot take ourselves thus to have explained the senses of (open) sentences of the form "*x* is the number of *F*s", sentences in which a name of a number and a free variable flank the identity-sign, and this would appear to imply that we cannot, by means of HP alone, "obtain... any satisfactory concept of Number" (Gl, §68). The concept "ξ is a number" can only be defined thus: For some *F*, ξ is the number of *F*s (Gl, §72). But we have just admitted that we have not explained what it is for "ξ is the number of *F*s" to be true of an object, but only what it is for certain very special sentences containing this concept-expression to be true.

Now, as said earlier, extensions of concepts, and the explicit definition of numbers, are introduced by Frege to resolve the Caesar problem, which prima facie is a philosophical problem, not a formal one: That Frege appeals to the explicit definition only in the derivation of HP suggests, further, that extensions and the explicit definition are introduced only to resolve the Caesar problem. If Frege could not abandon Basic Law V, it is therefore because he could not abandon the explicit definition; and if that is so, he could not abandon Basic Law V because he could not solve the Caesar problem without it. Of course, it is antecedently possible that Frege thought that there was some formal problem, a formal reflection of the Caesar problem, that could only be overcome with the help of the

---

[6] I have altered Austin's translation here. The German "*erkannten wir*" is translated by Austin "we found", which does not imply quite so strongly as does "we recognized" that we *correctly* found.

[7] I am deeply indebted to Dummett's discussion of this issue (Dummett, 1991b, pp. 178–9), as should be clear from what follows.

explicit definition: In fact, there is no such problem, at least within the theory of arithmetic, since axioms for second-order arithmetic can be derived from HP without appeal to the explicit definition. Since, as I shall shortly argue, Frege knew that the axioms could be derived from HP, he knew there was no such formal problem: no problem that arises within the formal theory for whose resolution the explicit definition was needed. Whether there is some meta-theoretic problem for whose resolution the explicit definition is required is another question, one to which we shall return.

4.2 The Importance of HP in Frege's Philosophy of Arithmetic

I have claimed that Frege's derivation of the axioms of arithmetic, in *Grundgesetze*, requires appeal only to HP, the only essential appeal to Basic Law V being in the proof of HP itself. Unless Frege himself knew as much, however, this fact is of little significance for the interpretation of his work. There is, however, good reason to think that Frege did know that neither the explicit definition nor Basic Law V was not needed for the proofs of the axioms. Moreover, there is good reason to think that he thought it significant that the explicit definition was needed only for the derivation of HP. Let me discuss the second point first.

In *Frege: Philosophy of Mathematics*, Michael Dummett argues that Frege's explicit definition of numerical terms is intended to serve just two purposes: To solve the Caesar problem, that is, to "fix the reference of each numerical term uniquely", and "to yield" HP (Dummett, 1991b, ch. 14). The explicit definition is in certain respects arbitrary, since numbers may be identified with a variety of different extensions (or sets, or possibly objects of still other sorts): There is, e.g., no particular reason that the number six must be identified with the extension of the concept "is a concept under which six objects fall"; it could be identified with the extension of the concept "is a concept under which only the numbers zero through five fall" or that of "is a concept under which no more than six objects fall".[8] Now, one might wonder how a definition that is arbitrary in this way could possibly be a definition of names of numbers, as we ordinarily understand them. Yet it is essential that it be just such a definition, for Frege's goal is not to show that some formal theory bearing but a syntactic relationship to arithmetic can be developed within logic: His goal is to show that arithmetic as we ordinarily understand it is analytic, that our concept of number is logical in character, and that the truths of arithmetic we take ourselves to know are analytic of our concept of number.

---

[8]See the Postscript for some reservations about this claim.

This point can be made more vivid if we consider the relationship between Frege's views about geometry and those about analysis, i.e., the
theory of the real numbers. Frege held not only that arithmetic, but analysis too, was analytic. Now, given a definition of ordered pairs (Gg, v.
I, §144), geometrical objects can be represented in the Cartesian plane
(more generally, in Cartesian n-space). Not only that: It is easy to prove
that the axioms of Euclidean geometry hold for such objects. Why, then, is
Frege not committed to the view, which he repeatedly rejects, that geometry is analytic (or, more generally, that it has the same epistemological
status as analysis)? Have we not explicitly defined what points are, what
lines are, and so forth, and produced a proof of the axioms of Euclidean
geometry?—I do not know exactly what Frege would have said about this
question, but the most natural reply, and the one I expect he would have
made, is as follows.[9] The fact that there is a definition of the fundamental geometrical concepts from which the axioms of Euclidean geometry
are provable does not imply that there is a *correct* definition from which
the axioms are provable. Or again: The representability of geometrical
objects in real 3-space does not necessarily yield a proof of the axioms of
Euclidean geometry, not if these axioms are supposed to have the same
content as the axioms as we ordinarily understand them.

To reject the analyticity of geometry, Frege must reject the identification of geometrical objects with parts of Euclidean 3-space. At the
same time, however, he must allow the identification of numbers with
the extensions of certain concepts. Something must thus distinguish the
explicit definition of (names of) numbers from this (alleged) explicit definition of (names of) points, lines, and so forth. The suggestion made by
Dummett, and hereby endorsed by me, is that, for Frege, an acceptable
explicit definition of names of Fs is one that immediately yields[10] an explanation of the senses of identity statements connecting names of Fs as
we ordinarily understand them. There is no obvious reason the definition must yield any verdict on the truth-values of statements of the form
"$t = x$", where t is our usual sort of name for an F and x apparently not
a name of an F. Hence, to be an explicit definition of numbers is immediately to yield HP, since it is in terms of HP that the senses of identity
statements connecting names of numerical terms are to be explained.[11]

What makes Frege's explicit definition a definition of numerical terms
is thus that it has HP as an immediate consequence: It is this that
constrains the explicit definition, that allows Frege to claim that, even

---

[9]Frege's two series of papers entitled "On the Foundations of Geometry" contain several
points very much along the lines of what I am about to say.

[10]Much more could be said about this notion of "immediately yielding" the explanation,
but I shall not pursue such questions here.

[11]One might wonder, however, if we have quite so much freedom in choosing the objects
with which we identify numbers. It seems likely, to me anyway, that our knowledge that
Caesar is *not* a number plays an important role in Frege's understanding of the problem.
This point is taken up in Section 5.3.

though his identification of numbers with extensions of concepts is required by nothing in our ordinary understanding of numerical terms, the definition is nonetheless a definition of numerical terms. Indeed, according to Dummett, it is HP that, according to Frege, "embodies the received sense of" numerical terms. If so, then, an explicit definition that yields HP thereby answers to everything in our ordinary understanding of names of numbers (Dummett, 1991b, p. 179).

Now, Frege's formal theory of arithmetic, constructed along the lines of *Grundgesetze*, is to contain an explicit definition of names of numbers. As said, however, different explicit definitions could be given, and each of these will yield a different class of theorems: So, one might ask, on what basis is it claimed that the logical consequences of Frege's explicit definition are analytic of the concept of number? The answer to this question should now be clear: Not all the theorems of any particular such theory are claimed to be analytic of the concept of number, only those whose proofs do not depend upon arbitrary features of the explicit definition. Since a given explicit definition is non-arbitrary precisely in so far as it implies HP, the theorems that are analytic of the concept of number are just those that follow from HP. It is therefore essential to Frege's philosophical project that the explicit definition, arbitrary as it is, should not figure in the proofs of any of the axioms of arithmetic—except, of course, that of HP itself. It is for this reason that the explicit definition must "immediately yield" HP, for the explicit definition must fall entirely out of consideration *once* it has yielded HP.

4.3   The Role of Basic Law V in Frege's Derivation of Arithmetic

Frege's views about arithmetic thus require him to derive the axioms of arithmetic in two stages: First, to derive HP from his explicit definition, and then to derive the axioms of arithmetic from HP, making no further appeal to his definition. This does not yet show, however, that Frege knew that no further appeal to Basic Law V was needed. Further argument is needed conclusively to establish this point.

Circumstantial evidence for this claim comes from the Introduction to *Grundgesetze*, in which Frege writes that "internal changes in my Begriffsschrift... forced me to discard an almost completed manuscript"; one of the most important of these changes was the introduction of value-ranges and Basic Law V (Gg, v. I, pp. ix–x).[12] It seems likely that this early

---

[12]It is unclear how much of the formal derivation Frege had when he wrote *Die Grundlagen*. That he had quite a bit of it is apparent from the letter from Stumpf to Frege, written in 1882 (PMC, pp. 171–2). Moreover, almost all of the interesting logical results needed in the derivation of arithmetic from HP are already present in *Begriffsschrift*, with the exception of *Grundgesetze*, Theorem 141. On the other hand, as argued in Chapter 3, there is a

derivation of the axioms of arithmetic from HP did not make any mention at all of extensions of concepts. Without Basic Law V, or something very much like it, Frege would have had no means for effecting reference to the extensions of first-level concepts.

One may feel some resistance to this claim, because one wants to ask: How could Frege have derived HP from the explicit definition without referring to the extensions of concepts, without some axiom like Basic Law V? How could he even have formulated the explicit definition? The most plausible answer, prima facie, is that, while he did not have terms for value-ranges in his system, Frege did have terms for the extensions of second-level concepts; and, furthermore, that the formal system contained some axiom, similar to Basic Law V, governing them. However, the extensions of second-level concepts are of much less general utility than are those of first-level concepts: It is obvious how one might make quite general use of the extensions of first-level concepts, since these are essentially (naïve) sets, and Frege does make quite general use of value-ranges in *Grundgesetze* (though, as mentioned, these more general uses are inessential). But it is harder to imagine how one might make general use of second-level extensions. So, even if Frege did have terms for the extensions of second-level concepts in his formal theory, they probably were not used in the derivation of the axioms of arithmetic from HP, even if they were used in the formulation of the explicit definition and the derivation of HP from it.

Further evidence is provided by Frege's remark, toward the end of *Die Grundlagen*, that he is not committed to the claim that the Caesar problem can be solved only by means of his explicit definition, that he "attach[es] no decisive importance even to bringing in the extensions of concepts at all" (Gl, §107). This would be a strange remark for him to make if his formal system contained an axiom governing extensions that was essential to the proofs of the axioms from HP.

The evidence thus far mentioned in not conclusive, however. Conclusive evidence is provided by a letter Frege wrote to Russell in 1902. Discussing whether it might be possible to do without value-ranges, or classes, in a logicist development of arithmetic, Frege writes:

We can also try the following expedient, and I hinted at this in my *Foundations of Arithmetic*. If we have a relation $\Phi\xi\eta$ for which the following propositions hold: (1) from $\Phi(a, b)$, we can infer $\Phi(b, a)$, and (2) from $\Phi(a, b)$ and $\Phi(b, c)$, we can infer $\Phi(a, c)$; then this relation can be transformed into an equality (identity), and $\Phi(a, b)$ can be replaced by writing, e.g., "$\S a = \S b$". . . . But the difficulties here are []^{13} the same as in transforming the generality of an identity into an identity of value-ranges. (PMC, p. 141)

---

subtle confusion in Frege's proof in *Die Grundlagen* which he could not have made had he formally worked out the whole proof. I discuss the matter further elsewhere (Heck, 2005).

^{13}At this point, the translation inexplicably contains the word "not", which is not found in Frege's original letter.

The suggestion is thus precisely that Basic Law V be abandoned and HP be taken as an axiom. The difficulty we must face if we choose this option is presumably the Caesar problem, which Frege has been discussing just prior to this passage. In any event, since it would be utterly pointless to replace the explicit definition with HP if value-ranges were going to be needed in the proof of the axioms of arithmetic from HP, Frege surely did know that the axioms of arithmetic could be derived from HP without the help of Law V. That is, he knew that, in so far as reference to extensions was required in his formal theory, such reference was required only in order to formulate the explicit definition and to derive HP from it.

4.4   Frege's Derivations of HP

Basic Law V, value-ranges, and the extensions of concepts play only a very limited formal role in Frege's derivation of the basic laws of arithmetic: They are required only for the derivation of HP. How, then, could the refutation of an axiom of such little formal import have had, in Frege's own estimation, such a devastating effect upon his philosophy of arithmetic? It is not easy to take this question seriously. It is easy to say that Basic Law V was obviously needed if Frege was to show arithmetic to be a branch of logic. But the question is why *that* is: why Frege himself regarded the use of Law V as indispensible to the logicist project.

What I wish to discuss, henceforth, is not this question, but a slightly different one, namely: What reason there is to suppose that the Frege of *Die Grundlagen* intended his explicit definition to be given within his formal theory in the first place? What reason is there to suppose that extensions were to play *any role at all* in his derivation of the laws of arithmetic? By answering this question, we can hope to understand a little better why Frege could not just reject Basic Law V, why it was essential to his philosophy of arithmetic. But, more importantly, the investigation of Frege's attitude toward Law V must, I think, begin with a proper understanding of the options he thought himself to have. For example, one might wish to know what Frege's position was *circa* 1884, because that would tell us what his position was before the discovery of Basic Law V. If Frege thought, at that time, that extensions had some important role to play in his formal derivation of arithmetic, if at that time he thought it necessary to appeal to some analogue of Law V governing the extensions of second-order concepts, then the "retreat" to an earlier view that he suggests to Russell would not have helped very much: The obvious analogue of Law V governing the extensions of second-order concepts[14] can be show to be inconsistent. But, if his earlier view were that extensions had no formal role to play, such retreat might have been an option for him.

---

[14] See below for the analogue.

Above, I raised the question, "How, *circa* 1884, could Frege formally have derived HP from the explicit definition?" If we accept the presupposition of the question, there can be no answer other than that he had, in his formal system, terms for the extensions of second-level concepts and an axiom governing them. But it is important to recognize that the question presupposes that, *circa* 1884, Frege envisioned, or would have required, a formal derivation of HP from his explicit definition (even if he did not then have such a derivation). We ought not just to assume that he would have. Our question is thus: Can it coherently be maintained that the Frege of *Die Grundlagen* did not intend HP to be derived, in the formal theory, from the explicit definition? that the explicit definition was not even to be stated in the formal theory? What makes this position seem so implausible is that, in *Die Grundlagen*, Frege sketches a *proof* of HP; during that discussion, he appeals directly to the explicit definition. It therefore cannot be denied that the explicit definition plays a role in this proof; the question is what role it plays.

Frege's sketch of the derivation of HP in *Die Grundlagen* begins as follows:

Our next aim must be to show that the Number which belongs to the concept *F* is identical with the Number which belongs to the concept *G* if the concept *F* is equinumerous with the concept *G*....

On our definition, what has to be shown is that the extension of the concept "equinumerous with the concept *F*" is the same as the extension of the concept "equinumerous with the concept *G*", if the concept *F* is equinumerous with the concept *G*.

By "our definition", Frege means his explicit definition. He continues:

In other words: it is to be proved that, for *F* equinumerous with *G*, the following two propositions hold good universally:

> if the concept H is equinumerous with the concept F, then it is also equinumerous with the concept G;

and

> if the concept H is equinumerous with the concept G, then it is also equinumerous with the concept F. (Gl, §73)

Frege here transforms the statement of the identity of the extensions of the concepts "equinumerous with the concept *F*" and "equinumerous with the concept *G*" into the statement that the concepts falling under the one are just the concepts falling under the other: He transforms the statement that their extensions are identical into the statement that the concepts themselves are co-extensional.[15] Frege thus seems to be appealing,

---

[15] In Chapter 3, Boolos and I note that Frege does mention extensions once in *Die Grundlagen* after he proves HP, in §83. Austin's translation of that passage is somewhat mis-

without mentioning that he is, to an axiom governing names of extensions of second-level concepts: This axiom, a natural analogue of Basic Law V, states that the extension of a second-level concept $\Phi_x(\phi x)$ is the same as that of $\Psi_x(\phi x)$—ϕ marks the argument-place—just in case, for every F, $\Phi_x(Fx)$ just in case $\Psi_x(Fx)$. Using this axiom, it is easy to derive HP from Frege's explicit definition.[16]

The proof of HP in *Grundgesetze* is in two parts. Frege proves in Part II, A(lpha), that, if F is equinumerous with G, then the number of Fs is the same as the number of Gs (Theorem 32). The central lemma in this proof is Theorem 32δ.[17] If we abbreviate "the concept Φ is equinumerous with the concept Ψ" as "$\mathrm{Eq}_x(\Phi x, \Psi x)$", then Theorem 32$\delta$ is:

$$\mathrm{Eq}_x(Fx, Gx) \to \forall H[\mathrm{Eq}_x(Fx, Hx) \equiv \mathrm{Eq}_x(Gx, Hx)]$$

That is: If F is equinumerous with G, then the concepts equinumerous with F are just the concepts equinumerous with G. The consequent is exactly what, in the passage from *Die Grundlagen* just cited, Frege says we must prove if we are to show that the number of Fs is the number of Gs.

Frege's proof of Theorem 32δ makes only inessential appeal to Law V and so may be reconstructed in second-order logic. The proof follows the sketch in *Die Grundlagen* closely. Theorem 32δ will follow easily once it has been shown that $\mathrm{Eq}_x(\Phi x, \Psi x)$ is an equivalence relation or, more precisely, that it is transitive and symmetric. If $\mathrm{Eq}_x(Fx, Gx)$, then $\mathrm{Eq}_x(Gx, Fx)$, by symmetry; hence, if $\mathrm{Eq}_x(Fx, Hx)$, then $\mathrm{Eq}_x(Gx, Hx)$, by transitivity. Similarly, if $\mathrm{Eq}_x(Gx, Hx)$, then $\mathrm{Eq}_x(Fx, Hx)$, again by transitivity. All the work in the proof thus goes into establishing that the relation of equinumerosity is transitive and symmetric.

Once he has established (32δ), Frege infers, via Basic Law V and the explicit definition, that the number of Fs is the number of Gs.[18] In fact,

---

leading. The relevant sentence is: "And for this, again, it is necessary to prove that this concept has an extension identical with that of the concept 'member of the series of natural number ending with d'". The German is: "Und dazu ist wieder zu beweisen, dass dieser Begriff gleichen Umfanges mit dem Begriffe 'der mit d endenden natürlichen Zahlenreihe angehörend' ist". What Frege says is that we must prove that one concept "is the same in extension" as another: One natural translation of "gleichen Umfanges mit... ist" would be is "is co-extensional with".

[16]For a proof of the inconsistency of this axiom, see footnote 2 of Chapter 3, on p. 71.

[17]By this I mean the intermediate result marked "δ" that occurs *during* the proof of (and so *prior* to the appearance of) Theorem 32.

[18]The second-level relation $\mathrm{Eq}_x(\Phi x, \Psi x)$ would be represented by its value-range, so, strictly speaking, we should thus have something like $\mathrm{Eq}(\grave{e}F\epsilon, \grave{e}G\epsilon)) \to \forall u[\mathrm{Eq}(\grave{e}F\epsilon, u) \equiv \mathrm{Eq}(\grave{e}G\epsilon, u)]$, where "$u$" ranges over value-ranges (as well as other objects). (Frege does not actually have a special term for our "$\mathrm{Eq}_x(\Phi x, \Psi x)$".) Basic Law V then yields: $\mathrm{Eq}(\grave{e}F\epsilon, \grave{e}G\epsilon) \to \grave{\alpha}\mathrm{Eq}(\grave{e}F\epsilon, \alpha) \equiv \grave{\alpha}\mathrm{Eq}(\grave{e}G\epsilon, \alpha)$. Since, by definition, $\mathcal{N}(x) = \grave{\alpha}\mathrm{Eq}(x, \alpha)$, we have: $\mathrm{Eq}(\grave{e}F\epsilon, \grave{e}G\epsilon) \to \mathcal{N}(\grave{e}F\epsilon) = \mathcal{N}(\grave{e}G\epsilon)$. The formal derivation, using extensions of second-level concepts, would proceed similarly.

however, the full strength of Basic Law V is not required for this inference.[19] What is required is obviously just the following:

$$\forall H[\mathrm{Eq}_x(Fx, Hx) \equiv \mathrm{Eq}_x(Gx, Hx)] \rightarrow \mathrm{N}x : Fx = \mathrm{N}x : Gx$$

That is: If the concepts equinumerous with F are just the concepts equinumerous with G, then the number of Fs is the number of Gs.

The other direction of HP—if the number of Fs is the number of Gs,
then the Fs are equinumerous with the Gs—is Theorem 49 of *Grundgesetze*, which is proven at the beginning of section B(eta). Frege's proof
of Theorem 49 is extremely peculiar, and it makes essential use of the
very specific way Frege defines the numbers.[20] I think he must have been
overcome by the slickness of that proof, as it would have been just as easy,
and ultimately more informative, to give a proof along the lines sketched
in *Die Grundlagen*. That proof needs only the converse of the principle
just mentioned, namely:

$$\mathrm{N}x : Fx = \mathrm{N}x : Gx \rightarrow \forall H[\mathrm{Eq}_x(Fx, Hx) \equiv \mathrm{Eq}_x(Gx, Hx)]$$

Given the reflexivity of $\mathrm{Eq}_x(\Phi x, \Psi x)$, the proof is then easy. Suppose that
$\mathrm{N}x : Fx = \mathrm{N}x : Gx$. By *modus ponens* and instantiation of "H" with "G",
$\mathrm{Eq}_x(Fx, Gx)$ iff $\mathrm{Eq}_x(Gx, Gx)$. Since $\mathrm{Eq}_x(\Phi x, \Psi x)$ is reflexive, $\mathrm{Eq}_x(Gx, Gx)$,
so $\mathrm{Eq}_x(Fx, Gx)$.

The derivation of HP therefore requires only what we may call *Frege's
Principle*:[21]

$$\mathrm{N}x : Fx = \mathrm{N}x : Gx \equiv \forall H[\mathrm{Eq}_x(Fx, Hx) \equiv \mathrm{Eq}_x(Gx, Hx)]$$

That is: The number of Fs is the number of Gs if, and only if, the concepts equinumerous with F are just the concepts equinumerous with G.
Or: The number of Fs is the number of Gs just in case the (second-
level) concept $\mathrm{Eq}_x(Fx, \Phi x)$ is co-extensional with the (second-level) concept $\mathrm{Eq}_x(Gx, \Phi x)$.

I conclude that Basic Law V and the explicit definition serve no essential purpose in Frege's proofs of the axioms of arithmetic in *Grundgesetze*

---

[19]On a slightly different note, Robert May has observed that only Basic Law Va is needed
for the proof of Theorem 32 from (32δ); it is used to infer (32ϵ). This is the 'safe' direction of
Law V, which Frege calls Law Va:

$$\forall x(fx = gx) \rightarrow \grave{\epsilon}f(\epsilon) = \grave{\epsilon}g(\epsilon)$$

That we only need Va here is not surprising, since (32) is the 'safe' direction of HP, i.e., the
one that has no consequences for the cardinality of the domain.

[20]Frege first derives Theorems 45 and 46α from the explicit definition. We can record
these as: $\mathrm{Eq}(w, z) \rightarrow w \in \mathcal{N}(z)$ and $w \in \mathcal{N}(z) \rightarrow \mathrm{Eq}(w, z)$. Given that equinumerosity is
reflexive (39, 42), Frege then uses (45) to get Theorem 48: $w \in \mathcal{N}(w)$. Identity then yields:
$\mathcal{N}(w) = \mathcal{N}(z) \rightarrow w \in \mathcal{N}(z)$, and Theorem 49 follows immediately from this and (46α).

[21]Something very much like this principle is discussed by Boolos under the name "Numbers" (Boolos, 1998b, p. 186).

other than to yield Frege's Principle. This is what I had in mind when I said earlier that Basic Law V plays a ludicrously meager formal role, and so this is what makes pressing the question why the truth, indeed the logical truth, of Basic Law V was so essential to Frege's philosophy of mathematics.

4.5   HP versus Frege's Principle

Our question is whether it can coherently be maintained that, *circa* 1884, Frege did not intend formally to derive HP from his explicit definition. As said, one of the chief reasons to think that it cannot is that Frege sketches a derivation of HP in *Die Grundlagen*: He presumably intended to derive it from something; hence, if we deny that he intended to derive HP from his explicit definition, we are left with the question from what he did intend to derive it. We have seen, however, that, in the first instance, Frege derives HP from Frege's Principle; and it is clear that, in *Grundgesetze*, anyway, he would have derived Frege's Principle from the explicit definition and Basic Law V. What dispute there might be concerns the status of Frege's Principle in *Die Grundlagen*. The standard view would be that Frege intended somehow to derive Frege's Principle from his explicit definition; an alternative view is that Frege's Principle itself was to be the fundamental axiom of the theory of arithmetic. Is this alternative view tenable?

The natural objection to it is that it would then be utterly unclear why Frege gives the explicit definition at all, what role it is supposed to play. More to the point, Frege introduces the explicit definition to resolve the Caesar problem: But Frege's Principle is no more immune to the objections he brings against HP than is HP itself; the one no more settles whether Caesar is a number than does the other. Hence, numbers must still be identified with the extensions of certain concepts. But, then, how it can be held both that Frege's Principle was to be the fundamental axiom of the theory of arithmetic and that numbers were to be identified with extensions? To put the point another way: What advantage, exactly, is Frege's Principle supposed to have over HP?

The natural suggestion is that the identification of numbers with extensions is not made in the formal theory but in the meta-theory. Consider again Frege's sketch of the proof of HP in *Die Grundlagen*. At the very beginning of that proof, he writes that the sentence "The concepts equinumerous with F are just the concepts equinumerous with G" states the identity of the number of Fs and the number of Gs, as he explicitly defines them, "in other words". Now, in *Grundgesetze*, the explicit definition of numbers as extensions is used in a formal proof of Frege's Principle; on the standard view, this is also what is envisaged in *Die Grundlagen*. On the alternative view, what we have here is (not a short formal proof but) an *informal justification* of Frege's Principle in terms of the explicit definition, which is itself being given in the meta-theory, not in the formal theory. Thus, Frege's Principle does have an advantage over HP, for it can straightforwardly be informally justified in terms of the explicit definition.

Why shouldn't the explicit definition be given in the meta-theory? The Caesar problem is not a formal problem, not, that is, a problem upon whose solution the successful execution of the formal part of the logicist program depends. The derivability of second-order arithmetic from HP shows that the Caesar problem is not a formal problem in this sense. Furthermore, the identification of numbers as value-ranges does not fully resolve the Caesar problem: A version of the Caesar problem arises again in *Grundgesetze*, the question, this time, being whether either of the two truth-values is a value-range and, if so, which value-ranges they are (Gg, v. I, §10). From our current perspective, what is interesting about Frege's solution to this problem is that *it is not incorporated into the formal theory*. As said, the Caesar problem is not a formal problem, so no solution to it needs to be incorporated into the formal theory: It is sufficient that it be resolved in the meta-theory, and that is precisely where Frege resolves it in *Grundgesetze*. According to the alternative view, Frege resolves it there in *Die Grundlagen*, also.

As yet, then, we have seen no reason to suppose that, *circa* 1884, Frege should not also have been satisfied to resolve the Caesar problem, for numerical terms, in the meta-theory: That is, we have discovered no objection to the claim that the explicit definition was to be given in the meta-theory, that Frege's Principle was to be justified informally in terms of it, and that Frege's Principle was to be the fundamental axiom of the formal theory of arithmetic.

4.6   Frege's Principle and the Explicit Definition

The philosophical value, for Frege, of formalizing arithmetic is that doing so affords insight into its epistemological status. The derivation of Frege's Principle from the explicit definition, and indeed the formulation of the explicit definition itself, could, however, be argued to involve an appeal to intuition: Perhaps an appeal to intuition is required to justify the transition from "the extension of F is the same as the extension of G" to "the Fs are exactly the Gs"; perhaps the notion of the extension of a concept is itself an intuitive one. One could attempt to answer such objections in a variety of ways. Presumably, objections similar to those Frege offers against the view that numbers are given in intuition could also be made against the view that the extensions of concepts are given in intuition: For example, the extensions of concepts, the objects falling under which are not given in intuition, are presumably also not given in intuition (Gl, §§14, 24). Conclusively to answer such an objection, however, Frege would have had to show that the theory of the extensions of concepts is itself part of logic; he would, that is, have had to show that the explicit definition could be formulated in logical terms and that Frege's Principle could be derived from it in accord with general logical principles.

If Frege is going to derive Frege's Principle from his explicit definition, then he must do so formally: To show arithmetic to be a branch of logic, the derivation of Frege's Principle must be carried out formally, so that the epistemological basis of the principles employed in that derivation can be uncovered. On the alternative view we have been discussing, Frege's Principle is to be derived from the explicit definition within the meta-theory, rather than within the formal theory; but it is still to be derived from the explicit definition. What the objection just considered shows is that this claim cannot stand: One cannot interpret Frege as deriving Frege's Principle from the explicit definition only informally. Since, as was said earlier, Frege does appear to derive Frege's Principle from something, the alternative view surely fails as an interpretation of Frege.

In fact, however, this argument is less conclusive than it seems. What we need to know is not just that Frege proposed to derive Frege's principle from the explicit definition. There might be many different reasons one would want to do that. What we need to know is that Frege proposed to *justify* Frege's Principle in terms of the explicit definition. The argument just given assumes that he did. Did he?

4.7   The Caesar Problem Revisited

There is an important difference between Frege's resolution of the Caesar problem in *Die Grundlagen* and his resolution of the corresponding problem in *Grundgesetze*. As said earlier, the problem that arises in *Grundgesetze* is whether either of the truth-values is a value-range and, if so, which value-ranges they are. Frege's solution to the problem consists of two parts (Gg, v. I, §10). The first part is a proof that, compatibly with Basic Law V, one may identify each of the two truth-values with the value-range of any distinct functions one chooses; more precisely, if a bit anachronistically, given any model of the theory at all, there is, for each function in the domain of the model, a model in which, say, Truth is the value-range of that function.[22] The second part is a stipulation that Truth is to be its own unit class; Falsity, its own unit class. Of course, this does not solve the Caesar problem in any generality: It tells us neither

---

[22]Of course, the full theory of *Grundgesetze* has no models, since it is inconsistent. But the first-order fragment of the theory does, and with respect to it this claim can be given a proper proof, though it needs some restriction: There are some very special pairs of concepts whose extensions cannot simultaneously be taken to be Truth and Falsity (Schroeder-Heister, 1987; Parsons, 1995b).

whether Caesar is a value-range nor, if so, which one he is. But this appears not to trouble Frege: He says simply that the formal system, as it stands, contains only terms that refer either to value-ranges or to truth-values and that, if terms referring to other sorts of objects are introduced, additional stipulations will have to be made.[23]

In *Die Grundlagen*, the Caesar problem is resolved by giving an explicit definition from which HP follows; in *Grundgesetze*, the corresponding problem is resolved by making assignments to (stipulations regarding the references of) certain terms, which assignments are then proven to be consistent with Basic Law V. One can imagine a similar strategy being employed in *Die Grundlagen*: We make some stipulation about what the referents of numerical terms are to be, a stipulation that would resolve the Caesar problem compatibly with HP; these stipulations could be made by means of the explicit definition itself. To show that the assignment was consistent with HP, it would be sufficient, though not necessary, to derive the latter within some theory in which a statement of the former assignment is taken as an axiom (which would then amount to a relative consistency proof). We probably would not find it natural then to say that our stipulations were made by means of a definition (except within the context of the relative consistency proof): Frege's calling it a definition[24] may be good reason to think that he did so intend it. Nevertheless, such a stipulation would resolve the Caesar problem, in the sense that it would fix the truth-values of identity statements of the form "the number of Fs is t", for every term t.[25]

If HP is to be justified in terms of the explicit definition, then the principles used in that derivation must be examined, if we are to be clear whether HP itself is thereby shown to be analytic, to be a truth of logic. On the other hand, suppose that HP is not to be justified in terms of the explicit definition, that the explicit definition is merely a tool for resolving the Caesar problem. Then the only requirement on the explicit definition is that it be consistent with HP. In that case, even if the definition made use, say, of geometrical notions, is it obvious that that would show HP not to be a logical truth? Some of our knowledge of numbers—of their identity with and distinctness from objects not given to us as numbers—would be infected with knowledge derived from intuition. But our knowledge of the basic laws of arithmetic would not be, since HP is not, on this view, to be to be thought of as justified by, or as known because it follows from, the

---

[23] Indeed, Frege does not even consider the possibility, in the main text, that additional primitive singular terms may be introduced into the language: He considers only the introduction of additional, primitive functions. It is natural to read the footnote, however, as concerning primitive singular terms.

[24] It is worth mentioning here that Austin translates both "erklären" and "definieren" as "define", and similarly for cognate expressions. Frege may use them interchangeably, but this is not clear to me. I have not, however, studied this question.

[25] At least, it will do so if the corresponding explicit definition does so, so the explicit definition has no advantage on this score.

explicit definition.[26]

Whether a stipulation of the sort under discussion would resolve the Caesar problem depends, of course, on what we take the Caesar problem to be.[27] The question we need to consider, at this point, is: Is there some reason that, in the case of numerical terms, the problem must be resolved not just by giving an explicit definition but by giving one in terms of which HP is supposed to be justified? It is important to remember that the Caesar problem is not a problem upon whose solution the proper development of the formal theory of arithmetic depends. As Frege presents the problem whether the truth-values are value-ranges, it is a formal problem, but explicitly a meta-theoretic one: The problem is to fix the truth-values of certain statements, so that every well-formed sentence of the theory will have a unique, well-determined truth-value (Gg, v. I, §§10, 30–1). The Caesar problem, as it arises in *Die Grundlagen*, on the other hand, is presented in nothing like this way: Rather, the problem is primarily philosophical in character. This difference is presumably of fundamental importance: The problem of *Grundgesetze*, which Frege thinks he can solve by stipulation, is presumably not the same problem as the problem of *Die Grundlagen*, a problem he would not (at least by *Grundgesetze*) have been willing to resolve by stipulation. But that means that we have come, after much wandering, back to where we began, for what we need now to understand is what objection, exactly, Frege raises against HP in *Die Grundlagen*. The question, once again, is: What exactly is the Caesar problem?

4.8   Closing

When he wrote *Die Grundlagen*, Frege might well have been prepared, if confronted with Russell's Paradox, to renounce appeal to extensions and to install HP as an axiom. Extensions, as we have seen, are introduced only to resolve the Caesar problem: Abandoning the explicit definition, and so use of extensions of concepts, would require Frege to resolve the Caesar problem in some other way. At the time he wrote *Die Grundlagen*, however, Frege thought that, in principle, the Caesar problem could be resolved in some other way (Gl, §107), though he may well have thought it probably could not be. Perhaps the Caesar problem could be resolved by giving a different explicit definition, which did not "bring[] in the exten-

---

[26]And there is, in a way, nothing novel about this way of looking at things. As said earlier, even if the explicit definition is given in the formal theory and is given in purely logical terms, not all of its consequences are going to be analytic of the concept of number, anyway, since some of its features are arbitrary.

[27]Similarly, whether an appeal to non-logical notions in a solution of the Caesar problem would undermine the claim of HP to be a truth of logic will depend upon whether our knowledge of the truths of arithmetic is supposed somehow to depend upon how the Caesar problem is solved.

sions of concepts"; perhaps it could be resolved by making certain sorts of stipulations, as discussed above. Perhaps it could even be shown not really to be a problem, HP, together with more general sorts of considerations, actually settling the truth-values of 'mixed' identity statements, as Wright (1983, §§xiv–xv) would have it.

But by 1903, Frege's position had changed. As he wrote to Russell:[28]

I myself was long reluctant to recognize value-ranges and hence classes; but I saw no other possibility of placing arithmetic on a logical foundation. But the question is, How do we apprehend logical objects? (PMC, pp. 140–1)

This question is closely related to that raised in §62 of *Die Grundlagen*: "How, then, are numbers to be given to us, if we can have no ideas or intuitions of them?" Logical objects are certainly given neither in intuition nor in perception.[29] To answer this earlier question, Frege had once suggested, it would be sufficient to "fix the sense of a numerical identity": The difficulty, however, was that HP alone did not suffice to do this. Frege continues:

And I have found no other answer to it than this, We apprehend them as extensions of concepts, or more generally, as value-ranges of functions. I have always been aware that there are difficulties connected with this, and your discovery of the contradiction has added to them; but what other way is there? (PMC, p. 141)

The reason that abandoning Basic Law V was not an option for Frege in 1902 was thus this: Only by identifying numbers with the extensions of certain concepts could he answer a certain epistemological question, first raised in *Die Grundlagen*, namely, how we apprehend numbers with the aid of neither intuition nor perception. Frege's logicist program thus did not fail, even by his own lights, because he could not derive the axioms of arithmetic from principles that have some claim to be logical in character. By his own lights, his program failed because he could not explain how we can apprehend the objects of arithmetic as logical objects; it failed because he was unable to resolve a particular philosophical problem, the Caesar problem, and it is the nature of this problem that we, as Frege's interpreters, must come to understand.

Postscript

Following Michael Dummett, I claim above that Frege's explicit definition of numbers is "in certain respects arbitrary", on the ground that various other definitions would also have allowed Frege to prove HP (see p. 95).[30]

---

[28] I have altered the translation slightly.

[29] The delicacy here is required by the fact that neither are directions given in intuition or perception, though they are not (purely?) logical objects.

[30] Similar claims were made in the original version of Chapter 6.

For example, one could take the number of Fs to be the class of all con-cepts that can be injected into F, or into which F can be injected, and there are plainly many other definitions that would also do.

This claim now seems to me to be over-stated,[31] though not in way that undermines the main arguments of the chapter. In particular, it now seems to me that there is a strong case to be made that the particular explicit definition Frege gives—assuming we are going to give an explicit definition—is almost completely forced.

As Frege makes clear in *Die Grundlagen*, he regards the case of num-bers as but one of a range of similar cases. Indeed, his discussion of the question whether HP can be taken as a complete explanation of names of numbers is actually carried out in terms of a different example, names of directions, and Frege mentions other examples, too (Gl, §64). So what is really at issue in the central sections of *Die Grundlagen* is a very general strategy for explaining names of, and so our ability to think about, ab-stract objects, and the Caesar problem is supposed to lead to the collapse of that strategy quite generally. What Frege needs, therefore, is not just a new way to explain reference to numbers but a new *strategy* for explain-ing reference to abstract objects, and that new strategy is first articulated with respect to directions: the direction of l is the extension of the concept $\xi \parallel l$.

So consider the matter quite generally. We have some equivalence relation $\xi R\eta$, and we want to define a function $\rho(\xi)$ in such a way as to validate the corresponding abstraction principle

$$\rho(x) = \rho(y) \equiv xRy$$

How, *in general*, can we do this? So far as I can see, the only general strategy that is available here is essentially the one Frege adopts: Take $\rho(x)$ to be x's equivalence class under R, that is, the extension of the concept $xR\xi$. To be sure, there is a kind of 'dual' definition: Take $\rho(x)$ to be x's 'anti-equivalence class', that is, the extension of the concept $\neg xR\xi$. But one might reasonably regard that as a trivial variant.

So, if we think of the explicit definition of numbers and its relation to HP not in isolation but as one instance of a general phenomenon, then the explicit definition is required not just to validate HP but also to instance a general strategy for defining functions that will validate abstraction principles. Seen in that light, the explicit definition is, as I said, almost completely forced, though there is a small, residual choice to be made.

It's worth noting that a similar point can be made about the definition of cardinailty in set theory. The standard definition of the cardinality of a set S is: $|S|$ is the least ordinal α that can be mapped one-one onto S. As

---

[31]I realized this when thinking through some of the issues raised in Tyler Burge's paper "Frege on Truth" (Burge, 2005c). Though Burge does not make the point I am about to make, he did make it clear to me how many different sorts of considerations might be brought to bear upon the correctness of such a definition.

should be obvious, however, there will be no such ordinal unless S can be well-ordered, which is to say that sets that cannot be well-ordered will not have a cardinal number. In the presence of the axiom of choice, this is not a problem, of course, but what if we do not have choice? Frege's definition does not depend upon choice, but it has a worse flaw, namely, that we can prove that $|S|$, which would then be the set of all sets equinumerous with S, does not exist, unless S is empty.[32] As was first noted by Dana Scott, however, there is a natural way of modifying Frege's definition. The modification, sometimes known as "Scott's trick", involves taking $|S|$ to be the set of the sets *of lowest rank*[33] that are equinumerous with S. Clearly, the strategy generalizes: We may take $\rho(x)$ to be

$$\{y : xRy \wedge \forall z(xRz \rightarrow y \leq z\}$$

where $y \leq z$ means: z's rank is at least that of y. One might well argue, therefore, that Scott's definition of $|S|$ is preferable to the standard one, *even if* we accept the axiom of choice: Scott's definition instances a general strategy for defining equivalence classes, a strategy of which the definition of cardinality is but one instance; the standard definition not only exploits special features of the case of cardinality but is, by comparision, *ad hoc*.

Fortunately, these observations do not undermine but rather reinforce the conclusion for which the claim about the arbitrariness of HP was meant to be evidence. That conclusion is stated at the end of Section 4.2: "What makes Frege's explicit definition a definition of numerical terms is... that it has HP as an immediate consequence...". The explicit definition of numbers is one instance of a general strategy for defining functions that satisfy abstraction principles. Although the explicit definition is *formally* prior to HP, then, it is HP that is *philosophically* fundamental: To know what explicit definition to give, we must antecedently know which abstraction principle we want to validate. The philosophical justification of the explicit definition therefore goes via the claim that, whatever (cardinal) numbers are, HP must be true of them.

---

[32] I am not sure exactly what the proof requires, but it will not be much. We need first to show that, if $S \neq \emptyset$, then $\cup|S|$ is the universal set; it is enough to show that, for any x, there is a set S_x that is equinumerous with S. By separation, using the formula $\xi \neq y$, we have a set T that is S without y; we may then put x in its place (pair T with $\{x\}$ and take the union). If we have a one-one mapping of S onto itself—I am not sure precisely what we need for that—then we can modify this mapping, again using separation, pairing, and unions, to give us one that maps S_x onto S. So, as said, $\cup|S|$ is the universal set, and now separation will give us the Russell set, in the usual way.

A different proof would use foundation: Show, in the same way, that $\cup|S|$ is a member of some set equinumerous with S; then $\cup|S| \in \cup|S|$. But it is worth avoiding foundation.

[33] That sets have well-defined rank depends upon the axiom of foundation.

5

Julius Caesar and Basic Law V

Much of Frege's philosophical and mathematical work is devoted to an attempt to show that, given appropriate definitions, all theorems of arithmetic can be proven from logical laws alone. In his *Grundgesetze der Arithmetik*, Frege presents formal proofs intended to show "that arithmetic is a branch of logic and need not borrow any ground of proof whatever from either experience or intuition" (Gg, v. I, p. 1). But the formal system in which Frege proves the basic laws of arithmetic is inconsistent, since Russell's Paradox is derivable from Frege's Basic Law V in (full) second-order logic.[1] Basic Law V is:

$$(\acute\epsilon(F\epsilon) = \acute\epsilon(G\epsilon)) \; = \; \forall x (Fx = Gx)$$

Law V governs the term forming operator "$\acute\epsilon(\phi\epsilon)$", from which terms standing for 'value-ranges' are formed: It states that the value-range of $F\xi$ is the same as that of $G\xi$ just in case F and G have the same values for the same arguments. Since, for Frege, the truth-values are objects, extensions of concepts are among the value-ranges.

Value-ranges are used throughout Part II of *Grundgesetze*, in which Frege proves the axioms of arithmetic and various related results. As argued in Chapter 2, however, with just two exceptions, Frege uses value-ranges only for convenience, to make certain parts of his proofs easier; most of his uses of them can be eliminated in a uniform manner. The two ineliminable uses occur in Frege's proofs of the two directions of HP: The number of objects falling under a concept $F\xi$ is the same as the number of objects falling under $G\xi$ if, and only if, the Fs can be correlated one-to-one with the Gs.

HP is stated by Frege in *Die Grundlagen*, and he there uses it in informal proofs of various fundamental facts about the natural numbers, including axioms for arithmetic. Famously, however, Frege derives HP, in *Die Grundlagen*, from an explicit definition of numbers as extensions of concepts—much as he derives it, in *Grundgesetze*, from a definition of numbers as value-ranges. But once again, although the use of extensions is necessary for the proof of HP from the explicit definition, Frege makes no essential use of extensions during the derivation of the axioms of arithmetic *from* HP. His proof-sketches therefore amount to an informal derivation of the laws of arithmetic from HP alone. Similarly, since,

---

[1] In fact, what one needs for the derivation is just Σ_1^1 comprehension (Heck, 1996).

in *Grundgesetze*, Basic Law V is used essentially only in the proof of HP, the proofs Frege there gives of the axioms of arithmetic amount to a formal second-order derivation of them from HP (*modulo* the inessential uses of Basic Law V).

As Frege Arithmetic—second-order logic, with HP taken as the sole 'non-logical' axiom—is equi-interpretable with second-order arithmetic (Boolos and Heck, 1998, appendix 2), it follows that Frege's formal proof of the axioms of arithmetic can be carried out within a (presumably) consistent sub-theory of the formal theory of *Grundgesetze*. Furthermore, Frege knew full well that the other uses he made of value-ranges were only for convenience (see Section 4.3). The point of our discussion up to this point is thus this: Frege knew that the basic laws of arithmetic could be derived from HP in that sub-system of his formal system that results from the exclusion of Basic Law V.[2] Formally speaking, then, there was no reason that, upon receiving Russell's famous letter, Frege could not have abandoned Law V, installed HP as an axiom, eliminated the inessential uses of value-ranges, and then have declared himself to have derived the axioms of arithmetic from HP, a principle arguably "analytic of the concept of number", as neo-Fregeans might put it. That would have been no mean feat.

Moreover, not only did Frege know that he could have substituted HP for Basic Law V, he explicitly considered doing so. In a letter written to Russell in 1902, discussing how he might avoid using Law V, Frege writes:

We can also try the following expedient, and I hinted at this in my *Foundations of Arithmetic*. If we have a relation $\Phi(\xi, \eta)$ for which the following propositions hold: (1) from $\Phi(a, b)$ we can infer $\Phi(b, a)$, and (2) from $\Phi(a, b)$ and $\Phi(b, c)$ we can infer $\Phi(a, c)$; then this relation can be transformed into an equality (identity), and $\Phi(a, b)$ can be replaced by writing, e.g., "$\S a = \S b$". If the relation is, e.g., that of geometrical similarity, then "a is similar to b" can be replaced by saying "the shape of a is the same as the shape of b". This is perhaps what you call "definition by abstraction". But the difficulties here are []³ the same as in transforming the generality of an identity into an identity of value-ranges. (PMC, p. 141)

The idea is indeed familiar from *Die Grundlagen*: If $\Phi(\xi, \eta)$ is an equivalence relation, we may take

$$\mathrm{fnc}(a) = \mathrm{fnc}(b) \equiv \Phi(a, b)$$

---

[2]And, strictly speaking, Basic Law VI, which governs the description-operator and is formulated in terms of value-ranges. Frege uses Basic Law VI only in his definition of the application-operator. It is therefore not needed once value-ranges are excluded from the system. The system remaining once Axioms V and VI are dropped is a version of axiomatic second-order logic, with comprehension formulated as a rule of substitution, for which see Frege's Rule 9 (Gg, v. I, §48).

[3]At this point, the translation contains the word "not", which is not found in the German edition. Thanks to Michael Kremer for originally pointing this out to me and to Thorsten Sander for reminding me again, more recently. Christian Thiel has confirmed that the German edition is faithful to Frege's original letter.

as a 'contextual definition' of the functional expression "fnc(ξ)" (as an axiom governing it, in the formal system).[4] HP, of course, is a somewhat different case: The relation of equinumerosity, though provably an equivalence relation, is one between *concepts*, not objects, so HP is a second-order abstraction principle. Writing "Eq$_x$($\Phi x, \Psi x$)" for any of the usual formalizations of "The Φs can be correlated one-one with the Ψs", and "N$x : \Phi x$" for "the number of Φs", HP may then be formulated as follows:

$$\text{N}x : Fx = \text{N}x : Gx \equiv \text{Eq}_x(Fx, Gx)$$

Thus, to adopt HP as a fundamental axiom is precisely to follow the suggestion Frege is making in his letter to Russell.

Our question is why Frege did not take his own advice: Abandon Basic Law V, install HP as an axiom, and make one's stand on the logical character of HP itself.

5.1   The Caesar Problem

In the cited letter to Russell, Frege remarks that there are certain difficulties connected with adopting HP as a primitive axiom, ones that are, in fact, the same as certain difficulties that afflict Basic Law V. I take it that Frege is not suggesting that HP is inconsistent.[5] What difficulties might he have in mind, then? He does not say explicitly, but it is natural to look for them in *Die Grundlagen*. Frege there discusses, at some length, the question whether a principle similar in spirit to HP can be taken as explaining the concept of direction. In this case, the principle is:

$$\text{dir}(a) = \text{dir}(b) \equiv a \parallel b$$

That is: The direction of a is the same as the direction of b if, and only if, a is parallel to b. Frege considers three objections to the claim that this principle explains the concept of direction, the first two of which he rebuts. In the end, though, he rejects the proposed explanation on the ground that it fails to decide the truth-values of what have come to be called *mixed identity-statements*, identity statements of the form "$t = \text{dir}(a)$", where t is a term not itself of the form "dir(x)". Frege's own example is "England is the direction of the Earth's axis" (Gl, §66). It is this problem, the so-called Caesar problem,[6] that prevents Frege from

---

[4]Frege omits the condition of reflexivity—$\forall x[(\exists y)(Rxy \lor Ryx) \to Rxx]$—but the other two conditions imply it, though they do not imply the stronger condition that Φ is *totally* reflexive—$\forall x(Rxx)$.

[5]Indeed, since Frege is talking about abstraction principles quite generally, such a suggestion would be absurd.

[6]So-called because Frege argues, a little earlier, that a familiar sort of inductive definition of names of finite numbers fails to decide whether Caesar is a number (Gl, §56). I discuss the relation between these two versions of the Caesar problem in Chapter 6.

regarding HP as explaining the concept of number—and so, within the formal theory, from adopting it as an axiom.

It is important to realize that the Caesar problem itself is *not* one upon whose solution the *formal* part of the logicist project depended— and that Frege knew as much. Frege knew that the axioms of arithmetic are derivable from HP, and it is simply obvious that the derivation does not require a solution to the Caesar problem.

Frege raises the Caesar problem as an objection to his 'contextual' explanation of the concept of direction, an explanation that is offered as part of an attempt to answer the famous question of *Die Grundlagen* §62: "How, then, are numbers to be given to us, if we cannot have any ideas or intuitions of them?" According to Frege, to answer this question, it is necessary (and apparently sufficient) to explain the senses of identity statements in which number-words occur; analogously, it is necessary (and apparently sufficient), in order to answer the question how directions are given to us, to explain the senses of identity statement in which names of directions occur. The suggestion Frege is considering, when he raises the Caesar problem, is that this explanation may be given by means of the abstraction principle considered above; the analogous suggestion, in the case of numbers, is that the senses of identity statements containing names of numbers may be explained by means of HP.

It is far from obvious, however, on what ground Frege concludes, from the failure of the relevant abstraction principle to decide whether England is a direction, that it fails as an explanation of the senses of identity statements containing names of directions. Of course, the Caesar problem does show that the abstraction principle, on its own, does not provide a sense for *all* identity statements containing names of directions, since it does not provide one for "England is the direction of the Earth's axis".[7] But why should that be thought a difficulty? We shall return to that question. At present, the important point is just that extensions of concepts— and later, value-ranges—are introduced by Frege in order to resolve the Caesar problem (Gl, §68). Rather than attempt to explain the senses of identity statements involving names of numbers by means of HP, Frege explicitly defines the number of *F*s as the extension of the concept "concept which can be correlated one-one with the concept *F*", "assum[ing] that it is known what the extension of a concept is" (Gl, §68, note). Frege then derives HP from this explicit definition and, as said, proves the axioms of arithmetic from HP, making no further use of extensions (nor any essential use of value-ranges).

As has been said, Frege knew that he could do without extensions *formally*. His abandonment of the logicist program is thus, in a certain sense, not the result of Russell's discovery of the contradiction. Russell of

---

[7] It is worth emphasizing that this is the objection. The objection is not just that the contextual definition fails to decide the *truth-value* of this sentence, but that it fails to give any clear sense to it at all.

course showed Basic Law V to be inconsistent, but this axiom plays a very limited role in Frege's proofs. What ultimately forces Frege to abandon his logicism is his inability to resolve the Caesar problem, his inability, without making reference to value-ranges, to answer the question how we apprehend[8] logical objects. Indeed, just before mentioning to Russell that HP might replace Law V, and alluding to the "difficulties" confronting this suggestion, he writes:

> I myself was long reluctant to recognize the existence of value-ranges and hence classes; but I saw no other possibility of placing arithmetic on a logical foundation. But the question is, How do we apprehend logical objects? And I have found no other answer to it than this, We apprehend them as extensions of concepts, or more generally, as value-ranges of functions. I have always been aware that there were difficulties with this, and your discovery of the contradiction has added to them; but what other way is there? (PMC, pp. 140–1)

The question how we apprehend logical objects is much the same question as that raised in §62 of *Die Grundlagen*: For logical objects are those our apprehension of which does not depend upon intuition or experience either of them or of objects by means of which they are identified.[9] Thus: Frege introduces extensions of concepts into his system to explain how we apprehend logical objects. It was because he could not otherwise explain how we apprehend logical objects that he could not do without extensions.

Frege's abandonment of the logicist program is thus the result of a failure to resolve not a *formal* problem but an *epistemological* one. If we are to understand Frege's logicism, we therefore must understand, first, what Frege meant by the question how we apprehend logical objects and, second, why the Caesar problem frustrates the attempt to answer this question by means of abstraction principles, such as HP.

5.2   The Caesar Problem in *Grundgesetze*

I have argued that Frege is unwilling to adopt HP as a fundamental axiom because he does not think he can solve the Caesar problem without making reference to value-ranges. In this section, we shall look at some of Frege's later discussions of the Caesar problem: It continued to haunt

---

[8]I shall use this term of Frege's throughout, though it obviously could use some explanation. In my own opinion, Frege is using the term to mean "refer to" or "think about", that is, "have cognitive access to". What is wanted is thus an account of how we might be able to refer to or think about certain objects though we have neither intuition nor experience of them.

[9]Care is needed here. It is tempting to say just that logical objects are those an apprehension of which requires neither intuitions nor ideas of them. As said below, however, directions are not *logical* objects, but their apprehension does not rest upon intuition or experience *of directions* (Gl, §64). Nonetheless, Frege would maintain that apprehension of directions does require intuition, since one must apprehend the direction as the direction of a given line. The question raised in §62 is thus more general than the question how we apprehend logical objects.

him long after he 'solved' it in *Die Grundlagen*. We shall see that, in fact, Frege was never able to solve the Caesar problem to his satisfaction.

Frege's 'solution' of the Caesar problem in *Die Grundlagen* consists in the identification of numbers with the extensions of certain concepts. As has often been remarked, however, this solution works only if we assume that we know how to resolve the Caesar problem for extensions themselves: The identification of numbers with extensions decides whether Caesar is a number only if it has already been decided whether Caesar is an extension and, if so, which one he is. Frege's remark that, in giving his solution, he is "assum[ing] it is known what the extension of a concept is" (Gl, §68, note; see also §107) is naturally interpreted as a recognition of this fact. The Caesar problem is, therefore, not so much solved by Frege's identification of numbers with extensions as it is relocated by it.

In *Grundgesetze*, reference to extensions is formalized as reference to value-ranges, and value-range terms are governed by Basic Law V, which bears a marked formal similarity to HP. It is therefore not surprising that Frege should raise the question, in §10 of *Grundgesetze*, whether either of the truth-values (Truth and Falsity) is a value-range and, if so, which value-ranges they are. Frege argues that, consistently with Basic Law V, Truth and Falsity may be identified with the value-ranges of any (extensionally distinct) functions. Frege chooses to identify each of them with its own unit class.

Thus, something much like the Caesar problem arises in *Grundgesetze*, and Frege resolves it by making a stipulation regarding the references of certain terms. Now, the domain of Frege's theory consists only of Truth, Falsity, and the value-ranges:[10] So, for whatever *formal* purposes Frege might have needed to resolve the Caesar problem—for whatever reason he might need to fix the truth-values of mixed identity-statements *of the formalism*—his stipulation may suffice. Nevertheless, the Caesar problem, as it is raised in *Die Grundlagen*, is surely not a problem Frege was prepared to resolve by a stipulation applicable only to such objects as are in the domain of the formal theory. A similar 'stipulative' solution would work just as well in the context of second-order logic augmented by HP: Identify Truth with 1; Falsity, with 0.

In a long footnote, Frege considers the question whether a general solution to the Caesar problem can be modeled upon the partial solution he offers in the case of the truth-values: "A natural suggestion is to generalize our stipulation so that every object is regarded as a value-range, viz., as the extension of a concept under which it and it alone falls" (Gg, v. I, §10). Frege's argument against this proposal is that it works only for such objects as are not "already given to us as value-ranges". Consider

---

[10]This claim contradicts the oft-made claim that, as a consequence of his so-called 'universalism', the domain of Frege's theory must always comprise all the objects there are. For further discussion of this matter, and of other issues related to those discussed in this paragraph, see my paper "*Grundgesetze der Arithmetik* I §10" (Heck, 1999).

$\grave{\alpha}(F\alpha)$, the value-range of the concept $F\xi$. Law V implies that the unit class of $\grave{\alpha}(F\alpha)$, i.e., $\grave{\epsilon}(\epsilon = \grave{\alpha}(F\alpha))$, is the same as $\grave{\alpha}(F\alpha)$ just in case the objects falling under $\xi = \grave{\alpha}(F\alpha)$ are exactly the objects falling under $F\xi$, i.e., that:

$$\grave{\alpha}(F\alpha) = \grave{\epsilon}(\epsilon = \grave{\alpha}(F\alpha)) \equiv \forall x(Fx \equiv x = \grave{\alpha}(F\alpha))$$

As Frege says, however, "Since this... is not necessary, our stipulation cannot remain intact in its general form": Not *every* object can be the same as its unit class; in particular, no class that does not have exactly one member can be its own unit class. The most natural emendation would be this: Objects *other than* value-ranges are to be the same as their unit classes. But that does not work either: If x were not a value-range, the stipulation would imply that it *was* a value-range, namely, its own unit class, whence it ought *not* to be identified with its own unit class. (A thing cannot both be and not be a value-range.) So we are forced to say instead that every object which is not *obviously* a value-range, which is not "already given to us as a value-range", is to be identified with its unit class. Thus intrude ways in which objects are given.

Frege's discussion of this proposal is reminiscent of his discussion in *Die Grundlagen* of the suggestion that Caesar is not a number because only such objects are numbers as are "introduced by means of" HP:[11]

If... we were to adopt this way out, we should have to be presupposing that an object can be given only in one single way; for otherwise it would not follow, from the fact that [an object] *was* not introduced by means of our definition, that it *could* not have been introduced by means of it. (Gl, §67)

The proposal under consideration in the footnote in section 10 of *Grundgesetze* is rejected on similar grounds:

...[I]t is intolerable to allow [the stipulation] to hold only for such objects as are not given us as value-ranges; the way in which an object is given is not an immutable property of it, since the same object can be given in a different way.

Thus, while Frege will make stipulations regarding the truth-values of certain identity statements, for whatever formal reason he might need to do so, he twice rejected attempts to model a *general* solution to the Caesar problem on such stipulations.[12]

---

[11]Frege is actually discussing directions here, but of course the discussion is meant to apply equally to the case of numbers.

[12]Frege discusses such 'stipulative' solutions to the Caesar problem in at least one other place. In one of his letters, Russell had expressed concern about the inference from "The members of u are the same as the members of v" to "u is the same as v". The inference holds, in Frege's system, only when both u and v are value-ranges: Non-value-ranges have no members, so all non-value-ranges have the same members. Russell therefore asks Frege how it can be known whether a given object is a value-range (PMC, p. 139). Unsurprisingly, Frege is unable to answer Russell's question in the general terms in which it is posed. So he says instead that the question may be answered piecemeal, for each sort of mathematical object, as it is introduced:

Frege thus had no general solution to the Caesar problem, there being no other proposed solutions he considers. If so, however, the Caesar problem posed an enormous threat to his position. According to Frege, the Caesar problem shows that the question how we apprehend numbers as logical objects cannot be answered by means of abstraction principles such as HP. Similarly, the analogue of the Caesar problem, for value-ranges, ought to show that our apprehension of value-ranges as logical objects cannot be explained in terms of Basic Law V alone. But the only thing Frege has to say about what value-ranges are is this:

I use the words "the function $\Phi(\xi)$ has the same *value-range* as the function $\Psi(\xi)$" generally to denote the same as the words "the functions $\Phi(\xi)$ and $\Psi(\xi)$ have always the same value for the same argument". (Gg, v. I, §3)

And that amounts to explaining what value-ranges are by means of a meta-linguistic version of Basic Law V. It follows that, since he was without a solution to the Caesar problem for value-ranges, Frege cannot explain, even to his own satisfaction, how we can apprehend *value-ranges* as logical objects. Since his view was that we apprehend all logical objects *as* value-ranges, he was therefore unable to explain, to his own satisfaction, how we can apprehend logical objects at all.

We have yet, however, to see just why Frege thought the Caesar problem such a threat, for we have yet to see why he thought it showed abstraction principles on their own to be explanatorily impotent.

5.3   The Caesar Problem and the Apprehension of Logical Objects

In the first section, I argued that what prevented Frege from adopting HP as a primitive axiom of his formal theory was his inability to resolve the Caesar problem without appealing to value-ranges; in the last section, I argued that Frege ought to have concluded, on similar grounds, that the appeal to value-ranges failed to accomplish what was required of it—and so that he had no account of our apprehension of logical objects. One might wonder, however, whether this can be right. For one thing, Frege seems perfectly willing to accept Basic Law V as a primitive axiom,

---

Now, all objects of arithmetic are introduced as value-ranges. Whenever a new object is to be considered which is *not introduced as a value-range*, we must at once answer the question whether it is a value-range, and the answer is probably always no, since it would have been introduced as a value-range if it was one. (PMC, p. 142, emphasis added)

Similarly, Frege writes in section 10 of *Grundgesetze* that we shall have to decide such questions as they arise, and that "...this can then be regarded...as a further determination of the value-ranges...". Plainly, Frege is not here offering a solution to the Caesar problem: A piecemeal 'solution' is not a solution to the problem but a recipe for side-stepping it.

though he appears to have known that he could no more solve the version of the Caesar problem that arises in connection with it than he could solve the version that arises in connection with HP.

Frege's fondness for Basic Law V should not be overstated, however. He was famously dissatisfied with it even before Russell's discovery of the contradiction. He writes, in a famous passage from the Introduction to *Grundgesetze*:

A dispute can arise, so far as I can see, only with regard to my basic law (V) concerning value-ranges, which logicians perhaps have not yet expressly enunciated, and yet is what people have in mind, for example, where they speak of the extensions of concepts. I hold that it is a law of pure logic. In any event the place is pointed out where the decision must be made. (Gg, v. I, p. vii)

But *what* decision must be made here? This remark is preceded by the following:

Of course the pronouncement is often made that arithmetic is merely a more highly developed logic; yet that remains disputable so long as transitions occur in proofs that are not made according to acknowledged laws of logic, but seem rather to be based upon something known by intuition. Only if these transitions are split up into logically simple steps can we be persuaded that the root of the matter is logic alone. (Gg, v. I, p. vii)

It seems to me that the dispute Frege envisions is not a dispute about the *truth* of Basic Law V but rather one about its epistemological status: a dispute about whether it is a law of logic.

Showing that it is possible to derive arithmetic from certain axioms, whatever they may be, cannot decide the question of arithmetic's epistemological status on its own. The point should be obvious: If the laws of arithmetic follow logically from Basic Law V, then they are laws of logic if Law V is; if the laws of arithmetic follow logically from HP, then they are laws of logic if it is. A dispute can always arise, in principle, concerning the logical character of one's axioms and rules of inference, and Frege knew as much. His most transparent statement of this point is in his 1897 paper "On Mr. Peano's Conceptual Notation and My Own":

I became aware of the need for a conceptual notation when I was looking for the fundamental principles or axioms upon which the whole of mathematics rests. *Only after this question is answered* can it be hoped to trace successfully the springs of knowledge upon which this science thrives. (PCN, op. 362, emphasis added)

If "the axioms upon which the whole of mathematics rests" were those of the formal theory of *Grundgesetze*, then the question of the epistemological status of Basic Law V would become the critical one, the one on which the epistemological status of arithmetic itself would turn. But, as Frege notes, "[a] dispute can arise" regarding it, a dispute he all but explicitly acknowledges he cannot resolve: Frege "hold[s] that it is a law of pure logic", but he has no convincing argument that it is.[13]

Frege was thus dissatisfied with Basic Law V for two sorts of reasons. On the one hand, he was unable to resolve the Caesar problem as it arose in connection with it; on the other, he had no defense of his claim that it is a law of logic. These two difficulties are not unrelated.

To understand the connection, it is useful to compare the case of the axioms of Euclidean geometry. Frege maintains that these axioms are non-logical truths because he holds that our knowledge of them depends upon intuition. More precisely, his view is that *apprehension of the objects of geometry* requires intuition of them and that our knowledge of the truth of the axioms is founded upon that intuition. Similarly, the question whether Basic Law V is a truth of logic is, for Frege, the question whether our recognition of its truth requires intuition or sense-experience. This question, in turn, Frege takes to reduce to the question how we apprehend the *objects* to which reference is made in Basic Law V: Only if we can apprehend value-ranges as logical objects can we recognize the truth of Basic Law V independently of intuition and experience; only then can we recognize Basic Law V as a law of logic rather than as a law of one of the 'special sciences'.

In the case of HP, Frege's asking how one can apprehend numbers as logical objects is—or so I am suggesting—a way of asking how one can know HP to be true. If HP is to be a truth of logic, we must be able to recognize its truth without relying upon either experience or intuition, whence we must be able to apprehend the *references* of numerical terms (i.e., numbers) without perceiving or intuiting them.[14]

It is important to remember that Frege raises the Caesar problem in the context of a certain argument: It is easy to be distracted by its generality, to forget that it is not raised in a vacuum, as if Frege were asserting that it is a requirement on *any* acceptable definition of a singular term that it decide the truth-values of all identity statements containing that term.[15] The Caesar problem is raised as an objection to the claim that HP completely explains (identity statements involving) names of numbers. Now, again, the question under discussion at this point in *Die Grundlagen* is how we apprehend numbers as objects. So the view to which the Caesar problem is raised as an objection is this: We apprehend numbers as the referents of names of the form "the number of *F*s", and our under-

---

[13] For further discussion of the issues raised in the last two paragraphs, see my paper "Frege and Semantics" (Heck, 2010).

[14] I am thus suggesting that Frege held that intuition in mathematics is primarily intuition of objects rather than intuition of truths. For the distinction, see Parsons's paper "Mathematical Intuition" (Parsons, 1980).

[15] This is a relatively common view of the Caesar problem. The demand that functions be defined for all arguments is, of course, characteristic of Frege's later writings. But I know of no real evidence that Frege held this view in *Die Grundlagen*, let alone evidence that the Caesar problem is simply a manifestation of this more general demand.

standing of these names consists entirely in our grasp of HP.[16]

Now, as has been said, it is broadly agreed that Frege's objection to this view is that

[HP] will not, for instance, decide for us whether [Caesar] is the same as the [number zero]—if I may be forgiven an example which looks nonsensical. Natu- rally, no one is going to confuse [Caesar] with the [number zero]; but that is no thanks to our definition of [number]. (Gl, §66; example changed)

But it is rarely mentioned, because it is not thought important, that Frege takes for granted that we *do* recognize that Caesar is not a number: Frege's objection is *not* that HP does not decide whether the singleton of the null set is the number zero (which, on Frege's explicit definition, it happens to be). The example Frege chooses is one about which he takes us to have strong intuitions: Whatever numbers may be, Caesar is not among them. Thus, there must be *more* to our apprehension of numbers than a mere recognition that they are objects that satisfy HP. Something explains *why* "no one is going to confuse [Caesar] with the [number zero]". Frege is thus insisting that any complete account of our apprehension of numbers as objects must include an account of how we recognize that Caesar is not a number. But HP alone yields no such explanation.[17]

The Caesar problem is not intended to show only that our apprehen- sion of numbers as *logical* objects cannot be explained in terms of our knowledge that they satisfy HP. Frege's discussion of the Caesar problem takes place, after all, in the context of a discussion of names of directions, and directions are surely not logical objects. The intended lesson of the Caesar problem, in the case of directions, therefore cannot be that our knowledge that directions satisfy the appropriate abstraction principle does not explain our apprehension of directions as *logical* objects. The lesson is supposed to be that our apprehension of directions *as objects at all* cannot be explained in terms of our knowledge that they satisfy the abstraction principle. For again: If we apprehended directions only by recognizing them to satisfy the appropriate abstraction principle, we would have no basis on which to claim that England is not the direction of the Earth's axis; but we all *do* recognize that England is not the direction of the Earth's axis, so there must be something about our apprehension of directions, something about our capacity to refer to them, that is not captured by the abstraction principle.

Because the Caesar problem, as here interpreted, concerns only mixed identity statements about which we have reasonably strong intuitions, it may not be the same problem as the one Frege raises in *Grundgesetze* §10, which concerns all sentences (of the formalism) in which value-range

---

[16]This view, I take it, is similar to that defended by Wright in his book *Frege's Conception of Numbers as Objects* (Wright, 1983).

[17]It is just this that Wright (1983, §xiv) proposes to deny: He thinks that HP does, in some way, decide whether Caesar is a number. See also the later discussion by Hale and Wright (2001c).

terms occur. To solve the Caesar problem in the form in which it is raised in *Die Grundlagen*, we must explain in virtue of what we recognize (e.g.) that Caesar is not a number. It is not obvious that doing so requires that we fix the truth-values of all mixed identities.

Moreover, even if Frege had been able to fix the truth-values of all mixed identity statements, not just any way of doing so would have solved the Caesar problem to his satisfaction. To see this, note that the truth-values of all such sentences could be fixed, in principle, by identifying numbers with *non*-logical objects. If we suppose, for the moment, that the only (cardinal) numbers are countable numbers, numbers may be identified with numerals. And if we take the same liberties that Frege did and suppose that it is already known what numerals are—that is, if we suppose we already know such things as whether Caesar is a numeral and, if so, which one he is—then this stipulation will decide the truth-values of mixed identity statements. Caesar, for instance, not being a numeral, is not a number, either.

Frege surely would not have accepted such a solution to the Caesar problem. Why not? Why should the identification of numbers with non-logical objects pose any threat to Frege's logicism? It is far from clear that an identification of numbers with numerals *need* pose any real threat to logicism. One might argue, for example, that the Caesar problem does not raise any questions about the truth of, or grounds for, HP: Its truth is established prior to, or independently of, any such identification, which serves only to fix the truth-values of certain statements (for whatever reason one might want to do that).[18] On this view, we apprehend numbers as the objects of which HP is true. HP, in turn, we know to be true independently of any intuition or experience because, say, we recognize it to be analytic of the concept of cardinal number. Thus, we apprehend numbers as logical objects because our recognition of the truth of HP requires neither intuition nor experience. Note the order of explanation: It is essential to the argument that our knowledge that HP is true does not depend upon any prior apprehension of numbers. So, to summarize: On this view, we apprehend numbers as the objects of which HP is true; if numbers are numerals and knowledge of HP requires neither intuition, nor experience, nor even *thought about* numbers, we can apprehend *numerals* as logical objects. That may be surprising, but no absurdity appears to be forthcoming: As Frege emphasizes time and again, objects may be given in different ways.

Why, then, would Frege have rejected the identification of numbers as numerals? What one would like to say is that, if numbers are numerals, our knowledge about numbers—in particular, our knowledge of HP and its consequences—would be corrupted by our knowledge of the relevant facts about numerals. Now, granted, *some* of our knowledge about numbers would not be logical knowledge; for example, our knowledge that

---

[18]There is a bit more on this matter in Chapter 4.

the number zero is round would not be logical in character. But there is no obvious reason that *all* one's knowledge about numbers would be so 'corrupted', and we have seen that it is possible to reject this claim in a principled fashion, by maintaining that our apprehension of the truth of HP does not require any prior apprehension of numbers themselves. I suggest, however, that for Frege the identification of numbers as objects of another kind (whether numerals, as in the example, or value-ranges) is to be part of an *explanation* of the truth of HP, part of an account of how it can be known to be true, and so part of an explanation of its epistemological status. If, with this in mind, we re-describe what the identification of numbers as numerals accomplishes, we see immediately why Frege would have rejected it: If numbers are identified with numerals, and if the truth of HP is explained in terms of the existence of a mapping from concepts to numerals, then our recognition of the truth of HP, so understood, depends upon our recognition that there is such a mapping and so that there are enough numerals to constitute the range of such a mapping (at least countably many). Such an explanation of the truth of HP would presumably not show it to be a truth of logic: It would, rather, show it to be a truth of whatever sort the relevant truths about numerals are; the epistemological status of HP would then depend upon the epistemological status of the relevant knowledge about numerals.[19]

Frege is thus maintaining that our knowledge of the truth of HP depends upon our knowledge about the objects with which numbers are identified. To put the point in more Fregean language, his view is not that we apprehend numbers as logical objects because we recognize HP to be a truth of logic; his view is that we recognize HP to be a truth of logic because we apprehend numbers as logical objects and recognize HP to be true of them.[20]

That this is Frege's view is clear from his discussion, in *Grundgesetze*, of explanations of terms by means of abstraction principles (for short, 'contextual explanations'). Regarding such explanations, Frege writes:

... [W]e may not define a symbol or word by defining an expression in which it occurs, whose remaining parts are known. For it would first be necessary to investigate whether—to use a readily understandable metaphor from algebra—the equation can be solved for the unknown, and whether the unknown is unambiguously determined. (Gg, v. II, §66)

If we were to take HP as a contextual explanation of names of numbers, two questions would arise: Whether there are *any* objects that satisfy HP, and whether there is any *unique* set of objects that satisfy it.

---

[19]For present purposes, it matters little what the epistemological status of our knowledge about numerals might be, other than that the relevant truths about numerals are not logical. But Parsons (1980) argues that numerals, as types, are objects of intuition.

[20]There are strong similarities between the point now being argued and Michael Dummett's view that terms explained by means of abstraction principles refer, though not in a manner suitable to a realist interpretation of them (Dummett, 1981a, ch. 14).

The latter problem is the focus of Frege's attention in *Die Grundlagen*, the question being, as it were, *which* objects the numbers are. The former problem arises naturally, however, from reflection on the Caesar problem. As said earlier, the Caesar problem is intended to show that we cannot explain our capacity to refer to numbers (to apprehend numbers as objects) solely in terms of our knowledge that they satisfy HP: If we take HP as a contextual explanation of names of numbers, we have no explanation of how we apprehend numbers as objects at all. But, if not, we are presumably without any defense of the claim that there are such objects.[21] The question to which the Caesar problem leads is thus: What distinguishes HP, if taken as a primitive logical law, from a 'creative definition', where to give a creative definition of an object is to specify an equation it must satisfy and then to stipulate that there is to be an object satisfying that equation (Gg, v. II, §§143–4)?[22] By the time he wrote *Grundgesetze*, at least, Frege not only thought he could not answer this question, he thought it unanswerable: HP, construed as a contextual explanation, simply is a creative definition.

Now, as I have emphasized, Frege was also without a solution to the Caesar problem as it arises in the case of value-ranges. Indeed, his claim that the Caesar problem, in general, can be satisfactorily resolved only by identifying numbers (or directions, or what have you) with value-ranges threatens to make it impossible to resolve the Caesar problem, as it arises for value-ranges *themselves*: It is obviously useless to attempt an identification of value-ranges with value-ranges, and Frege has foresworn any other sort of solution. The question therefore arises whether Basic Law V is not *itself* a creative definition:

... [S]omebody might indicate that we ourselves have nevertheless constructed new objects, viz. value-ranges (vol. I, §§3, 9, 10). What, then, did we do there? or, rather, in the first place, what did we not do? We did not enumerate properties and then say: we construct a thing that is to have these properties. (Gg, v. II, §146)

What is important here is Frege's conception of how Basic Law V differs from a creative definition. First, he insists that he is *not* just stipulating that there are to be objects that satisfy Basic Law V. Then he argues that what he really said—in *Grundgesetze* §3, quoted above—was:

If a (first-level) function (of one argument) and another function are such as always to have the same value for the same argument, then we may say instead that the value-range of the first is the same as that of the second. We are then *recognizing something common* to the two functions, and we call this the value-

---

[21]One could avoid this conclusion, as said above, if one denied—as I take it Wright does—the implicit assumption that the explanation of our knowledge of HP depends upon an explanation of our apprehension of numbers as objects, rather than conversely.

[22]The discussion of 'postulationism' in §§92ff of *Die Grundlagen* is remarkably similar, both in general spirit and, at times, in detail to the cited discussion in *Grundgesetze*.

range of the first function and also the value-range of the second function. (Gg, v. II, §146, my emphasis)

That is: In making the transformation from the thought that $\forall x(Fx = Gx)$ to the thought that $\acute{\epsilon}(F\epsilon) = \acute{\epsilon}(G\epsilon)$, we are "recognizing something common" to the functions $F\xi$ and $G\xi$. Now, it is extremely tempting to read Frege's words "we may say instead" as suggesting that this transition is merely verbal—or, perhaps, merely conceptual—and so that we require no justification to make that transition. But we have already seen that Frege must deny that our recognition (or apprehension) of value-ranges can be explained in terms of our recognition that they satisfy Law V: The Caesar problem prevents explanations of that form. The explanation must instead proceed the other way:

We must regard it as a fundamental law of logic that we are justified *in thus recognizing something common to both*, and that *accordingly* we may transform an equality holding generally into an equation (identity). (Gg, v. II, §146, my emphasis)

Thus, it is to be a law of logic[23] that we may *recognize* something common to co-extensional functions: It is *because* we can so recognize (apprehend) value-ranges that the inference from the co-extensionality of functions to the identity of their value-ranges is permitted, which inference is then formalized in Basic Law V. Our recognition of the truth of Law V is thus to be grounded in our apprehension of value-ranges as objects: Its status as a law of logic therefore requires that we be able to apprehend value-ranges as logical objects, that is, apprehend them without relying upon intuition or experience.

But, as we have seen, Frege has no argument that value-ranges are logical objects. And he has no argument that there are any such objects.

5.4   Closing

Why was Frege unwilling to abandon Basic Law V, install HP as a primitive axiom, and derive the laws of arithmetic from it? The answer at which we have arrived is this: Without the explicit definition of numbers in terms of extensions, Frege could not solve the Caesar problem as it arises for numbers; he could not explain how we apprehend numbers as objects—logical or otherwise—and so could not explain how we know HP to be true. On the other hand, however, Frege was also without a solution to the Caesar problem as it arises for value-ranges: He was unable to explain how we apprehend value-ranges as objects—logical or otherwise—and so was unable to explain how we know Basic Law V to be

---

[23]Frege would seem to be using the word "logic" here in the broad sense in which it was used in German philosophy in the nineteenth century, so that it included parts of what we would now call epistemology and metaphysics.

true. What we have seen is thus that the objections to treating HP as a primitive axiom can also be raised against Frege's treatment of Basic Law V as a primitive axiom: The situations seem exactly parallel, and Frege regarded them as parallel. Why then was he willing to accept Basic Law V but not HP as a primitive axiom?

To this question, there is a dissatisfyingly simple answer. Though he has no argument that extensions are logical objects, Frege does not expect to meet much opposition on this point. Most of his admissions that he is unable to produce such an argument are immediately followed by remarks like the following:

Logicians have long since spoken of the extension of a concept, and mathematicians have used the terms set, class, manifold; what lies behind this is a similar transformation [from the generality of an identity to an identity of value-ranges]; for we may well suppose that what mathematicians call a set (etc.) is nothing other than the extension of a concept, even if they have not always been clearly aware of this. (Gg, v. II, §147; see also v. I, p. vii)

Dialectically, the supposition that there are value-ranges, and that they satisfy Basic Law V, was not an unreasonable one for Frege to make.

The dialectical situation with respect to HP, however, could not be more different. To suggest that we regard it as a fundamental law of logic that we are justified in recognizing something common to two *equinumerous* concepts, and that accordingly logic allows us to transform a statement of equinumerosity into an identity of numbers (see Gg, v. II, §146, again), would blatantly beg the question whether arithmetic is a branch of logic. To derive arithmetic from Basic Law V, and then to suggest that Law V be taken as a fundamental law of logic, is to make substantial dialectical progress, progress rightly characterized as showing "where the decision must be made" (Gg, v. I, p. vii). To derive arithmetic from HP, and then merely to remark that *it* must be regarded as a fundamental law of logic,[24] is to make no such progress.

That does not imply that HP is not *in fact* the right place for the decision to be made.

---

[24] Let me make it clear that contemporary neo-logicists do not merely make such a remark.

6

The Julius Caesar Objection

Recent research has revealed three important points about Frege's philosophy of arithmetic. First, his attempt to derive axioms for arithmetic from principles of logic does not require Frege to appeal to his Basic Law V, the axiom which gives rise to Russell's Paradox. The proofs sketched in *Die Grundlagen der Arithmetik* depend only upon HP: The number of Fs is the same as the number of Gs just in case there is a one-one correspondence between the Fs and the Gs. Formally, the relevant result is that, if a formalization of HP is added as an axiom to standard, axiomatic second-order logic, second-order arithmetic can be interpreted in the resulting theory. Second, this theory—which may be called *Frege Arithmetic*—is itself interpretable in second-order arithmetic and so, presumably, is consistent (Boolos and Heck, 1998, appendix 2). And third, Frege's own formal proofs of axioms for arithmetic, given in his *Grundgesetze der Arithmetik*, do not depend upon Basic Law V essentially. Indeed, Frege himself knew that he did not require any more than HP, this being essential if he is to draw certain of the philosophical conclusions he wishes to base upon his formal results.[1]

All of this having been said, the question arises why, upon receiving Russell's famous letter, Frege did not simply drop Basic Law V, install HP as an axiom, and claim to have established logicism anyway. The question is not only of historical interest. Though Frege did not himself adopt it, this position has seemed to some a worthy heir to Frege's logicism: Crispin Wright, for example, has suggested that HP embodies "an explanation of the concept of cardinal number in general", whence, even though it is not a principle of logic, it might yet have a similarly privileged epistemological position (Wright, 2001b, p. 310). In attempting to evaluate this view, I for one would very much like to know why Frege did not adopt it.

The historical question is made pressing by the fact that, in a letter to Russell, Frege explicitly considers adopting HP as an axiom, remarking only that the "difficulties here" are the same as those plaguing Basic Law V (PMC, p. 141).[2] Frege says nothing else about these "difficulties", but surely does not mean to suggest that HP might similarly be inconsistent.

---

[1] See Chapters 2–4 for defense of these claims.

[2] The English translation has Frege saying that the difficulties are *not* the same, but the negation is not found in the German edition. Thanks to Michael Kremer for pointing this out to me originally and to Thorsten Sander for reminding me later. Christian Thiel has confirmed that the German edition is faithful to Frege's original letter.

Rather, he presumably had in mind the "third doubt" discussed in §66 of *Die Grundlagen*. It is this that forces Frege to abandon HP, understood as a free-standing explanation of names of numbers, and to replace it with an explicit definition from which HP may be derived. Transposed from the context of his discussion of an analogous principle governing names of directions, the worry is this:

In the proposition "the number of Fs is the same as the number of Gs", the number of Fs plays the part of an object, and our definition affords us a means of recognizing this object as the same again, in case it should happen to crop up in some other guise, say as the number of Gs. But this means does not provide for all cases. It will not, for instance, decide for us whether Julius Caesar is the same as the number zero—if I may be forgiven an example that looks nonsensical. Naturally, no-one is going to confuse Caesar with the number zero; but that is no thanks to our definition of number. That says nothing as to whether the proposition "the number of Fs is identical with q" is to be affirmed or denied, except for the one case where q is given in the form of "the number of Gs". What we lack is the concept of number; for if we had that, then we could lay it down that, if q is not a number, our proposition is to be denied, while if it is a number, our original definition will decide whether it is to be affirmed or denied. (Gl, §66, adapted)

This is *the Caesar objection*. As said, it is Frege's inability to answer this objection that forces him to give an explicit definition in §68, which in turn requires reference to extensions and so (something like) the disastrous Basic Law V.

To understand why Frege could not treat HP as an axiom, we must understand the Caesar objection. But this has proved far from easy. I myself have come to the conclusion that the Caesar objection does not pose any single problem but at least three different, though related, ones. I cannot discuss all of these here, so let me simply indicate two of them, if only to set them aside.

The first problem is epistemological. Frege raises the Caesar objection against a proposed answer to the famous question of §62, "How, then, are numbers to be given to us, if we cannot have any ideas or intuitions of them?" Frege takes it that, to answer this (Kantian) challenge, it is necessary and sufficient to explain the senses of identity statements in which number-words occur, this claim being underwritten by the context principle (Gl, p. x and §107). The suggestion Frege is considering when he raises the Caesar objection is that this may be done by means of HP. So the view against which the Caesar objection is offered is this: We recognize numbers as the referents of names of the form "the number of Fs", and our understanding of these names consists (wholly) in our grasp of HP. Frege's objection to this view is, once again, that HP "will not, for instance, decide for us whether [Caesar] is the same as the [number zero]..." (Gl, §66). He concludes, since he is unable to answer the objection, that HP fails as an explanation of the senses of identity statements containing names of numbers. Now, the Caesar objection does seem to show that HP, on its own, does not provide a sense for all identity statements containing names of numbers.[3] But why should that be thought a difficulty?

It is rarely mentioned that Frege takes for granted we *do* recognize that Caesar is not a number. If, as is often said, his objection were that HP does not decide the truth-values of *all* 'mixed' identity statements, then our intuitions about the truth-value of this particular 'mixed' identity statement would be quite irrelevant. But it is important to remember that Frege raises the Caesar objection in the context of a particular argument: One should not be so distracted by its apparent generality that one imagines it raised in a vacuum, so that it could only depend upon some general requirement that every well-formed sentence must have a truth-value. The specific objection Frege raises is not, say, that HP does not decide whether the singleton of the null set is the number zero (which, on Frege's explicit definition, it happens to be). The example Frege chooses is one about which he takes us to have strong intuitions: Whatever numbers may be, Caesar is not among them. Thus, one might think, there must be *more* to our apprehension of numbers than a mere recognition that they are the references of expressions governed by HP. Any complete account of our apprehension of numbers as objects must include an account of what distinguishes people from numbers. But HP alone yields no such explanation. That is why Frege writes: "Naturally, no one is going to confuse [Caesar] with the [number zero]; but that is no thanks to our definition of [number]" (Gl, §62).[4]

The second problem raised by the Caesar objection is semantical, but it has obvious epistemological overtones. HP purports to explain *names* of numbers, expressions that must be treated, semantically, as purporting to refer to objects, the numbers. Only if expressions of the form "the number of *F*s" are so understood, as purporting to refer to objects, can our capacity to refer to numbers be explained in terms of our grasp of HP. But why is "the number of *F*s is the number of *G*s", as explained by HP, not just an idiomatic rendition of "there is a one-one correlation between the *F*s and the *G*s"? On what ground is it claimed that the former has the sort of *semantic*, as opposed to merely orthographic, structure that it needs to have? If this explanation of 'identity statements' involving 'names of numbers' really does license us so to treat those statements, then the understanding conveyed by HP must enable us to understand such predicates as "ξ is the number of *G*s", these being true or false of *objects*. These predicates are formed, after all, merely by omitting a se-

---

[3] Note that the objection is not so much that HP fails to decide the truth-value of "Caesar is zero", but that it fails to give any sense to it at all.

[4] One might want to object that there is no reason to suppose that our recognition that Caesar is not a number must be explained in terms of the resources employed in an explanation of our apprehension of numbers as objects. But it is hard to see how one's recognition that Caesar is not a number could flow from anything but one's understanding of numerical terms. In any event, Frege does not discuss this assumption; for present purposes, let us just record it.

mantic constituent from the sentences so explained. To put the point differently, if "the number of *F*s" is a semantic constituent of such sentences, it must be replaceable by a variable: It must be an intelligible question whether the open sentence "*x* is the number of *G*s" is true or false of any particular object, *independently of how it might be given to us* (Gl, §67). But HP does not even appear to explain sentences of the form "*x* is the number of *G*s", but only statements of the form "the number of *F*s is the number of *G*s".[5] The best we seem able to do is to understand such questions as whether the open sentence is true when a term of the form "the number of *F*s" is substituted for the variable. But that is to invite the question whether our understanding of quantification over numbers is not merely substitutional; if so, it would seem that our capacity to refer to numbers (at least, as objects independent of our ways of thinking of them) has not been explained.

To summarize: To meet the Caesar objection, one must meet at least two challenges. The first is to show how, on the basis of the understanding of names of numbers captured by HP, one can come to understand questions of 'trans-sortal identification' and, in particular, to know that numbers are of a sort different from people and other such objects.[6] The second is to explain how, on the same basis, one can arrive at an understanding of such predicates as "ξ is the number of *G*s". These two challenges apply not only to the explanation of names of numbers embodied in HP, but to *any* 'abstractionist' explanation of names, e.g., to the analogous explanation of names of directions considered in *Die Grundlagen*. Since Frege raises the Caesar objection both against this explanation of names of directions and against that of number in terms of HP, some quite general problems must be raised by the Caesar objection.[7] One should not conclude, however, that the Caesar objection does not also raise quite specific problems in the case of numbers. Such an aspect of the Caesar objection is what I wish to discuss here.

The Caesar objection is first raised in a different context. In §55 of *Die Grundlagen*, Frege considers an 'inductive' definition of cardinal numbers, the two important clauses of which are:

The number 0 belongs to a concept, if the proposition that *a* does not fall under that concept is true universally, whatever *a* may be.

The number $n + 1$ belongs to a concept *F*, if there is an object *a* falling under *F*

---

[5]This is why Frege mentions sentences of the form "*q* is the direction of *a*". I take it that a nice, rhetorical way to make this point is to emphasize that the definition gives us no purchase on the question whether this open sentence is true or false of England. To the best of my knowledge, this point, like many others relevant to the present topic, was first made by Charles Parsons (1995a).

[6]This is what Wright attempts to do in §xiv of *Frege's Conception* (Wright, 1983). The use of the term "sort" is intentional. Adopting the view discussed below, that numbers are simply a different *sort* from people, will not relieve one of the obligation to explain the origins of speakers' knowledge of this fact.

[7]These general problems are discussed in Chapters 8 and 9.

and such that the number n belongs to the concept "falling under F, but not a".

Among the objections Frege makes to these definitions is that

...we can never—to take a crude example—decide by means of our definitions whether any concept has the number Julius Caesar belonging to it, or whether that same familiar conqueror of Gaul is a number or not. (Gl, §56)

It is implausible that this occurrence of the Caesar objection should not be closely related to that in §66. Part of my goal here is to explain how these two objections are related and so to throw light on them both.

The remainder of this chapter consists of four sections. In Section 6.1, I shall discuss a version of the Caesar objection that Frege raises in *Grundgesetze*. Explaining why Frege is compelled to answer the question whether Truth is a 'value-range' will motivate my treatment of the Caesar objection, as it arises in *Die Grundlagen*. In Section 6.2, I shall argue that the Caesar objection arises, in §56, as a manifestation of a technical obstacle to the development of arithmetic on the basis of the inductive definitions considered in §55. Furthermore, the later occurence of it, in §66, is connected to a similar technical obstacle to the development of arithmetic on the basis of HP, and the most obvious way of overcoming this obstracle forces Frege to provide a sense for such sentences as "Caesar is the number zero". In Section 6.3, I shall argue that the technical obstacle can be overcome in an *un*obvious way, and so that there is a way of founding our knowledge of arithmetic on (an analogue of) HP that sidesteps this aspect of the Caesar objection. The resulting conception of the genesis of our knowledge of arithmetic has, I think, independent virtues, some of which I shall mention in the closing Section 6.4.

6.1   Why the Caesar Objection Has To Be Taken Seriously

One common view about the Caesar objection is that the demand that a sense be provided for "Caesar is the number zero" is a consequence of Frege's general "requirement as regards concepts that, for any argument, they shall have a truth-value as their value..." (CO, op. 20).[8] It follows that, if "$F(0)$" has a truth-value, every sentence resulting from the replacement of "0" by some other (referential) proper name must also have a truth-value. But there is no indication that, at the time of writing *Die Grundlagen*, Frege subscribed to this 'requirement of complete determination'; this interpretation of the Caesar objection thus reads post-1891 doctrines back into *Die Grundlagen* without independent justification.[9]

---

[8]I have argued elsewhere that this requirement itself results from Frege's desire to provide his formal theory with a classical semantics (Heck, 2010).

[9]In recent years, I have become convinced that almost *none* of Frege's mature views are present before about 1881, and that their development is extremely complicated. How

Indeed, even though Frege does subscribe to the requirement of complete determination in his mature period, and even though it does lead to the Caesar objection, Frege does not raise analogues of the Caesar objection, even in his mature period, solely because he subscribes to this requirement. In §10 of *Grundgesetze*, Frege considers the question whether truth-values—which, for him, are the referents of sentences—are value-ranges, and if so, which value-ranges they are; that is, whether such sentences as "Truth is the extension of the concept *non-self-identical*" are true or false. Frege shows, by means of the so-called 'permutation argument', that the semantical stipulations he has made up to that point do not decide such questions.[10] Frege eventually stipulates that Truth and Falsity are to be their own unit-classes and then argues, in §§30–1, that this stipulation, together with those made earlier, suffices to provide every expression of the theory with a unique referent.[11]

Obviously, however, the requirement of complete determination will force Frege to provide the sentence "Truth is the value-range of the concept *non-self-identical*" with a sense *only if it is well-formed*. Once made, the point is obvious: The version of the Caesar objection discussed in §10 would not arise if Frege did not treat truth-values as objects, sentences as singular terms. His insistence that every well-formed sentence must have a truth-value cannot, on its own, then, explain why this *or any other* instance of the Caesar objection should arise. The question is why Frege insists upon treating sentences as singular terms when doing so causes so much trouble. Answering this question will help us to understand the answer to the analogous question concerning the instance of the Caesar objection discussed in *Die Grundlagen*.

It is almost cliché to remark that Frege's texts are conspicuously thin on argument for the claim that sentences are proper names. His argument, such as it is, is that sentences are 'saturated', like proper names; hence, sentences cannot be functional expressions, and so must be proper names. But this argument is unpersuasive: Why shouldn't there be more than one kind of saturated expression, just as there are different kinds of 'unsaturated' expressions?[12] Given just how poor this argument is,

---

these views develop is the central theme topic of the various papers Robert May and I have written over the last few years.

[10]Schroeder-Heister has shown that the permutation argument can be so reconstructed that it does indeed show, with respect to the first-order fragment of Frege's theory, what is here claimed (Schroeder-Heister, 1987). The argument can be modified to establish a similar claim about the predicative second-order fragments of the theory. Both of these are consistent (Parsons, 1995b; Heck, 1996).

[11]One of the interesting facts about the stipulation is that it is not embodied in the axioms of the formal theory. The reason, one might conjecture, is that the problem under discussion in §10 is a *semantical* problem; that Frege appeals to his stipulation only during §31 also suggests this, for §31 is a (flawed) attempt to prove the consistency of Frege's theory by a semantical argument. But that is another paper (Heck, 1998a).

[12]Ricketts suggested to me that perhaps the point of Frege's 'argument' is just that, since sentences are saturated, allowing them to stand where names stand is legitimate. Doing so

I strongly suspect that Frege is not concerned to establish that truth-values 'really are' objects at all, but rather that they can be so treated, should that be convenient.[13] So perhaps it would be best to inquire why one might *want* to treat truth-values as objects.

Michael Dummett has suggested that doing so simplifies Frege's formal system by simplifying its ontology. If we treat truth-values as objects, then we need not distinguish between concepts (that is, functions whose values are always truth-values) and functions more generally. Nor need we distinguish between one-place functions whose arguments are always truth-values and others (Dummett, 1981a, pp. 183–5). Now, I have no quarrel with Dummett's claim that Frege was motivated to treat truth-values as objects because doing so somehow simplifies his system, but I am not sure Dummett has properly identified the simplification the identification effects.

Frege's Basic Law V governs terms that (purport to) stand for value-ranges. As it is often formulated, Basic Law V is:

(Vc) $\qquad (\hat{x}Fx = \hat{x}Gx) \equiv \forall x(Fx \equiv Gx)$

Thus, the *extension* of the concept F is the same as that of the concept G just in case these concepts are co-extensive. But Frege's formulation Basic Law V would better be written:

(Vf) $\qquad (\grave{\epsilon}f\epsilon = \grave{\epsilon}g\epsilon) \equiv \forall x(fx = gx)$

Thus, the *value-range* of the function f is the same as that of the function g just in case these functions have the same value for every argument. Now, in the sort of theory with which we are most familiar—namely, one that distinguishes the logical types of sentences from those of proper names—one might well want to take both (Vc) and (Vf) as axioms.[14] In Frege's theory, however, since sentences are of the same logical type as proper names, one-place (first-level) functions are of the same logical type as one-place (first-level) concepts. So, for Frege, (Vf) includes (Vc) as a kind of special case, and Basic Law V is in fact stated in the form:

(V) $\qquad (\grave{\epsilon}f\epsilon = \grave{\epsilon}g\epsilon) \ = \ \forall x(fx = gx),$

since material equivalence is then just identity of truth-value.

---

will not, at least, leave unfilled argument places. This is plausible, but the matter could no doubt use more attention.

[13]Burge (2005c) has argued that there are good, philosophical reasons both for Frege to identify truth-values as objects and for him to identify them with the value-ranges with which he does. His arguments merit a reply, but I cannot discuss them here. Let me say, though, that *if* we assume Frege has reason to identify truth-values with value-ranges, Burge gives as good an account as one can of why Frege chooses the value-ranges he does.

[14]Until, of course, one realized that (Vc) is inconsistent and (Vf) implies that there is exactly one object. We shall, however, abstract from these problems with (Vc) and (Vf) here. One could transpose much of the present discussion to the context of predicative second-order logic, where (Vc) and (Vf) raise no such problems.

Frege uses (V), in large part, so that he may speak of the value-ranges of one-place functions, instead of speaking of the functions themselves.[15] But it also allows him to speak of the value-ranges of two-place functions, which he calls 'double' value-ranges. Consider, for example, the function "$\xi + \eta$". Fix one of its arguments, say the second, and consider the function "$\xi + 2$". The value-range of this function, $\grave{\epsilon}(\epsilon + 2)$, is the graph of the function whose value, for a given argument x, is $x + 2$. Suppose the second argument is now allowed to vary; the resulting value-range, $\grave{\epsilon}(\epsilon + n)$, will be the graph of the function whose value, for given argument x, is $x + n$. What, then, is the value-range of the function $\grave{\epsilon}(\epsilon + \eta)$? It is the *double* value-range $\grave{\alpha}\grave{\epsilon}(\epsilon + \alpha)$, the graph of the function whose value, for argument y, is the value-range $\grave{\epsilon}(\epsilon + y)$.[16] This double value-range Frege uses as if it were the value-range of the two-place function itself. And it is easy to see that $\grave{\alpha}\grave{\epsilon}(f\epsilon\alpha) = \grave{\alpha}\grave{\epsilon}(g\epsilon\alpha)$ if, and only if, $\forall x \forall y(fxy = gxy)$.[17]

By means of this lovely construction,[18] Frege manages to do without special axioms governing the value-ranges of two-place (and similarly, many-place) functions. Not that it would have been a difficult matter to formulate such axioms, but the use of double value-ranges certainly does simplify Frege's system. The same construction enables Frege to utilize the double value-range $\grave{\alpha}\grave{\epsilon}(R\epsilon\alpha)$ of a relation R as if it were its extension. And it is at this point that the utility of Frege's identification of truth-values as objects becomes apparent. One can see this by considering what would happen were we to try to mimic Frege's use of double value-ranges in a theory that did not treat truth-values as objects. Consider, first, a theory containing only Basic Law (Vc). The 'double extension' term "$\hat{y}\hat{x}(x < y)$", which one might have supposed could serve as a term denoting the extension of the relation $\xi < \eta$, is not even well-formed. Extension terms are formed by prefixing "$\hat{y}$", say, to a one-place *predicate*, the argument-place of which is then filled by "y". But "$\hat{x}(x < \eta)$" is not a predicate at all; it is a functional expression. That this expression is not

---

[15]For discussion of how value-ranges are used in *Grundgesetze*, see Section 2.1.

[16]For Frege, ordered pairs are defined in terms of value-ranges, so value-ranges are not sets of ordered pairs. But if we pretend that they are, the points made in the text can be put as follows. The value-range of the function $\xi + 2$ is the set of ordered pairs $\{<\epsilon, \epsilon + 2>\}$. The value-range of $\xi + n$ will be the set of ordered pairs $\{<\epsilon, \epsilon + n>\}$. And the value-range of the function $\grave{\epsilon}(\epsilon + \eta)$ is the set of ordered pairs $\{<\alpha, \grave{\epsilon}(\epsilon + \alpha)>\}$, or $\{<\alpha, \{<\epsilon, \epsilon + \alpha>\}>\}$.

See the Postscript to Chapter 2 for some speculation about the relationship between these two treatments.

[17]Frege does not prove this result, since he does not need it. Theorems 2 and 3 of *Grundgesetze* do the necessary work.

[18]In effect, then, Frege is treating every function, of however many arguments, as a one-place function, with values possibly themselves being functions. So, for example, a two-place function is here treated as being a one-place function whose value is a one-place function. This treatment, I am told, is characteristic of combinatory logic and is credited to Moses Schönfinkel. In his introductory note to the reprinting of Schönfinkel's paper in the van Heijenoort *Sourcebook* (1967), Quine notes that Frege "anticipated" Schönfinkel's construction in his discussion of double value-ranges.

of the correct type for the formation of extension terms is a consequence of the fact that truth-values are of a different logical type from extensions: Only if truth-values are of the same logical type as extensions will concepts—i.e., functions from objects to truth-values—be of the same logical type as functions from objects to extensions.

Now, it is true that one does not have to treat truth-values as objects to effect something like Frege's reduction of the extensions of relations to those of concepts. One remedy is to add Basic Law (Vf) to the theory and write "$\acute{\alpha}\hat{x}(x < \alpha)$" instead of "$\hat{y}\hat{x}(x < y)$"; the extension of $\xi < \eta$ is then the *value-range* of the function $\hat{x}(x < \eta)$. Really to see what Frege's treating truth-values as objects amounts to, however, one needs to consider a different remedy, which involves neither identifying truth-values as objects nor taking both (Vc) and (Vf) as axioms.[19]

We continue to suppose that we are working in a language that distinguishes the logical types of names and sentences. Suppose, now, that we limit ourselves to Basic Law (Vf). We can then speak, easily enough, of the single and double value-ranges of functions. Can we also be speak of the extensions of concepts? A familiar trick will enable us to do that, too: We can employ *characteristic functions*.[20] To do so, we require a description-operator "ι"[21] and two arbitrary objects—which we denote "$\bot$" and "$\top$". We then define the extension of the concept $\Phi\xi$ as follows:

$$\hat{x}(\Phi x) = \grave{\epsilon}\{\iota x[(\Phi\epsilon \wedge x = \top) \vee (\neg\Phi\epsilon \wedge x = \bot)]\}$$

Thus, the extension of $\Phi\xi$ is the value-range of the function whose value, for given argument x, is $\top$ if x is Φ; $\bot$, if x is not Φ. It is easy enough to show that, so defined, "$\grave{\epsilon}(\Phi\epsilon)$" satisfies Basic Law (Vc).

In fact, one could go yet further and refuse to make serious use of anything *but* characteristic functions. Instead of introducing primitive relations into the system, for example, one would introduce their characteristic functions. One could even introduce the logical constants in terms of *their* characteristic functions: Expressions such as "$(2+2 = 4) \vee (1+1 = 3)$" would then be names, not sentences. To be able to form sentences (and

---

[19]Thanks to Michael Glanzberg for asking the question that made what follows clear to me.

[20]In set-theory, the characteristic function associated with a set S is the function $\phi_S(\xi)$ whose value, for argument x, is 1 if $x \in S$; 0, otherwise.

[21]Frege has a description operator in the formal system of *Grundgesetze*, but it is applied to value-range terms, not to predicates. What is needed here is description operator of the more usual sort. The axiom

$$a = \iota x(Fx) \equiv \forall x \forall y(Fx \wedge Fy \rightarrow x = y)$$

will serve as an analogue of Frege's Basic Law VI. A pure analogue of Basic Law VI would be:

$$a = \iota x(x = a)$$

but this will not imply the above without an axiom asserting the extensionality of concepts. Such an axiom is not present in Frege's system (and is not usually present in axiomatic second-order logic), though his semantics for second-order logic does justify it.

so make any assertions), one would need to have at least one real predicate in the language, and the most natural choice for such a predicate would be one that meant: ξ is identical with $\top$. This predicate would be remarkably like Frege's horizontal, which means: ξ is identical with Truth; indeed, one might wonder whether he had something like this construction in mind when he wrote, in *Begriffsschrift*, that the system has only one predicate (Bg, §3). Thus, although Frege could have distinguished truth-values from objects and still been able to make do with Basic Law (Vf), he does not take this course: What he effectively does is identify concepts with their characteristic functions. Once the identification is made, the truth-values become the objects $\top$ and $\bot$, in terms of which the characteristic functions were defined. Nothing could be more natural, mathematically speaking.[22]

Treating sentences as of the same logical type as proper names thus has substantial technical advantages in the context of Frege's system. And once one sees that it amounts to identifying concepts with their characteristic functions, it should not seem all that perplexing. Nonetheless, Frege's making this move imposes certain obligations on him. As he understands the notion of logical type, two expressions are of the same logical type only if they are intersubstitutable *salva significatione*. Thus, if there is any sentence of the form "...t..." that has a sense, and there is a corresponding sentence of the form "...u..." that does not, then t and u cannot be of the same logical 'Sort'.[23] Hence, if truth-values are of the same Sort as value-ranges, identity statements such as "$\top = \grave{\epsilon}(\epsilon \neq \epsilon)$" simply *must* have a sense, since "$\grave{\epsilon}(\epsilon \neq \epsilon) = \grave{\epsilon}(\epsilon \neq \epsilon)$" most certainly does.

6.2   The Caesar Objection and the Feasibility of the Logicist Project

We have seen that the version of the Caesar objection Frege considers in *Grundgesetze* arises because he takes value-ranges and truth-values to be of the same logical Sort. More generally, a version of the Caesar objection will arise whenever one makes such suppositions about the Sorts of objects of apparently different kinds. For example, if one supposed that (names of) directions had to be of the same logical Sort as (names of)

---

[22]Moreover, we can see now why Frege thought himself free to stipulate which value-ranges the truth-values are: For the selection of objects in terms of which to define the characteristic functions is arbitrary, constrained only by the resources available in the language. One might wonder, though, how Frege might have proven that there are two objects in the domain. Treating the truth-values as objects does resolve this problem (as well as the more general problem of the non-emptiness of the domain). But Frege thought he had a philosophical argument that the truth-values exist: "These two objects are recognized, if only implicitly, by everybody who judges something to be true—and so even by a skeptic" (SM, op. 34).

[23]To emphasize that I am using the term, on Frege's behalf, in this sense, I shall speak of expressions as being of the same logical Sort, rather than of the same logical type.

the lines whose directions they were, "$\ell = \mathrm{dir}(\ell)$" would have to have a sense. For to say that "ℓ" and "$\mathrm{dir}(\ell)$" are of the same logical Sort is just to say that they are intersubstitutable *salva significatione*, whence this sentence must have a sense, since "$\mathrm{dir}(\ell) = \mathrm{dir}(\ell)$" does.

Why, then, does Frege raise the question whether Caesar is a number in *Die Grundlagen*? One might suggest that it is because he is assuming that all objects are of a single logical Sort: That would certainly give rise to the Caesar objection. But Frege does not argue for this claim, which is not so much as mentioned. A similar answer could, of course, be given to the question why Frege raises the Caesar objection in *Grundgesetze*. But, even though Frege did then hold that all objects belong to a single Sort, he had a specific, technical reason to suppose that truth-values and value-ranges were of the same Sort. One might wonder, therefore, whether there is not, in the case of HP, too, a similarly specific reason Frege needed to suppose that numbers were of the same Sort as objects of other kinds.

Imagine a language, devoid of mechanisms for reference to numbers, into which names of numbers are to be introduced by means of HP. Prior to the definition of names of numbers, the speakers of this language may be supposed to understand names of, and predicates true or false of, objects of various other kinds; among these 'basic objects', one might suppose, will be such things as people and trees and rocks and rivers.[24] HP will, in the first instance, explain (or define) terms of the form "the number of Fs", where F is a predicate true or false of basic objects. Once speakers have understood the abstractionist explanation of these terms, they will understand terms that refer to numbers, such as "the number of Roman emperors", and predicates true or false of numbers, such as "ξ is a number less than 5".[25] Now, if names of numbers, so explained, were of the same Sort as names of basic objects, a version of the Caesar objection would arise: The question whether Caesar is a number, and, if so, which one he is, would have to be provided with an answer. But *why* should Frege suppose that numbers are of the same Sort as basic objects?

This supposition is not gratuitous; nor need it based upon some prior assumption that all objects are of the same Sort. The need for the supposition is connected with Frege's oft-repeated insistence that numbers can themselves be counted.[26] One way to explain the force of this idea would be to observe that, once names of numbers have been introduced by means of HP, no further explanation of such expressions as "the num-

---

[24]For simplicity, I assume that these are of a single Sort. Just how unimportant this assumption is will be clear by the end of the chapter.

[25]It is exactly at this point that the second problem raised by the Caesar objection, discussed on page 129, becomes pressing.

[26]Thus, Frege writes in *Die Grundlagen*: "The truths of arithmetic govern all that is numerable. This is the widest domain of all; for to it belongs... everything thinkable" (Gl, §14). And, in "Formal Theories of Arithmetic": "...[W]e can count just about everything that can be an object of thought: ... even numbers can in their turn be counted" (FTA, op. 94).

ber of numbers less than 5" would appear to be required. One would expect speakers immediately to understand such expressions and to know, for example, that "the number of Roman emperors is the same as the number of numbers less than 5" is true if, and only if, there is a one-one correlation between the Roman emperors and the numbers less than 5. It would appear to follow that the predicates "ξ was a Roman emperor" and "ξ is a number less than 5" must be of the same Sort, for the names explained by HP contain predicates of a single Sort, namely, those of the same Sort as "ξ was a Roman emperor". But predicates of this Sort are predicates true or false of objects of the same Sort as basic objects (since the Sort of a predicate is determined by the Sorts of its acceptable arguments). Hence numbers must be of the same Sort as basic objects; hence the question what to make of "Caesar is the number zero".[27]

Frege's insistence that numbers can be counted is also important in the context of his logicism. The counting of numbers—the use of terms of the form "the number of numbers less than 5"—is essential to the logicist project as Frege envisions it. Frege proves that every number has a successor by showing that every natural number n is succeeded by the number of numbers less than or equal to n; the proof thus makes essential use of numerical terms containing predicates true or false of numbers. Hence—to reprise the above—if expressions such as "the number of numbers less than 5" are supposed to have been explained by HP, names of numbers apparently must be of the same Sort as names of persons.

There is a bad misunderstanding of what has just been argued that must be avoided here. One might well remark that the use of terms of the form "the number of numbers less than or equal to n" is not essential to Frege's proofs, since he could equally well have used terms formed from predicates true or false of objects of other sorts. Suppose, for example, that Basic Law V is in place and that predicates true or false of value-ranges are substitutable into HP (i.e., that value-ranges are 'basic objects'). Consider the sequence of value-ranges: $\grave{\epsilon}(\epsilon \neq \epsilon)$, $\grave{\alpha}(\alpha = \grave{\epsilon}(\epsilon \neq \epsilon))$, etc., i.e., the sequence beginning with the empty value-range that is determined by taking, at each step, the singleton of that at the previous step. It is not difficult to show that, for each natural number n, there is a concept true of exactly the first n members of this series; since the series is unending, there will be a concept F under which fall those value-ranges and the next one in the series.[28] It will then be an easy matter to show that the number of Fs is the successor of n.

---

[27] It is more usual to make this argument by concentrating on the Sort of object over which the bound variables appearing in HP range.

[28] This is done, of course, by inductively defining a map from the numbers into this series. The techniques for doing so were known to Frege (Heck, 1995a, 1998b). The same construction can be carried out given various consistent axioms weaker than Basic Law V, for example, Boolos's New V (Boolos, 1998j). An analogous construction can be carried out given the ordered pair axiom and another axiom asserting that there are at least two objects. Infinity is cheap, formally speaking.

As a technical observation, this is plainly correct: If Frege had been willing to make such an appeal to Basic Law V, or if he had been willing to appeal to an axiom of infinity, he could have done without the claim that numbers are of the same Sort as basic objects. But appeal to a special axiom of infinity is precisely what Frege is trying to avoid, since part of his purpose is to explain the genesis of our knowledge that there are infinitely many objects. Of course, HP, as it is usually formulated, is an axiom of infinity; but that fact is uninteresting, since it is completely obvious that any principles from which the truths of arithmetic can be derived will imply the existence of infinitely many things. To object to Frege's use of HP simply on the ground that it is an axiom of infinity is to object that his premises imply his conclusion. What is interesting about HP is not so much that it implies the existence of infinitely many things, but that the infinitely many things whose existence it implies are (or at least are intended to be) *precisely the natural numbers*.[29],[30] Even if one does not go on to claim that HP is a conceptual truth, or an explanatory principle, or any other such thing, it is still one thing to rest one's development of arithmetic upon a principle implying that there are infinitely many natural numbers and something else entirely to presuppose that there are infinitely many objects of some other, possibly even non-mathematical, kind (as do both Dedekind and Russell). This is why I said, above, that Frege wants to avoid appeal to a *special* axiom of infinity. If so, then Frege would have had reason to regard the claim that numbers are of the same Sort as basic objects as essential to the *kind* of proof of the infinity of the number-series required by his philosophical purposes.[31]

The argument given so far shows merely that the proofs Frege gave of the axioms of arithmetic require the assumption that numbers are of the same Sort as basic objects: It does not show that Frege would have had reason to suppose that those axioms cannot be proven without that assumption. Consider, then, a formulation of HP in a two-sorted language. There are 'basic' individual variables, "x", "y", and the like, and 'numeric' individual variables, "x", "y", and the like. There are also basic and numeric predicate variables, "F", "G", and "F", "G", respectively. And there are relation variables of various kinds, the logical types of which may be indicated by subscripts: "R_{bn}" is a relation variable whose first argument must be basic and whose second must be numeric; "R_{bb}", one both

---

[29]Another way to put this point is as follows: Any (reasonably typical, non-modal, etc.) development of arithmetic will imply the existence of the natural numbers, but Frege's has the virtue of implying the existence of nothing else.

[30]Actually, HP does imply the existence of the first transfinite cardinal, though that just points to a difference between Frege's project and the usual one nowadays: Frege is looking for a general theory of cardinal number. See Chapter 11 for more on this matter. There is further discussion in the postscript to Chapter 10.

[31]I owe special thanks to Bill Tait for discussions relating to the preceding few paragraphs.

of whose arguments must be basic; and so on.[32] As for identity, there is
not one identity-sign in this language, but two, which we might write as
"$=_b$" and "$=_n$"; no 'mixed' identity statement will be well-formed.

A version of HP may then be formulated as follows:[33]

$$\mathbf{N}x{:}Fx =_n \mathbf{N}x{:}Gx \equiv$$
$$\exists R_{bb}\{\forall x\forall y\forall z\forall w(R_{bb}xy \wedge R_{bb}zw \rightarrow x =_b z \equiv y =_b w)\wedge$$
$$\forall x[Fx \rightarrow \exists y(Gy \wedge R_{bb}xy)]\wedge$$
$$\forall y[Gy \rightarrow \exists x(Fx \wedge R_{bb}xy)]\}$$

That the "N" is boldface indicates that terms governed by HP are numeric. Note that, in this language, there is no well-formed term such as
"$\mathbf{N}x{:}(x =_? \mathbf{N}y{:}(y \neq y))$", since the terms appearing on either side of "="
are of different Sorts.

What will one be able to prove in a theory whose sole non-logical axiom
is this version of HP? The answer may be somewhat surprising. Frege's
definition of predecession can be formulated as follows:

$$\mathbf{Pmn} \overset{df}{\equiv} \exists F\exists y[\mathbf{n} =_n \mathbf{N}x{:}Fx \wedge Fy \wedge \mathbf{m} =_n \mathbf{N}x{:}(Fx \wedge x \neq y)]$$

His definitions of zero and of the concept of natural number can also be
stated without difficulty:[34]

$$\mathbf{0} =_n \mathbf{N}x{:}(x \neq_b x)$$

$$\mathbf{Nn} \overset{df}{\equiv} \forall \mathbf{F}[\mathbf{F0} \wedge \forall \mathbf{x}\forall \mathbf{y}(\mathbf{Fx} \wedge \mathbf{Pxy} \rightarrow \mathbf{Fy}) \rightarrow \mathbf{Fn}]$$

Now, where "$\mathbf{P}^*$" denotes the strong ancestral of $\mathbf{P}$, the axioms of arithmetic, as Frege formulates them, may be stated as follows:[35]

---

[32]Here and below, the comprehension axioms for the theory are to be of the usual, impred-
icative sort. That is to say, where "F" and "$\mathbf{F}$" are not free in "$A(x)$" and "$A(\mathbf{x})$", respectively,
all formulae of the forms:

$$\exists F\forall x(Fx \equiv A(x))$$
$$\exists \mathbf{F}\forall \mathbf{x}(\mathbf{Fx} \equiv A(\mathbf{x}))$$

and so forth are to be comprehension axioms.

[33]I apologize for the ugliness of this formula, but the only way to ensure that this can all
be done without our violating any sortal restrictions is to be completely pedantic.

[34]I ignore here some of the complexity inherent in Frege's actual definition of "ξ is a
natural number". See the end of Section 2.4 for discussion.

[35]The strong ancestral of a relation R is defined as follows:

$$R^*ab \overset{df}{\equiv} \forall F[\forall x(Rax \rightarrow Fx) \wedge \forall x\forall y(Fx \wedge Rxy \rightarrow Fy) \rightarrow Fb]$$

Intuitively, a bears the strong ancestral of R to b if there is a non-empty, but finite, sequence
of R-steps connecting a to b. Thus, an object does not, in general, bear the strong ancestral
of a relation to itself, but only to its 'proper' ancestors.

The second axiom requires that there be no 'loops' in the sequence of natural numbers.

1. $\forall x \forall y \forall z (\mathbf{Pxy} \wedge \mathbf{Pxz} \rightarrow y =_n z)$

2. $\neg \exists x (\mathbb{N} x \wedge \mathbf{P}^* xx)$

3. $\forall x (\mathbb{N} x \rightarrow \exists y (\mathbf{Pxy}))$

4. $\mathbb{N} n \equiv \forall \mathbf{F} [\mathbf{F0} \wedge \forall x (\mathbb{N} x \wedge \mathbf{F} x \wedge \mathbf{Pxy} \rightarrow \mathbf{F} y) \rightarrow \mathbf{F} n]$

Frege's own proofs of Axioms (1) and (2) may simply be carried over into this new theory; Axiom (4) follows almost immediately from the definition of "$\mathbb{N}$".[36] That one cannot prove Axiom (3), however, may be easily shown: Simply consider the model in which the basic domain contains only Caesar, in which the numerical domain contains only 0 and 1, and in which terms of the form "$\mathbb{N} x{:}Fx$" are interpreted in the obvious fashion. It follows that all of Frege's axioms for arithmetic, other than the third, may be proven in this theory.[37]

The foregoing was known to Frege. It is part of what Frege is arguing, when he first raises the Caesar objection in *Die Grundlagen*: Without the assumption that numbers are of the same Sort as basic objects, the logicist program is impossible.

The inductive definition of cardinal numbers that Frege considers in §55 of *Die Grundlagen* may be formalized as follows:[38]

$$\exists_0 x (Fx) \overset{df}{\equiv} \neg \exists x (Fx)$$

$$\exists_{n+1} x (Fx) \overset{df}{\equiv} \exists y [Fy \wedge \exists_n (Fx \wedge x \neq y)]$$

Frege writes that "these definitions suggest themselves so spontaneously in the light of our previous results that we shall have to go into the reasons why they cannot be reckoned satisfactory" (Gl, §56). He makes three objections. The first is that "we can never... decide whether any concept has the number Julius Caesar belonging to it, or whether that same familiar conqueror of Gaul is a number or not". The second is that "we cannot by the aid of our suggested definitions prove that, if the number a belongs to the concept F and the number b belongs to this same concept, then necessarily $a = b$". The third is that "[i]t is only an illusion that we have defined 0 and 1; in reality we have only fixed the senses of the

---

For a discussion of Frege's axioms for arithmetic, see my paper "Definition by Induction" (Heck, 1995a). It is perhaps worth noting, too, that the second axiom is specially emphasized in Frege's discussion in *Die Grundlagen* (Gl, §83) and is discussed at some length in *Grundgesetze* (Gg, v. I, §108).

[36] Less immediately than is often thought, however: The proof of Axiom (4) requires impredicative comprehension. See the discussion of this matter in Chapter 12 and in my paper "Ramified Frege Arithmetic" (Heck, 2011).

[37] A similar result holds for the Dedekind-Peano axioms. That zero has no predecessor and that predecession is one-one can be proved, but not the existence of successors.

[38] For reasons to be mentioned shortly, the second definition must be understood schematically.

phrases 'the number 0 belongs to' and 'the number 1 belongs to'...". Our task here is to understand the nature of these objections.

The "previous results" that "suggest" the inductive definition are two: First, "that the individual numbers are best derived...from the number one [better, zero] together with increase by one" (Gl, §18); Second, "that the content of a statement of number is an assertion about a concept" (Gl, §46). Now, the 'inductive' character of the definitions certainly embodies the first of these results. What embodies the second is the fact that, as the numbers are here defined, they are second-level concepts: 'properties of concepts'. That this view is Frege's target is confirmed by his pointing out, at the beginning of §57, that, contrary to what one might have thought, it does not follows from the fact that a statement of number ascribes a property to a concept that "a number [is] a *property* of a concept". It is by disposing of this view that Frege means to motivate his alternate view, that numbers are objects.[39]

The discussion in §§55–6 is ostensibly non-technical, but I suggest that what is really under discussion here is not just how the numbers should be defined but how they must be defined *if we are to have any hope of proving the axioms of arithmetic*. By this point in *Die Grundlagen*, Frege takes himself already to have motivated the view that arithmetic is analytic. Earlier parts of the book have disposed of the alternative that it is synthetic, either *a posteriori* (Mill) or *a priori* (Kant). So, at the beginning of chapter IV, we are already searching for a purely logical proof of the axioms of arithmetic from definitions of arithmetical concepts (Gl, §3), and the question here, again, is how the arithmetical concepts must be defined, if such a proof is to be possible. So we're going to try defining them as second-level concepts and see how far we can get.

There is no difficulty with the definitions of the fundamental notions. The number zero may be defined as $\exists_0 x(\Phi x)$. The 'inductive' part of the definition may be recast as a definition of the relation of predecession:[40]

$$P_\Phi(Qx(\Phi x), Rx(\Phi x)) \overset{df}{\equiv} \forall F[Rx(Fx) \equiv \exists y(Fy \wedge Qx(Fx \wedge y \neq x))]$$

That is to say, the second-level concept $Rx(\Phi x)$ will be the successor of the second-level concept $Qx(\Phi x)$ just in case: A concept F falls under $Rx(\Phi x)$ if, and only if, there is an object y that falls under F and the concept $F\xi \wedge y \neq \xi$ falls under $Qx(\Phi x)$. Finally, then, the concept of a natural number may be defined by means of a fourth-order analogue of Frege's definition of the ancestral—the formal statement of which I am happy to omit.

Perhaps the most obscure of the three objections Frege raises against this treatment is the second: that "we cannot by the aid of our suggested

---

[39]This is how Dummett (1991b, ch. 9) reads these sections. My own interpretation owes heavily to his, as well as to Wright's (1983, §vi).

[40]Note that "Φ", on the left-had side, is just a bound (second-order) variable.

definitions prove that, if the number a belongs to the concept F and the number b belongs to this same concept, then necessarily $a = b$" (Gl, §56). As Dummett (1991b, ch. 9) notes, if we assume that 'numbers' here are numerically definite quantifiers, then this remark is worse than wrong, since the proof is then quite easy. But, contrary to what often seems to be assumed, the inductive definition does *not* tell us that numbers are numerically definite quantifiers. Numbers, at this point, are being regarded, as was said earlier, as second-level concepts, and the problem to which Frege is pointing is that there are always going to be *lots* of second-level concepts under which any given first-order concept falls.

Frege lists the need "never to lose sight of the distinction between concept and object" as one of the three fundamental principles on which *Die Grundlagen* rests (Gl, p. x). But he had not always recognized that distinction himself (Heck, 2005; Heck and May, 2012). In particular, Frege does not distinguish concept from object in *Begriffsschrift* in anything like the way he does later in his career. Consider, for example, this remark:

Since the sign Φ occurs in the expression $\Phi(A)$ and since we can imagine that it is replaced by other signs, Ψ or X, which would then express other functions of the argument A, we can also regard $\Phi(A)$ as a function of the argument Φ. (Bg, §10)

Frege simply does not say here what familiarity with his mature views would lead one to expect him to say: that, when we so regard $\Phi(A)$, the function is not just A itself but the second-level concept $\phi(A)$, ϕ being the argument. If that had been his view, it would have needed a great deal of explanation, but it gets none. So Frege's view in *Begriffsschrift* was instead the obvious one: that a sentence like "Caesar coughed" is composed of two parts, "Caesar" and "to cough", both of which can be regarded either as argument or as function, with "to cough" being the function if "Caesar" is the argument and "Caesar" being the function if "to cough" is the argument.[41]

In 1879, then, Frege regarded Caesar as both an object and a second-level concept. Of course, he would not then have put it that way, but that is the view: We may think of Caesar as a function that, in this case, takes coughing as argument. If so, then it is surprisingly easy to see what it might mean to say that the number Julius Caesar belongs to a concept F: It simply means that Caesar is F! And then it is equally clear why we cannot prove that every concept has only one number: Any concept under which Caesar falls will have the number Julius Caesar as well as whatever 'ordinary' number it might also have.[42]

This may seem silly. Surely the obvious thing to say is that Julius Caesar is not a number, in particular, that the second-level concept ϕ(Caesar) is not a number, since it is not numerically definite. But it is important to

---

[41]The sloppiness about use and mention in which I am about to indulge is Frege's own. His official view in *Begriffsschrift* that functions are *expressions*, but he does not always speak that way.

[42]Exercise: What is the successor of Julius Caesar?

appreciate where this discussion occurs in *Die Grundlagen*. At this point in the book, nothing whatsoever has been said about equinumerosity and one-one correspondence. If the right response to the second objection is to insist that numbers are numerically definite quantifiers—and surely that is the right response—then the second objection directs us towards a concern with equinumerosity and so helps to prepare the way for the introduction of HP in §63.

What is true, then, is that there is a stronger view than the one Frege introduces in §55 and criticizes in §56, one that would couple the inductive definition with a characterization of numbers as numerically definite quantifiers. So we shall assume henceforth that the 'numbers' are numerically definite quantifiers.[43]

We can now formulate versions of Frege's axioms of arithmetic, and we will be able to prove many of them. The first axiom, stating that the relation of predecession is functional, is a trivial consequence of the extensionality of (second-level) concepts: If we have both $P_\Phi(Qx(\Phi x), Rx(\Phi x))$ and $P_\Phi(Qx(\Phi x), Tx(\Phi x))$, then:

$$Rx(Fx) \equiv \exists y(Fy \wedge Qx(Fx \wedge y \neq x)) \equiv Tx(Fx)$$

and so

$$\forall F(Rx(Fx) \equiv Tx(Fx))$$

which means that $Rx(\phi x)$ and $Tx(\phi x)$ are 'identical' in the sense appropriate to second-level concepts. The fourth axiom, induction, can be proven from the definition of natural number in the usual way. Moreover, it is immediate that every number has a successor. For $Qx(\Phi x)$ will have a successor just in case some second-level concept $Rx(\Phi x)$ satisfies:

$$\forall F[Rx(Fx) \equiv \exists y(Fy \wedge Qx(Fx \wedge y \neq x))]$$

But the sentence stating that there is such a concept is just an instance of (third-order) comprehension. The interesting case, of course, is that in which there are exactly n objects in the domain, for some natural number n. In this case, the successor of $\exists_n(\Phi x)$ will be the *empty* second-level concept $\emptyset x(\Phi x)$, which is false of every concept: If there are exactly n objects, then there are *no* concepts F for which there is an object y such that $F\xi \wedge y \neq \xi$ falls under $\exists_n(\Phi x)$. Similarly, $\emptyset x(\Phi x)$ succeeds itself.

---

[43] If we restrict ourselves to the case of *finite* numbers, we can actually prove that, if a concept has a number, it has only one, since we can prove that the finite numbers are all numerically definite quanitifers. And in some sense, Frege is concerned, in §§55–6, only with finite numbers. If he had been concerned here with cardinal number in general, one would have expected him to point out that, if the numbers are defined in accord with the inductive definition, one cannot prove that each concept has *at least* one number. The inductive definition gives us only the second-level predicates: $\exists_0 x(\Phi x)$, $\exists_1 x(\Phi x)$, $\exists_{1+1} x(\Phi x)$, and so on. It therefore does not provide an answer for every question of the form "How many Fs are there?" and, in particular, does not do so if there are infinitely many Fs. But Frege does not even raise this objection: He is here concerned only with finite numbers. (Thanks to Ori Simchen for this observation.)

Thus is the third of Frege's axioms for arithmetic proved. But now it is clear that the second axiom, which states that no natural number ancestrally follows itself, is not provable. As just noted, if the domain is finite, $\emptyset x(\Phi x)$ will be a 'natural number' that immediately, and so ancestrally, succeeds itself. Where, then, would an attempt to mimic Frege's proof of this axiom fail? As an examination of the proof will show, the only arithmetical facts upon which it depends are that zero has no predecessor and that different numbers have different successors (see Section 2.5). But the latter is not provable. Suppose that the domain contains exactly Caesar. Then the concept $\xi \neq \xi$ will be the only concept falling under $\exists_0 x(\Phi x)$; the concept $\xi = \xi$ will be the only concept falling under $\exists_1 x(\Phi x)$; and there will be *no* concept falling under $\exists_{1+1} x(\Phi x)$, nor any falling under $\exists_{1+1+1} x(\Phi x)$; and so on. $\exists_{1+1} x(\Phi x)$ will therefore be the same second-level concept as $\exists_{1+1+1} x(\Phi x)$, namely $\emptyset x(\Phi x)$; hence, the successor of $\exists_1 x(\Phi x)$ will be the same as that of $\exists_{1+1} x(\Phi x)$. What is needed if we are to prove that $\exists_{1+1} x(\Phi x)$ is different from $\exists_{1+1+1} x(\Phi x)$ is a proof that there are at least two objects. What is needed to prove that $\exists_n x(\Phi x)$ is different from $\exists_{n+1} x(\Phi x)$ is a proof that there are at least n objects. What is needed if we are to prove that all such quantifiers are distinct is a proof that, for each n, there are at least n objects. And what Russell and Whitehead do in *Principia Mathematica* (Whitehead and Russell, 1925) is simply to assume this claim: It is their axiom of infinity. Frege's objection to this would have been that to adopt an axiom of infinity is to abandon the epistemological project of accounting for our knowledge of the basic laws of arithmetic, for our knowledge of the infinity of the series of natural numbers, in particular. It would be more honest simply to assume that different numbers have different successors (a proposition to which the axiom of infinity is equivalent in *Principia*).[44]

There is another way to proceed. Suppose we insist that $Rx(\Phi x)$ is a number only if it is non-empty, that is, if there is a concept of which it is the number. In that case, $\emptyset x(\Phi x)$ is not a number at all and the existence of successors can no longer be proven. The one-one-ness of predecession can be, however, and the proof is very close to the one Frege gives himself. As a kind of technical lemma, we need first to prove that all natural numbers are numerically definite, which we can prove by the obvious induction. Now suppose that $P_\Phi(Qx(\Phi x), Tx(\Phi x))$ and $P_\Phi(Rx(\Phi x), Tx(\Phi x))$. Then we have:

$$Tx(Fx) \equiv \exists y(Fy \land Qx(Fx \land y \neq x))$$
$$Tx(Fx) \equiv \exists y(Fy \land Rx(Fx \land y \neq x))$$

---

[44]The axiom of infinity is introduced at *120.03 and is discussed on the following two pages. The equivalence mentioned is proved at *125.14. Interestingly enough, the axiom of infinity is also equivalent to the claim that $\emptyset x(\Phi x)$ is not a natural number: see *125.13. For a more sympathetic discussion of the role of the axiom of infinity in the theory of types, see Boolos's "The Advantages of Honest Toil over Theft" (Boolos, 1998a).

for all concepts F. And, crucially, *there will be* a concept F such that $Tx(Fx)$. So, for some a and b, we will have:

$$Fa \wedge Qx(Fx \wedge a \neq x)$$
$$Fb \wedge Rx(Fx \wedge b \neq x)$$

All we need then to show is that $F\xi \wedge a \neq \xi$ and $F\xi \wedge b \neq \xi$ are equinumerous, which is the core of Frege's own proof, and then it will follow that $\forall F(Qx(Fx) \equiv Rx(Fx))$.

Frege's own proof of the existence of successors is not even formulable in this context, for one cannot, given the inductive definitions alone, so much as make sense of such predicates as "$\xi = 0 \vee \xi = 1$", which is what is required if "$\exists_{1+1}x(x = 0 \vee x = 1)$" is to be proven. We are unable to make sense of such predicates because, as Frege points out in his third objection, the inductive definitions do not so much as define the expressions "0" and "1", but only the orthographically complex expressions "$\exists_0 x(\Phi x)$" and "$\exists_1 x(\Phi x)$".

What Frege is insisting upon in his third objection, then, is the very difference he had ignored in *Begriffsschrift*: the one between objects and second-level concepts. In this case, the crucial difference is between the concept $\exists_1 x(\Phi x)$ and a mere "element in the predicate" (Gl, §57), the proper name "1". It is but the barest hunch, but I suspect that, as so often with Frege, the real target of his criticisms here is his former self. If one could somehow ignore the difference between $\exists_1 x(\Phi x)$ and 1, then maybe the existence of successors could be proven, or maybe it would at least seem as if it could be proven.[45] Perhaps, that is, it was Frege's *own* first attempt to prove the axioms of arithmetic that was based upon the inductive definitions. A real defense of that claim is probably impossible, without rediscovery of the *Nachlass*, but the possibility is intriguing, and it would be worth exploring whether there is some way of trying to prove the infinity of the natural numbers in a framework in which the distinction between objects and second-level predicates is somehow elided.[46]

However that may be, though, what we have seen is that concerns about how the infinity of the natural numbers might be proven lie just beneath the surface of the discussion in §§55–6. So I conclude, as promised earlier, that what Frege is really arguing in these sections is that, if we are to have any hope of proving the infinity of the numbers, then we must regard numbers as objects. But, as pointed out earlier, if we do not regard numbers as being of the same Sort as basic objects, then the mere claim that numbers are objects will not help us: "$Nx:(x = Nx:x \neq x)$" will still be ill-formed in that case. So Frege would have had good reason to

---

[45] Compare the discussion in Section 6.4 of why Frege does not even need the assumption that numbers are objects.

[46] See my Boolos lecture (Heck, 2005) for some ideas about how the logic itself might be formulated. Of course, the resulting system is likely to be inconsistent.

suppose that, if the axioms of arithmetic are to be proven, numbers have not only to be objects but also to be of the same Sort as basic objects.

In fact, however, that turns out not to be true.

6.3   Avoiding the Caesar Objection

Our knowledge of arithmetic begins with our understanding of attributions of number to concepts true or false of basic objects. At this early stage, our use of names of numbers is formalizable by the higher-order theory discussed in the last section, numerals being understood merely as orthographic parts of numerically definite quantifiers. What further is required if one is to achieve an understanding of names of numbers sufficient to ground our knowledge of the infinity of the number-series? A very natural suggestion, compatible with Frege's general outlook, is that one must come to know that numbers are objects. But clearly that is not sufficient: If one treats numbers as objects of a Sort different from that of basic objects—and so in a way formalizable by means of the two-sorted theory discussed in the last section—then one will have no way of proving that there are infinitely many numbers. So, it would seem, if anything like Frege's explanation of the genesis of our knowledge of the axioms of arithmetic is to succeed, we must confront the question *with what right* we suppose that numbers are of the same Sort as basic objects.[47] One way to press this point is to note that no sense has been given to such statements as "Caesar is the number zero": That such statements do make sense is precisely the content of the claim that numbers are of the same Sort as basic objects, that is, that numbers and basic objects are members of a single domain of quantification. Now, I am not going to discuss the question whether there is some general way of giving sense to such statements. It would surely be better if Frege's position did not require us to broach the issue.

Consider a speaker who has mastered only attributions of number to concepts true or false of basic objects. What distinguishes our mature understanding of names of numbers from hers is, most fundamentally, that we know that numbers can themselves be counted. What this comes to is that we understand terms of the form "the number of numbers satisfying such-and-such a condition". How is our understanding of such terms to be explained? Plainly, the two-sorted form of HP discussed in the last section utterly fails to explain terms of this form; equally plainly, the familiar single-sorted version of HP does explain such terms. If such terms can be explained only in terms of the single-sorted version of HP, then Frege's project really does require numbers to be of the same Sort as basic objects. What I am going to argue, however, is that there is another

---

[47]The importance of this question was first suggested to me by Parsons's paper "Intuition and Number" (Parsons, 1994, pp. 150–1).

way of explaining these terms, one equally in the spirit of Frege's project, one that does not lead to a problematic instance of the Caesar objection.

One tempting suggestion is that an analogue of HP may be introduced, one that specifically governs terms of the form "the number of numbers that are F". This proposal may formalized within a many-sorted, higher-order language. There are to be variables of various types, the types indexed by numerals: Thus, variables of type 0 are written "x_0", "y_0", etc.; variables of type 1, "x_1", "y_1", etc. Similarly, there will be predicate variables and relation variables of various types, the types of whose arguments are indicated by subscripts: thus, "F_0", "G_0", "F_1", etc.; "$R_{0,0}$", "$R_{0,1}$", "$R_{5,2,3}$", and so forth. The language also contains countably many term-forming operators "$N_{n+1}v_n{:}\Phi(v_n)$", which form terms of type $n+1$ from formulae containing a free variable of type n. HP, in its original form, may then be formalized as:[48]

$$N_1x_0{:}F_0x_0 = N_1x_0{:}G_0x_0 \equiv$$
$$\exists R_{0,0}\{\forall x_0\forall y_0\forall z_0\forall w_0[R_{0,0}x_0y_0 \wedge R_{0,0}z_0w_0 \rightarrow x_0 = z_0 \equiv y_0 = w_0]\wedge$$
$$\forall x_0[F_0x_0 \rightarrow \exists y_0(G_0y_0 \wedge R_{0,0}x_0y_0)]\wedge$$
$$\forall y_0[G_0y_0 \rightarrow \exists x_0(F_0x_0 \wedge R_{0,0}x_0y_0)]\}$$

The suggested analogue of HP, required for 'counting numbers', may then be formalized as:

$$N_2x_1{:}F_1x_1 = N_2x_1{:}G_1x_1 \equiv$$
$$\exists R_{1,1}\{\forall x_1\forall y_1\forall z_1\forall w_1[R_{1,1}x_1y_1 \wedge R_{1,1}z_1w_1 \rightarrow x_1 = z_1 \equiv y_1 = w_1]\wedge$$
$$\forall x_1[F_1x_1 \rightarrow \exists y_1(G_1y_1 \wedge R_{1,1}x_1y_1)]\wedge$$
$$\forall y_1[G_1y_1 \rightarrow \exists x_1(F_1x_1 \wedge R_{1,1}x_1y_1)]\}$$

Thus, the number of *type 1* numbers satisfying a given condition is a *type 2* number. And, more generally, the number of *type n* numbers satisfying a given condition will be a *type n+1* number:

$$N_{n+1}x_n{:}F_nx_n = N_{n+1}x_n{:}G_nx_n \equiv$$
$$\exists R_{n,n}\{\forall x_n\forall y_n\forall z_n\forall w_n[R_{n,n}x_ny_n \wedge R_{n,n}z_nw_n \rightarrow x_n = z_n \equiv y_n = w_n]\wedge$$
$$\forall x_n[F_nx_n \rightarrow \exists y_n(G_ny_n \wedge R_{n,n}x_ny_n)]\wedge$$
$$\forall y_n[G_ny_n \rightarrow \exists x_n(F_nx_n \wedge R_{n,n}x_ny_n)]\}$$

Similarities to the theory of types are, of course, non-accidental. In this theory, as in the two-sorted theory discussed earlier, it will be possible, at each type $i > 0$, to formulate versions of the Frege's axioms for arithmetic

---

[48]I apologize again for the near unreadability of these formulae, due to the presence of so many subscripts. But it is again crucial that everything being done here can be done without violating any type-restrictions, and the only way to establish this point is to be as rigorous as possible.

and to prove all of them except that every number has a successor, for the same reason discussed above in connection with the two-sorted theory.

But is this treatment really compatible with Frege's observation that numbers can themselves be counted? Does this theory really capture our mature understanding of names of numbers? I think not. Consider the question what, on this view, is to be said about the use of numerals in ordinary language (or, for that matter, in informal mathematics). Presumably, when speakers make remarks such as "The number of numbers greater than 5 but less than 15 is 9",[49] this is to be taken as 'systematically ambiguous'. The speaker should be understood not as making some specific claim of the form "The number$_{n+1}$ of numbers$_n$ greater than 5_n but less than 15_n is 9_{n+1}", but rather as making an ambiguous claim: Roughly speaking, she claims that, for all permissible assignments of types, the resulting sentence is true.[50]

Whatever the prospects of this sort of manoeuver elsewhere, it will not work here. When one asks oneself, say, how many numbers there are that are less than 5, one supposes that the answer will be a *number*, an object that (logically speaking) *could be*, though it is not, one of the numbers less than 5; the type-theoretic treatment has it, however, that the answer is a 'number' of a different type, a different Sort. To put the point differently, one supposes that the statement "The number of numbers less than 5 is 5" is at least well-formed. The type-theoretic view can, of course, concede this point and suggest that what is really meant is something like: "The number$_{n+1}$ of numbers$_n$ less than 5_n is 5_{n+1}". But even if this re-interpretation of the claim should be deemed acceptable, the trick does not work more generally. Consider the statement "There is a number that is the number of numbers less than it", formally: $\exists x(x = Ny\!:\!y < x)$. This claim admits of no acceptable assignment of types:[51] "$\exists_n x[x_n = N_{n+1}y_n\!:\!y_n <_{n,n} x_n]$" is not well-formed, since terms of different types appear on the two sides of the identity-sign. Or again, consider the (Boolosese) claim "There are some numbers the number of which is one of them", formally: $\exists F(F(Nx\!:\!Fx))$. This claim is not only well-formed but true. But there is no permissible assignment of types in this formula.

What do these examples show? Not, admittedly, that there is no type-theoretic way to handle such statements. I am not about to make such a prediction, given the malleability of formal methods and the general

---

[49]It is frequently noted in such contexts that number theory does concern itself with the number of prime numbers less than a given number. This is, indeed, the example Dummett uses in his discussion of this matter.

[50]On the notion of systematic ambiguity, see Parsons' paper "Sets and Classes" (Parsons, 1981). If one were to insist that such statements involve definite assignments of types, that would only make matters worse.

[51]One might suggest that $<$ could be allowed to hold between numbers of different types. But, if it does, then why shouldn't identity hold between numbers of different types? And we could define identity in terms of $<$.

cleverness of logicians.[52] What the examples show is that, though we
have no strong intuitions regarding whether numbers and people are of
a single Sort, we *do* consider all *numbers* to be of a single Sort. That is
to say, we regard terms of the form "the number of Roman emperors" and
"the number of numbers less than 5" as being of the same Sort. Of course,
if we so regard them, we must be prepared to face an instance of the
Caesar objection. But this instance of the Caesar objection is innocuous.
For consider an identity statement linking terms of these two kinds, say:
The number of Roman emperors is the same as the number of numbers
less than 5. We discussed this statement earlier, and whether numbers
are of the same Sort as people is irrelevant to its truth-conditions: The
sentence will be true just in case there is a one-one correlation between
the Roman emperors and the numbers less than 5. More generally, a
statement of the form "The number of basic objects that are F is the same
as the number of numbers that are G" will be true just in case there
is a one-one correlation between the Fs and the Gs. This instance of
the Caesar objection thus has a simple, obvious, and obviously correct
answer.

Should we conclude from this that numbers must be of the same Sort
as basic objects? Such an inference would be completely fallacious. The
fact that "the number of basic objects that are F" is of the same Sort as
"the number of numbers that are G" says nothing whatsoever about the
Sorts of F and G. That functional expressions have *values* of the same
Sort does not imply that they have *arguments* of the same sort. What
we need, then, is to formulate a theory that distinguishes the Sorts of
numbers and basic objects, but that treats terms of the form "the number of Fs" and "the number of Fs" as being of the same sort. Perhaps
surprisingly, this turns out to be quite straightforward.

We revert to the two-sorted language employed in the last section.
The version of HP for concepts true or false of basic objects is one we
have already seen:

$$(\text{HP}_{bb}) \quad Nx{:}Fx =_n Nx{:}Gx \equiv$$
$$\exists R_{bb}\{\forall x \forall y \forall z \forall w (R_{bb}xy \land R_{bb}zw \to x =_b z \equiv y =_b w) \land$$
$$\forall x[Fx \to \exists y(Gy \land R_{bb}xy)] \land$$
$$\forall y[Gy \to \exists x(Fx \land R_{bb}xy)]\}$$

A version of HP that governs concepts true or false of numbers themselves
may then be formulated as:

---

[52] Still, there is reason to suppose that no such treatment will be forthcoming. The reason
is simple: If there is some way of construing such claims so that they are well-formed and
have the correct truth-conditions, then "If n is finite, the number of numbers less than or
equal to n is the successor of n" will have to come out true. But then Frege's project goes
forward (Heck, 1993a, p. 231).

(HP_{nn}) $\mathbf{Nx{:}Fx} =_n \mathbf{Nx{:}Gx} \equiv$

$\qquad \exists R_{nn}\{\forall\mathbf{x}\forall\mathbf{y}\forall\mathbf{z}\forall\mathbf{w}(R_{nn}\mathbf{xy} \wedge R_{nn}\mathbf{zw} \rightarrow \mathbf{x} =_n \mathbf{z} \equiv \mathbf{y} =_n \mathbf{w}) \wedge$

$\qquad \forall\mathbf{x}[\mathbf{Fx} \rightarrow \exists\mathbf{y}(\mathbf{Gy} \wedge R_{nn}\mathbf{xy})] \wedge$

$\qquad \forall\mathbf{y}[\mathbf{Gy} \rightarrow \exists\mathbf{x}(\mathbf{Fx} \wedge R_{nn}\mathbf{xy})]\}$

Note that terms of the form "$\mathbf{Nx{:}Fx}$" are to be of the same Sort as those of the form "$\mathrm{N}x{:}Fx$". The version of HP that governs mixed identities is then:[53]

(HP_{bn}) $\mathrm{N}x{:}Fx =_n \mathbf{Nx{:}Gx} \equiv$

$\qquad \exists R_{bn}\{\forall x\forall\mathbf{y}\forall z\forall\mathbf{w}(R_{bn}x\mathbf{y} \wedge R_{bn}z\mathbf{w} \rightarrow x =_b z \equiv \mathbf{y} =_n \mathbf{w}) \wedge$

$\qquad \forall x[Fx \rightarrow \exists\mathbf{y}(\mathbf{Gy} \wedge R_{bn}x\mathbf{y})] \wedge$

$\qquad \forall\mathbf{y}[\mathbf{Gy} \rightarrow \exists x(Fx \wedge R_{bn}x\mathbf{y})]\}$

There is no formal obstacle to formulating such a principle, even if basic objects and numbers are assumed to be of different Sorts.

Call the two-sorted, second-order theory whose (non-logical) axioms are (HP_{bb}), (HP_{nn}), and (HP_{bn}) *Two-sorted Frege Arithmetic* (2FA). Axioms for arithmetic are provable in 2FA. What is required for Frege's proof of the infinity of the number-series is that we should be able to prove that the number of numbers less than or equal to n is the successor of n (if n is a natural number). As was seen above, if this is to be proven, all that is required is that "the number of numbers less than or equal to n" be of the same Sort as "n" itself. But, in 2FA, this is so. More formally, suppose we simply drop all reference to basic objects from 2FA. What remains is a single-sorted theory whose sole (non-logical) axiom is (HP_{nn}). But this axiom, *modulo* the boldface, is just HP in its simple form: So 2FA is an extension of single-sorted Frege Arithmetic and so proves whatever it does.[54]

To see how insignificant it ultimately is that numbers and basic objects belong to different Sorts, consider the fact that not just objects but also concepts and functions can be counted.[55] Not only is it a sensible question how many (Fregean) concepts there are that are true of only

---

[53] Identity statements of the form "$\mathbf{Nx{:}Gx} =_n \mathrm{N}x{:}Fx$" may either be governed by another form of HP, whose formulation should be obvious at this point, or simply be stipulated to be equivalent to those of the form "$\mathrm{N}x{:}Fx =_n \mathbf{Nx{:}Gx}$".

[54] Frege's actual proofs of the axioms of arithmetic, that is to say, can simply be carried out by boldfacing all his object- and concept-variables and adding the subscript "nn" to all relation-variables.

[55] Frege makes this claim in at least two different places. First, in his letter to Marty of August 29, 1882, already quoted above: "Everything is enumerable, ... even concepts..." (PMC, p. 100). And again in "Formal Theories of Arithmetic": "...[W]e can count just about everything that can be an object of thought: ... concepts as well as objects..." (FTA, op. 94). Frege's insistence in *Die Grundlagen* that "everything thinkable" can be counted (Gl, §14) must presumably include concepts, too.

Caesar, the answer is obvious: One. Nor does the fact that one is counting concepts rather than objects imply that the answer is not a number of the usual kind. Nor is there any difficulty in saying under what circumstances the number of concepts F falling under some second-level concept $\Phi_x \phi x$ is the same as the number of objects x falling under some first-level concept $P\xi$: The numbers will be the same just in case there is a one-one correlation between the concepts and the objects. Where "Σ_z" ranges over relations between concepts and objects and "Φ_z" ranges over second-level concepts, this thought can be formalized as follows:

$$\mathbf{N}F : \Phi_x Fx = \mathbf{N}x : Px \equiv$$
$$\exists \Sigma [\forall x \forall F \forall y \forall G [\Sigma_z(x, Fz) \wedge \Sigma_z(y, Gz) \rightarrow x = y \equiv \forall x (Fx \equiv Gx)] \wedge$$
$$\forall x [Px \rightarrow \exists F(\Phi_z Fz \wedge \Sigma_z(x, Fz)] \wedge$$
$$\forall F [\Phi_z Fz \rightarrow \exists x (Px \wedge \Sigma_z(x, Fz)]\}$$

There is thus no formal obstacle even to the formulation of a version of HP governing mixed identity statements of this kind. Indeed, it should now be obvious that one can formulate versions of HP to govern terms of the form "the number of entities satisfying such and such a condition", in both pure and mixed identity statements, no matter what sorts of entities might be in question.[56] So HP may now be thought of as schematic—as specifying the truth-conditions of pure and mixed identity statements containing terms ascribing number to concepts (whose arguments may be) of any level or Sort.

6.4   Closing

The fundamental epistemological question of the philosophy of arithmetic is: What is the basis of our knowledge of the infinity of the series of natural numbers? On the position developed here, the acquisition of such knowledge occurs in three stages.[57]

1.   One must come to understand ascriptions of number to concepts true or false of basic objects, that is, of the various objects to which one is capable of referring before acquiring a capacity to refer to numbers.

---

[56]The higher-order theory containing as (non-logical) axioms all the infinitely many such principles is provably (formally) consistent. I first sketched a proof in a critical review of *Frege: Philosophy of Mathematics* (Heck, 1993a, p. 231), but it was a bit sloppy. A better proof observes that the theory in question is interpretable in ZFC+GCH and so is consistent if ZFC+GCH is. But of course ZFC+GCH is consistent if ZFC is. Theories weaker than ZFC will also do. And clearly, GCH is not really required. Quite possibly, however, this extremely general version of HP does impose some constraints. I do not know what these might be.

[57]These three stages correspond to the three axioms of 2FA: The first and second correspond to the axioms governing the 'pure' identity statements involving terms ascribing number to concepts true or false of basic objects and numbers, respectively; the third, to the axiom governing mixed identity statements.

2.   One must come to understand ascriptions of number to concepts true or false of numbers themselves.

3.   One must come to understand the conditions under which a number ascribed to a concept true or false of basic objects will be the same as one ascribed to a concept true or false of numbers.

At that point, one will know that terms of the form "the number of numbers that are **F**" and "the number of basic objects that are F" are of the same logical Sort and so will have acquired knowledge from which the infinity of the series of natural numbers may be inferred. I find this philosophical 'reconstruction' of the genesis of our knowledge of arithmetic compelling.

This position is subtly but importantly different from that mentioned at the beginning of Section 6.3. On that view, there is, in place of the second and third steps, a single step at which the speaker comes to understand 'that numbers are objects', that is, that numbers are objects of the same Sort as basic objects. If one formulates what a speaker must come to know in this way, the Caesar objection looms; the question with what right we suppose ourselves to know that numbers are of the same Sort as basic objects may well be unanswerable. However, though one could acquire 'knowledge' from which the axioms of arithmetic are derivable by coming to 'know' that numbers are of the same Sort as basic objects, such 'knowledge' is not necessary: What one needs to know is that numbers ascribed to concepts true or false of numbers are of the same Sort as numbers ascribed to concepts true or false of basic objects. It is quite fallacious to infer from this that numbers must be of the same Sort as basic objects.

Moreover, we are now in a position to see that Frege's proofs of the axioms of arithmetic do not even require the claim that numbers are objects. What is required is that expressions of the form "the number of numbers less than 5" should be of the same logical Sort as those of the form "the number of Roman emperors". So far as the proofs of the axioms are concerned, it matters not one bit whether this Sort is that of basic objects, of objects of some other kind, or of second-level concepts. What are required are axioms similar to those of 2FA.

Write "$\mathrm{Eq}_{vw}(\Psi(v), \Psi(w))$" for "$\Psi$ is equinumerous with Φ", where the variables v and w may be of any appropriate level. The first axiom is then:[58]

$$(\mathrm{HP}_{1,1})\quad \forall H\{\mathbf{N}_{xy}(Fx)(Hy) \equiv \mathbf{N}_{xy}(Gx)(Hy) \equiv \mathrm{Eq}_{xy}(Fx, Gy)$$

This says that, for each concept F, the second-level concept $\mathbf{N}_{xy}(Fx)(\phi y)$ is (co-extensive with) the concept $\mathbf{N}_{xy}(Gx)(\phi y)$ if, and only if, F and G are equinumerous. What we now require is an analogue of this for *third-*level concepts, i.e., for concepts true or false of the 'numbers', which are

---

[58]The subscripts mark the level of the concepts that occur as arguments to N.

the second-level concepts $N_{xy}(Fx)(\phi y)$. Let "Ψ" and "Φ" be variables for third-level concepts; "χ" and "θ", for second-level. Then the needed axiom is:

$$(\text{HP}_{2,2}) \quad \forall H\{N_{\chi y}(\Psi\chi)(Hy) = N_{\chi y}(\Phi\chi)(Hy) \equiv \text{Eq}_{\chi\theta}(\Psi\chi, \Phi\theta)$$

What this says is that, for every third-level concept Ψ, there is a second-level concept $N_{\chi y}(\Psi\chi)(\phi y)$ that is (co-extensive with) another second-level concept $N_{\chi y}(\Phi\chi)(\phi y)$ if, and only if, Ψ and Φ are equinumerous.[59] It is easy to see that the resulting system is consistent and proves axioms for arithmetic.[60] Frege's proofs can simply be mimicked, using $(\text{HP}_{2,2})$.[61]

It does not follow, of course, that there are no objections to a form of logicism based upon the derivability of axioms for arithmetic in 2FA. In particular, I have not addressed the 'bad company' objections to the claim that HP has a favored epistemological status—objections that derive from the observation that formally similar principles are inconsistent, are (though consistent) inconsistent with HP, or are in some other way naughty.[62] But let us not forget, in our haste to evaluate such a version of logicism, that the attractions of the genetic story told at the beginning of this section do not depend upon the claim that the various instances of HP are logical truths, analytic truths, or any such thing.

---

[59]The obvious third axiom would govern mixed identities:

$$(\text{HP}_{1,2}) \quad \forall H\{N_{xy}(Fx)(Hy) = N_{\chi y}(\Phi\chi)(Hy) \equiv \text{Eq}_{x\theta}(Fx, \Phi\theta)$$

But it is not actually needed here.

[60]$(\text{HP}_{2,2})$ says, in effect, that we can inject the set of cardinalities of things in the third-level domain into the second-level domain. This will require the domain to be infinite: If there were but n second-level concepts, then there would be $n + 1$ cardinalities of third-level concepts.

It is crucial here that $(\text{HP}_{1,1})$ and $(\text{HP}_{2,2})$ say nothing about what *kinds* of second-level concepts the numbers are. Thus, $N_{xy}(x \neq x)(\phi y)$ might be the second-level concept: $\varphi(\text{Caesar})$. One could proceed differently and define "$N_{xy}(Fx)(\phi y)$" as: $\text{Eq}_{xy}(Fx, \phi y)$, whence $(\text{HP}_{1,1})$ will be provable. But, though one can also define "$N_{\chi y}(\Psi\chi)(\phi y)$" as "$\text{Eq}_{\chi y}(\Psi\chi, \phi y)$", one will not then be able to prove $(\text{HP}_{2,2})$. The reason is that there will be concepts (such as the second-level concept true of all first-level concepts) that are equinumerous with no first-level concept, so these will end up being the empty second-level concept. We have no way to guarantee that all the concepts that are equinumerous with no first-level concept are equinumerous with each other, without assuming both that the first-order domain is infinite and the generalized continuum hypothesis. Only then will we be able to prove $(\text{HP}_{2,2})$. That is not to say that $(\text{HP}_{2,2})$ implies GCH. It doesn't. But it may well have implications concerning such matters.

[61]This is not to say there are not objections to a theory based upon $(\text{HP}_{2,2})$. There are plenty. Since the second-level concepts in question here are extensionally individuated, if there are κ objects, then there are 2^{2^κ} second-level concepts. But 2^{2^κ} is infinite only if κ is, so the theory implies that there are infinitely many objects. Since numbers here are not objects, the theory implies that there are infinitely many objects... of what kind, exactly? Compare here the discussion of the "bad misunderstanding" on page 138.

It would be worth thinking about the fate of this sort of theory if one took the 'numbers' here to be *in*tensionally individuated.

[62]The bad company objection originates with Boolos (1998k) and Dummett (1991b, pp. 187–9). For further discussion, see Chapter 10 and references therein.

Frege's most fundamental thought—that our knowledge of the truths of arithmetic is founded upon our knowledge of HP—could well be true, even if it does not have the epistemological implications he hoped it would.

7

Cardinality, Counting, and Equinumerosity

It is by now widely recognized that Frege's *Foundations of Arithmetic* pursues its philosophical goals, at least in part, by pursuing mathematical ones. Indeed, the centerpiece of the book is what we now know as Frege's Theorem: Given appropriate definitions of arithmetical notions, the basic laws of arithmetic can be derived from a single (non-logical) axiom, HP, which says that the number of Fs is the same as the number if Gs if, and only if, the Fs are in one-one correspondence with the Gs. This beautiful result has been the object of a good deal of study, and its inner workings are well understood. It is far less clear, however, what philosophical significance it might have. Crispin Wright, who was the first to recognize the Theorem's philosophical potential, has suggested that it supports a reformed logicism, according to which the basic laws of arithmetic are, though not logical truths, still analytic or conceptual truths in so far as they are logical consequences of an analytic or conceptual truth, HP.

Almost every aspect of Wright's proposal has been the subject of controversy. When it has not just been met with the Incredulous Stare Objection, the claim that HP is a conceptual truth has been vehemently disputed. There have also been doubts about the epistemological neutrality of the 'logic' needed to derive arithmetic from HP: The principles one needs include impredicative instances of second-order comprehension; one might wonder whether the class of conceptual truths is closed under deduction of this kind. But suppose we accept that HP is a conceptual truth and that analyticity is preserved by second-order deduction. Must we also agree that the basic laws of arithmetic are analytic? Counterexamples to the inference one is tempted to make abound. To use one I have used before, Frege held that analysis, as well as arithmetic, was analytic. He did not, however, regard all of mathematics as analytic, since he agreed with Kant that Euclidean geometry is synthetic *a priori*. But the truths of Euclidean geometry can be proven in analysis, given suitable definitions. Were Frege's views, inconsistent, then? Since he knew analytic geometry himself, were his views simply incoherent? I take it they were not.

Similarly, we know from Gödel that one can 'do syntax' in arithmetic: Given appropriate definitions, one can prove the basic principles of syntax, for any of a wide class of languages, in arithmetic. Is it, then, an *arithmetic* truth, then, that there are infinitely many sentences of the language of arithmetic? or that "$0 = S$" is not one of them? I would suppose not. The truths of arithmetic, it seems to me, are quite independent of the existence of even one linguistic expression.

What Gödel showed is that syntax is *interpretable* in arithmetic: One can map sentences of the language of syntax onto those of the language of arithmetic in such a way that the axioms of syntax go over into theorems of arithmetic and derivations in syntax go over into (typically more complicated) derivations in arithmetic; for certain purposes, therefore, one can think of arithmetic as if it contained syntax as a sub-theory. (Similarly, Euclidean geometry is interpretable in analysis.) But nothing follows about the epistemological status of the laws of syntax. What one proves when showing that arithmetic interprets syntax are not the laws of syntax but their 'translations' into arithmetic, and the notion of translation that is in play here is a purely formal one.

What Frege's Theorem shows, most fundamentally, is that arithmetic (by which I mean second-order arithmetic as standardly axiomatized, that is, second-order PA) is interpretable in Frege Arithmetic (FA, by which I mean second-order logic plus HP). Even if HP is analytic and second-order logic is rightly so called, it does not follow that arithmetic is analytic. Even granting these assumptions, it follows only that arithmetic is interpretable in an analytically true theory; it does not follow that its truths are provable in such a theory. To draw that conclusion, we need to know that, when we 'translate' sentences of FA into sentences of arithmetic *via* the 'definitions' of arithmetical notions Frege bequeathed to us, we are actually *translating*. We need to know, that is, that sentences of FA express the very propositions about numbers we pre-philosophically employ: We need to know, for example, that the sentence of FA that translates "Every number has a successor" *says* that every number has a successor, or at least something in the same ballpark.[1]

I am not going to discuss that problem directly, however. I mention it only to fix our attention upon the relationship between HP and our pre-theoretic concept of number, for it is that relationship that I want to discuss here. We can approach it by considering some remarks George Boolos made in connection with Wright's suggestion that HP embodies an explanation of the concept of number:[2]

[HP is] a biconditional whose right limb is a formula defining an equivalence relation between concepts F and G and whose left limb is a formula stating [that] the cardinal numbers of F and G are the same. Since the sign for cardinal number does not occur in the right limb, can one not appropriately say that HP *explains*

---

[1] Dummett (1991a) takes a similar route to this same conclusion. My own interest in this topic was inspired, many years ago, by hearing Dummett make similar remarks during lectures on Frege he delivered in Oxford.

[2] I've omitted a footnote. Boolos is responding to Wright's paper "The Philosophical Significance of Frege's Theorem" (Wright, 2001d).

the concept of cardinal number by saying what it is for two cardinal numbers, both referred to by expressions of the form "the number of...", to be identical? Certainly. HP states a necessary and sufficient condition for an identity [concerning numbers] to hold. ...So if one wants merely to sum up this state of affairs by saying that HP explains the concept of a cardinal number, I would not object. However, it is hard to avoid the impression that more is meant.... (Boolos, 1998d, p. 310)

And more of course is meant. Wright intends the claim that HP embodies an explanation of the concept of number to imply that it is *analytic of* the concept of cardinal number—and so that it is an analytic or conceptual truth, in much the way a definition would be. Boolos wants to block this argument:

It is certainly true that *one* of the ways in which HP can be used is to fix the character of a certain concept. Here's how: lay [HP] down. ...But [HP] is no different in this regard from any other statement that we might choose to take as an axiom. The axiom of choice fixes the concept of set in a similar manner. ...The principle of mathematical induction fixes the character of the natural numbers. The statement that bananas are yellow fixes the character of the concept of a banana. So nothing is said when it is said that one of the roles of HP is to fix the character of the concept of a cardinal number. (Boolos, 1998d, p. 311)

Boolos's claim is that, in whatever sense HP fixes the concept of cardinal number, *all* truths help "to fix the character" of the concepts they concern. Note especially the remark about bananas: No-one anymore would think it analytic that bananas are yellow.

What is behind these remarks, obviously, is a Quinean skepticism about the analytic–synthetic distinction, one that can do any philosophical work, anyway. Boolos does misplay his hand a bit, since Wright's claim is not that HP is "a super-hard truth", as Boolos suggests: The issue is not certainty or unrevisability. But Wright is claiming that HP is a special sort of truth: It is supposed to be *the* fundamental truth about cardinality, *the* truth that fixes the character of our concept of cardinality and so, in that sense, *the* truth on which all other truths about cardinality rest. In particular, HP is supposed to be more fundamental, in some sense, than the axioms isolated by Dedekind and popularized by Peano. So Boolos's Quinean skepticism is entirely appropriate: I was taught (by Boolos himself, among others) that Quine's real point is that there is no distinction among truths period that will do any philosophical work. But one does not have to accept this extraordinarily strong claim to be worried, and Boolos was sometimes given to press his point in another way.[3]

---

[3]One can find indications of what follows in many of Boolos's writings on this topic, all of which are collected in *Logic, Logic, and Logic* (Boolos, 1998e). But the formulation is mine and is based loosely on memories of conversations. MacFarlane has recently expressed a similar worry in his paper "Double Vision" (MacFarlane, 2009, see §2). Hale and Wright devote most of their reply (Hale and Wright, 2009a) to responding to it. (Those papers post-date this one.)

What reason do we have to think that HP is the 'fundamental' truth about cardinality, whatever exactly that may mean? In particular, what reason do we have to think HP is more fundamental than the Dedekind-Peano axioms? It is true, of course, that the axioms of PA can be proven from HP and Frege's 'definitions'; but, as I shall explain below, it is also true that (something very like) HP can be proven from (something very like) the axioms of PA and those same 'definitions'. Those are the mathematical facts, and Boolos would never have suggested that they do not profoundly illuminate the concept of number. But Wright's view is that one of these mathematical facts is much more, or at least quite differently, illuminating than the other. Wright thinks that the interpretability of Peano arithmetic in Frege arithmetic is more significant, epistemologically, than the (approximate) converse. But why? Waive the Quinean skepticism. Allow that talk of foundations makes sense. Allow that the question what follows from HP might be of more epistemological significance than what follows from the Peano axioms. Why *should* it be thought more significant? We all know what the Fregean answer is: It is that the concept of number is fundamentally bound up with the notion of one-one correspondence. But the question Boolos wants answered is what reason we have to suppose that it is any more bound up with the notion of one-one correspondence than it is with the facts about the structure of the natural numbers that the axioms of PA codify.

Two sorts of responses to this question come immediately to mind. The first we might call metaphysical. It would hold that among the truths themselves, as it were, some are more fundamental than others, and that HP is, as it happens, more fundamental than the axioms of PA. One does sometimes find intimations of such a conception in Frege, when, for example, he speaks of "the dependence of truths upon one another" (Gl, §2), but I find such ideas obscure and suspect I'm not the only one who does.[4]

The other sort of reply is epistemological: On this view, instead of grounding arithmetic itself, HP somehow grounds our *knowledge* of arithmetic. That may seem more promising, but is it? After all, one's first skeptical thought might be, it is not at all clear that HP is even known by ordinary, unreflective masters of the concept of number, let alone that it grounds their knowledge, e.g., that $7 + 5 = 12$. Of course, it will quickly be said, no-one is claiming that our ordinary knowledge of arithmetic is really derived from some prior knowledge of HP. Something like the distinction between 'context of discovery' and 'context of justification', which we do of course find in Frege (Gl, §3), must be invoked if the epistemological conception of HP's centrality is to be defensible. But how should that distinction be put to use? One way would be to deny that our ordinary knowledge of arithmetic is even at issue. Perhaps the question should be

---

[4]Burge (2005a; 2005b) has struggled mightily to make sense of this aspect of Frege's position. Much of the dissatisfaction Burge expresses with Frege's conception of apriority has its source, or so it seems to me, in its metaphysical, or Leibnizian, aspects.

not how we do come by knowledge of arithmetic but how someone *could* do so. Even if no one ever does, the idea might be, still someone could acquire a concept of cardinal number by receiving HP as an explanation. If she did, HP would then be analytic of her concept of cardinal number and hence would be a conceptual truth for her. So, if she had powers of thought and reasoning similar to ours, she would then be in a position to define the basic arithmetical notions in Frege's way and to prove the basic laws of arithmetic—the axioms of PA, in particular—from HP. So (assuming questions set aside above to have been answered), she would have analytic, *a priori* knowledge of those laws. Hence... well, *what* exactly? It is tempting to want to say: So arithmetical truths are analytic. But what does the claim that the laws themselves are analytic add to the claim that someone *could* have analytic, and so *a priori*, knowledge of the laws of arithmetic?

Indeed, of what interest is this latter claim anyway?[5] Even if we could earn a right to it, we would have been given no reason to suppose that *we* have, or even *could* have, analytic, or even *a priori*, knowledge of any arithmetical truths at all. And, I take it, when Kant claimed that arithmetical knowledge is synthetic *a priori*, he meant to be making a claim about our knowledge of arithmetic, not God's: As far as I can see, that is, a Kantian could just respond that it is impossible for possessors of finite minds to acquire the concept of cardinal number in the way that Wright suggests. And even if that objection could be met, it would remain an open question what our actual knowledge of arithmetic has or might have to do with HP, what the actual relationship is between our capacity for reasoning (Kant's 'faculty of understanding') and our arithmetical knowledge.

What follows will, I hope, contribute to the study of those (ultimately empirical) questions by clarifying the relationship between our concept of number—the one we employ in ordinary, unreflective applications of arithmetic—and the notion of one-one correspondence. I shall argue that the relation is close and that there is a sense in which any master of the concept of cardinal number must know something akin to HP: In that sense, something akin to HP is a conceptual truth. I shall, however, draw no conclusions about the relation between our knowledge of HP and our knowledge of the properties of the natural numbers enshrined in the axioms of PA. Even if there is no significant such relationship, though, a non-epistemological account of the philosophical significance of Frege's Theorem may be available, one that still requires a close relationship between HP and our ordinary concept of number. Unfortunately, I shall only be able to gesture at a possible view, since my own thought on the matter remains in flux.

---

[5]There is further discussion of this set of issues in Section 1.3.

7.1   Technical Preliminaries

Before we begin that discussion, however, I need to make a few remarks about technical matters and to make one further clarification about the scope of the present inquiry.

Stated in a form that is useful for comparison with Frege arithmetic, the Dedekind-Peano axioms are (the universal closures of) the following:

1. $\mathbb{N}0$

2. $\mathbb{N}x \wedge Pxy \rightarrow \mathbb{N}y$

3. $\mathbb{N}x \wedge Pxy \wedge Pxz \rightarrow y = z$

4. $\mathbb{N}x \wedge \mathbb{N}y \wedge Pxz \wedge Pyz \rightarrow x = y$

5. $\mathbb{N}x \rightarrow \neg Px0$

6. $\mathbb{N}x \rightarrow \exists y(Pxy)$

7. $F0 \wedge \forall x \forall y(\mathbb{N}x \wedge Fx \wedge Pxy \rightarrow Fy) \rightarrow \forall x(\mathbb{N}x \rightarrow Fx)$

Here, obviously, "0" is supposed to denote zero, and "$\mathbb{N}$" is to be true of exactly the natural numbers. "P" is a relational predicate that does the work of the more familiar functional expression "S":[6] "Pxy" means: x (immediately) precedes y, or: y is a(n immediate) successor of x. **PA** is the (full) second-order theory with non-logical axioms (1)–(7). HP is:

$$\mathbf{N}x : Fx = \mathbf{N}x : Gx \equiv \mathbf{Eq}_x(Fx, Gx)$$

where, of course, "$\mathbf{Eq}_x(Fx, Gx)$" abbreviates one of the usual formulae that define: the Fs and Gs are in one-one correspondence. HP, in this familiar form, is impredicative, in the sense that terms of the form "$\mathbf{N}x : Fx$" are terms of the same logical type as the first-order variables that appear in the formula displayed and, indeed, are of the same type as those bound by the cardinality operator itself. So e.g. "$\mathbf{N}x : (x = \mathbf{N}y : y \neq y)$" is well-formed.[7]

FA is (full) second-order logic, with HP the sole non-logical axiom.[8] The definitions Frege uses to interpret PA in FA are:

DZ $0 = \mathbf{N}x : x \neq x$

DP $Pmn \equiv \exists F \exists y[n = \mathbf{N}x : Fx \wedge Fy \wedge m = \mathbf{N}x : (Fx \wedge x \neq y)]$

---

[6]Using the relational expression allows us to state the assertions that every number has a successor and that no number has more than one successor as non-logical axioms, namely, as (6) and (3): In the more familiar formulation, using a function symbol, these are consequences of conventions that govern the use of all function symbols.

[7]See Chapter 6 for discussion of this feature of HP.

[8]It is convenient to take second-order logic here to include a principle guaranteeing the extensionality of operators of the type of "$\mathbb{N}$" (Boolos, 1998h, pp. 278ff).

DN $Nx \equiv P^{*=}0x$

where, in general, $R^{*=}$ is the weak ancestral of R, defined thus:

$$R^{*=}ab \equiv \forall F[Fa \land \forall x \forall y(Fx \land Rxy \to Fy) \to Fb]$$

Frege's Theorem (in one form) is that (1)–(7), the axioms of **PA**, are provable in **FA**, *via* the definitions (DZ), (DP), and (DN), which we shall collectively call **FD**. (We call them 'definitions', but we shall treat them as axioms.) No converse is possible, however: It can be shown that HP is *not* provable in **PA+FD**, though, of course, **FA** is relatively interpretable in **PA** and so the two are equi-consistent. HP is therefore stronger than it needs to be for the proof of Frege's Theorem. But now let "$\mathrm{Fin}_x(Fx)$" abbreviate some second-order formula defining the notion of (simple, not Dedekind) finitude and consider HPF:

$$\mathrm{Fin}_x(Fx) \lor \mathrm{Fin}_x(Gx) \to Nx\!:\!Fx = Nx\!:\!Gx \equiv \mathrm{Eq}_x(Fx, Gx)$$

HPF says that HP holds so long as at least one of F and G is finite: It leaves it open what the standard of equality is when both F and G are infinite. The axioms of **PA** are provable from HPF and **FD**. More importantly, for our purposes here, HPF itself is provable in **PA+FD**, given the additional axiom that every predecessor of a natural number is a natural number:[9]

PAF $Nx \land Pyx \to Ny$

For the most part, however, I won't fuss about the distinction between HP and HPF or between **PA** and **PA+(PAF)**.

The study of Frege's Theorem has revealed a good deal about HP and how it 'gets its strength', as Boolos puts it. Axioms (1) and (2) are consequences of **FD** alone, in fact, of (DN) alone, no comprehension being required. Axiom (7) follows from (DN), too, but its proof needs impredicative comprehension (Linnebo, 2004; Heck, 2011). Axioms (3), (4), and (5) are derivable from HP and **FD** using only predicative comprehension; moreover, Frege's definition of natural number—which is the only really controversial definition—is not needed for those proofs. Nor need one make any impredicative application of the cardinality operator in these derivations. To put this more precisely: Suppose we formulate HP in a two-sorted language, so that terms of the form "$Nx\!:\!Fx$" are *not* of the same logical type as the first-order variables bound by "N", in which case terms such as "$Nx\!:\!(x = Ny\!:\!y \neq y)$" are not well-formed. If we take the

---

[9] See Chapter 11 for discussion of this result and some suggestions about its philosophical implications.

I do not know how much induction is needed for this argument, but I believe at least Π_2^1 comprehension is required, because the formula defining finitude is Π_2^1. If so, then we need more comprehension for this direction, PAF $\to$ HPF, than we do for the other, HPF $\to$ PAF, since, as will be noted below, only Π_1^1 comprehension is needed for the latter.

logic of the theory to be predicative second-order logic, then all axioms of **PA**, other than Axioms (6) and (7), are provable in this theory, *via* **FD**.

Frege's proof of Axiom (6), on the other hand, appeals to instances of impredicative comprehension and requires impredicative applications of "N".[10] What is surprising, though, is that Axiom (6) turns out to be provable *from axiom* (3) *alone*, given **FD**. Moreover—and this point, unlike those I've been making thus far, has not been made before—one needs only Π_1^1 (equivalently, Σ_1^1) comprehension for this argument.[11]

The axioms of **PA** thus divide into three groups. Axioms (1), (2), and (7) follow from Frege's definitions of arithmetical notions, in fact, just from the definition of number. There are all kinds of issues here, but these are quite independent of the foundational status of HP, so I'll set them aside for now. Axiom (6) is often taken to be the crux of the matter: But it follows from Axiom (3) and Frege's definitions. There are big issues here, too, involving particularly the validity of 'impredicative' applications of the cardinality operator. But I want to set these aside, as well, since they too are independent of the foundational status of HP.

The issue thus concerns the status of the remaining Axioms, (3), (4), and (5), which are derivable from a predicative version of HP using only predicative comprehension and whose proofs do not involve the definition of natural number at all. One might think that, if we regard HP as a conceptual truth, we must regard these axioms too as conceptual truths, since they follow logically, by just about anyone's lights, from HP. But we saw above that matters aren't so straightforward. For one thing, the derivation of the axioms depends upon the definitions of zero and predecession: One might wonder whether the definitions really *define* those notions. More importantly, we might wonder whether HP is in any way more fundamental than these three axioms. That is the question I aim to discuss here.

7.2   Frege and Husserl

We may approach the problem by considering one of the many disagreements between Frege and Husserl.[12] Husserl famously argued that one should not explain the concept of number in terms of equinumerosity (or one-one correspondence), but should explain equinumerosity in terms of sameness of number, which in turn should be characterized in terms of

---

[10]It turns out that axiom (6) can be derived from HP using only predicative comprehension (Heck, 2011). The proof follows Frege's quite closely but is necessarily somewhat different from his. For further remarks on the strength of the logic needed for Frege's proof, see Chapter 12.

[11]Of course, one will only get as much induction as one puts into the comprehension axioms.

[12]The Husserl of whom I speak here is Husserl as seen through Frege's eyes. I hereby confess that I have not actually read *Philosophie der Arithmetik*.

the practice of counting. Frege quotes Husserl as having written: "The simplest criterion of equality of number is just that the *same* number results in counting the sets to be compared" (RevHus, op. 319). Husserl is thus claiming that Frege ought to have said, not that the number of *F*s is the number of *G*s just in case *F* and *G* can be put in one-to-one correspondence, but that the number of *F*s is the number of *G*s just in case you get the same number when you count the *F*s as when you count the *G*s. Frege considers a similar objection concerning (what he regards as) the analogous case of directions: The objection in this case is that parallelism ought to be explained in terms of sameness of direction, not conversely. Frege says, in response, that parallelism is more fundamental because all geometrical notions must originally be given in intuition (Gl, §64). But it is difficult, to say the least, to see how to apply such an observation to the case of numbers and equinumerosity.[13] Frege's response, in his review of *Philosophie der Arithmetik*, is that "counting itself rests on a one-one correlation, namely of the numerals 1 to *n* and the objects" to be counted. That there are *n* such objects follows, Frege intimates, only from the fact that, since the objects in question are in one-one correspondence with the numerals between 1 and *n*, the number of such objects is the same as the number of numerals between 1 and *n*, which number is *n* (RevHus, opp. 318–19).[14] Note that this argument appeals to (an instance of) HP.

Husserl, so far as I know, never responded to Frege. But there is a response to be had, and it has been forcefully developed by Charles Parsons (1994). Parsons concedes (as one obviously must) that counting does in fact establish a one-one correspondence between the objects to be counted and some numerals. But simply to rest with that observation is to miss Husserl's point, which concerns conceptual priority. Frege's observation, right though it may be, concerns, as we might put it, what someone who is counting *accomplishes*, not what she is *doing*: Frege gives us no reason to suppose that one must set out to establish a one-one correspondence when one counts. Of course, to be able to count,[15] one must in some sense realize that each object is to be counted once, and only once. But one can grasp that rule without thinking of oneself as establishing a one-one correspondence: One can have an 'operational' grasp of the rule of counting.

For consider: Suppose that, instead of spoken numerals, we had a stack of sticky notes with numerals on them and that counting involved literally tagging objects with sticky notes. It would be a rule that one was to tag each object once, and only once. But it's obviously one thing to be able to check whether all the objects in a given group have sticky notes on them and whether any object has more than one sticky note

---

[13]Hale (2001) discusses this issue at length.

[14]This remark dovetails nicely with certain of the results Frege proves in the later parts of the first volume of his *Grundgesetze der Arithmetik*, which was published in 1893, just a year before the review appeared. I have discussed these results elsewhere (Heck, 1998b).

[15]To count 'transitively', as it is sometimes put—that is, to count a group of objects—as opposed to counting 'intransitively', which is just to recite the numerals in order.

on it. It's quite another thing to have a general conception of one-one correspondence and to conceive of oneself, in tagging objects with sticky notes, as establishing such a relation between them. Parsons himself develops a treatment of counting that is, in relevant respects, similar. Counting, he suggests, can be conceived on a demonstrative model. Think of "1" as meaning something like: the first. So when one counts, one is saying, "the first, the second, the third, the fourth", where it is a rule of the game that one is supposed to 'tag' each object exactly once and then, when done, report how many objects there are by converting the last ordinal used to a cardinal. It is clear, once again, that one can grasp the 'once and only once' rule without thinking of oneself as establishing a one-one correspondence between the objects and the numerals. Indeed, there is a more powerful point in the vicinity, namely, that *one need not conceive of the numerals as objects in their own right* in order to count. To miss this point is simply to confuse use and mention: The numerals are not mentioned in counting—they are not treated as objects to be correlated with baseball players, for example—but are used. One no more correlates the numerals one-one with the members of the Boston Red Sox when one counts the players than one correlates the players with their names when one lists the members of the team.

A conversation I had with my daughter Isobel, when she was about three, illustrates this point. She was, by then, pretty good at counting. So, sometime neo-Fregean that I am, I decided to investigate how well she understood the connection between number and one-one correspondence and set up a little experiment. We had some Barbies and some hats and put them on the table. "How many Barbies are there?" I asked her. "One, two, three, four. Four Barbies!" she said proudly. And then we spent some time with the hats. We saw that we could put a hat on each Barbie—just one—there not being any left once each Barbie had a hat. "Just enough hats for the Barbies!" she said. So now the question: How many hats are there? No amount of prompting would elicit the inference: Four Barbies; one hat for each; so four hats.

As it happens, this sort of phenomenon is well-known to psychologists. In their important early study *The Child's Understanding of Number*, Gelman and Galistel write:

The preschooler's normal principle for determining whether two sets are numerically equal is "Count them and see". ... [T]he child's procedure actually presupposes the establishment of a one-to-one correspondence. In counting, the child establishes a one-to-one correspondence between the elements in his count sequence and the elements in the set being counted. From a logical point of view, the child's procedure for deciding numerical equivalence depends upon the fact that the numerosities of both sets can be placed in a relation of one-to-one correspondence with the same set of counting tags. But the child does not ordinarily take cognizance of the transitivity of one-to-one correspondence. He ignores or is indifferent to the fact that the cardinal numerons representing two equally numerous sets are identical precisely because both sets have been placed in one-to-one correspondence with that cardinal numeron. (Gelman and Galistel, 1978, pp. 198–9)

I take Gelman and Galistel's observation to show that Parsons is right, and not just about the abstract conceptual possibilities: The conceptual structures of very young children are just as Parsons describes. They know how to count and, to that extent, can answer questions of the form "How many?" and "Are there just as many?" But such children have no, or only a very minimal, understanding of one-one correspondence and its bearing upon questions of cardinality.[16]

If we look at the situation formally, we might come to a similar conclusion on different grounds: It turns out, when one looks closely, that HP plays a very limited role in the proofs of axioms (3)–(5). Consider, for example, the case of axiom (5), which says that zero has no predecessor. HP allows us to prove that a concept is 'empty'—that it is true of no objects— just in case it has the number zero. That is, HP allows us to prove that Zero-Concepts are Empty:

ZCE $\neg \exists x(Fx) \equiv Nx : Fx = 0$

From (ZCE), axiom (5) then follows almost immediately. Applying Frege's definitions to axiom (5), we get

$$\neg \exists y \{ Ny \land \exists F \exists x [0 = Nz : Fz \land Fx \land y = Nz : (Fy \land x \neq z)] \}$$

which (ZCE) immediately implies. Similarly, in the cases of axioms (3) and (4), HP allows us to prove that Adjunction Preserves Cardinality and that Removal Preserves Cardinality:

APC $Nx : Fx = Nx : Gx \land \neg Fa \land \neg Gb \rightarrow$
$$Nx : (Fx \lor x = a) = Nx : (Gx \lor x = b)$$

RPC $Nx : Fx = Nx : Gx \land Fa \land Gb \rightarrow$
$$Nx : (Fx \land x \neq a) = Nx : (Gx \land x \neq b)$$

Again, axioms (3) and (4) follow nearly immediately, given **FD**.

There is a case to be made that these Three Principles, as I shall call them, are absolutely fundamental to the notion of cardinal number. Consider what (ZCE) adds to the definition of "0": If there is an F, the number of Fs is not zero; or conversely, if the number of Fs is zero, there aren't any Fs. That's pretty basic. If you don't realize that there being zero books on the table means that there aren't *any* books on the table,

---

[16]It is, of course, possible that children do grasp the connection between cardinality and one-one correspondence, even from this young age, but that, for various reasons, that grasp is not revealed in the sorts of experiments so far devised. If so, then wonderful. But it is important to understand that the issue being discussed here concerns children's *conscious* knowledge of the relation between number and cardinality. Though it may be true, it would not advance the argument I am making to claim, say, that the connection is tacitly known. See footnote 22 for more on this matter.

then you simply don't understand what "zero" means—or, to put the point without mention of language, you don't have the concept *zero*. Or consider (APC): What it says is that, if you have the same number of Fs as Gs, and you adjoin an object to the Fs and an object to the Gs, then you will still have the same number. That too is pretty basic. If you have the same number of dolls as hats, and if you go get another doll and another hat, but you don't then realize that you *still* have the same number of dolls as hats—then, or so it seems to me, you don't really understand what it means to say that you've got the same number of dolls as hats. You don't have the concept *sameness of number*. And similarly for (RPC), whose converse is the natural direction: If you don't have the same number of dolls and hats, getting another doll and another hat isn't going to solve your problem.

So the Three Principles are pretty basic: Arguably, one could not have even the most primitive conception of cardinality without at least tacitly grasping them. Is HP so basic? It seems easy to imagine someone's grasping the Three Principles, and applying them in their dealings with numbers, while not grasping HP. Husserl, Parsons, and Isobel at age three give us models for just this possibility. Further, since sameness of number is now being conceived in terms of the results of counting, and counting itself is done with a collection of numerals, features of the numerals with which one counts can't help but be reflected in properties of the numbers. And that is what we have here: The Three Principles simply reflect features of the numerals. That (APC) holds, for example, is a reflection of the fact that there is always a definite 'next' numeral. For assume the antecedent of (APC), that the number of Fs is the same as the number of Gs; that is, assume that, in counting the Fs and the Gs, I end with the same numeral. Now suppose I add an object to each pile. To count what I have now, I need only continue the previous counts: Hence, the only way I could end up with different numerals now is if the next numeral might be different in the two cases. But the structure of the numerals makes that impossible.[17]

Furthermore, one might argue, the Three Principles implicitly characterize sameness of number.[18] They do not quite fix the extension of "$Nx : Fx = Nx : Gx$": They will not, in particular, tell us whether the number of evens is the number of odds. But it is easy to see that the Three Principles *do* fix the extension of "$Nx : Fx = Nx : Gx$" on finite concepts; it is only a little harder to see that they fix its extension so long as at least

---

[17]The case of (RPC) is somewhat more complicated, but that it holds reflects the fact that, if one says a numeral just before n, there is a definite numeral one always says just before n. For (RPC) to fail, it would have to be that, though in counting the Fs and the Gs, I ended up at the same numeral, still, if I'd stopped just before counting the last object, I would have ended up at different numerals. What Gelman and Galistel (1978, ch. 9) call the principle of order irrelevance is important here.

[18]In the mathematical sense, they do not do so, for the reason about to be mentioned. I thus use the term 'implicitly' in a non-technical sense here.

one of F and G is finite.[19] So, on this view, that sameness of number is (at least as concerns finite concepts) one-one correspondence is, far from being the fundamental fact about cardinality, a *theorem*, a consequence of the more fundamental Three Principles.

I think that one or another version of the view I've just outlined underlies much of the opposition to Fregean accounts of the source of our knowledge of (certain of) the axioms of arithmetic. Consider, for example, structuralist views. The analogue of the idea that the numbers are given *via* the numerals is that the numbers are given, in the first instance, as elements of a certain sort of structure, one that is also exhibited by the numerals. The analogue of the idea that the finite cardinals are characterized in terms of counting is the idea that they are characterized in terms of one-one functions onto initial segments of such structures. The structures in question are, of course, completely characterized (up to isomorphism) by the axioms of **PA**. The question whether our knowledge of those axioms depends upon our grasp of the notion of one-one correspondence thus becomes the question whether our grasp of such structures and our ability to employ them in thought, including our ability to employ them in counting, requires grasp of the concept of one-one correspondence. And what we have just seen is that there is at least some reason to doubt it does: It appears that a grasp of the rules that govern the practice of counting—which presupposes some sort of grasp of the structure of at least one ω-sequence, that of the numerals—does not require a grasp of the notion of one-one correspondence. If not, then it would seem that one could have a perfectly serviceable concept of cardinality without even having the concept of one-one correspondence.

7.3   Counting and Cardinality

It can seem like so much common sense that a child who has learned to count, and so has learned to answer questions like "How many cookies are on the table?" has acquired a concept of cardinality. But that is wrong: A child who can, *in that sense*, answer questions of the form "How many?" need have no idea what the answers to such questions mean—or, indeed, what the questions mean. Such a child need have no concept of cardinality.[20]

---

[19]The proof is, in essentials, just the proof of HPF in **PAF+FD**, which I discuss in detail in Chapter 11. The technique of that proof can be applied here, since the three principles are, as noted above, the result of applying Frege's definitions to Axioms (3)–(5) and doing some simple manipulations. The Three Principles can therefore play the role these axioms play in the original proof.

[20]Actually, what I will be arguing is simply that such young children are not *deploying* a concept of cardinality in answering "How many?" questions. They may, of course, have such a concept but not yet have connected it with counting and such. That is an empirical question.

Something like this point is, again, well-known to psychologists. Here are a couple illustrations of it. Recall the story about my daughter Isobel. Asked how many hats we had, her impulse was to count them, despite already knowing that we had four dolls and one hat for each doll. I didn't say before, though, that, if I'd asked her again how many dolls there were, *she would not just have told me*: What children of that age do when asked how many Fs there are is to count, even if they've just finished counting those same objects. In fact, barring boredom, children of that age will count a group of objects as many times as one asks them "How many?" It's almost as if they understand "How many dolls?" as meaning: Would you please count the dolls?

Such children also do not appear to understand what Frege called "ascriptions of number". If I had asked Isobel to give me three hats, her response would have been to grab some hats and hand them over. Whether she gave me three hats would have been a matter of chance. She wouldn't have given me one hat: Children of that age do understand the difference between one and not one. But it seems not to occur to them to count out three hats. Or again, if I'd shown Isobel two cards and asked her which had three balloons on it, she would have responded at chance—unless one of the cards had just one balloon on it, since children of that age do know one from not one. Perhaps one of the most amusing observations here is that children at this level of development will occasionally count with sequences other than the conventional numerals: For example, "*a, b, c*", or even "Monday, Tuesday, Wednesday" (Carey, 1995). Such children may well understand the numerals as mere tags, having no independent significance. For them, "There are four hats on the table" really does mean something like: I ended with "four" when I counted the hats. But these children seem to have no grasp at all of the point of such "ascriptions of number".

Think of the matter from the child's point of view. I have been taught how to 'count': I have, that is to say, been taught to point at objects and say certain words. I have to say the words in a certain order. I'm supposed to count each of the objects once and only once. And most importantly, when I get to the end, I'm supposed to say the last of the words loudly and proudly. But what all of this has to do with anything, who knows? The grown-ups seem to like it, and they're forever fawning over me when I do it right. That's more than enough reason for me to play.

The moral of the story is that mastery of the practice of (transitive) counting is compatible with one's having no concept of cardinality at all. One can be able to answer questions such as "Are there as many Fs as Gs?" in the terms allowed by one's mastery of the practice of counting, while yet having no idea what the significance of an answer to the question might be or even what the question means. And that is precisely the situation in which one will find oneself if one understands the question "How many Fs are there?" as meaning: With which numeral does one end when one counts the Fs? and understands the question whether there are as many Fs as Gs as meaning: Does one end with the same numeral when one counts the Fs as when one counts the Gs? Very young children simply illustrate this conceptual point.

So what *is* involved in understanding "How many?" questions as how many questions? What *is* the point of ascriptions of number? Well, suppose one knows that there are five cookies. There's not a lot one can do with that knowledge by itself. But suppose one also knows there are four children. Now one has knowledge one can use, for one can infer that there are more cookies than children. But what does that mean? Or suppose one knew that there were five children. Then one could infer that there were just as many cookies as children. But what does that mean?

There is a familiar answer to this question, namely: To know that there are just as many cookies as children is to know that there is a one-one correspondence between the cookies and the children. That is a bad answer. It is far from obvious that knowing what "just as many" means requires knowing what a one-one correspondence is, in general: The notion of a one-one correspondence is very sophisticated; it is far from clear that five year olds, who do seem to grasp the concept *just as many*, have any general grasp of one-one correspondence.

Another option would be to take the concept *just as many* to be characterized by analogues of the Three Principles. Let "$\mathrm{JAM}_x(Fx, Gx)$" be read: There are just as many Fs as Gs, and consider the following principles:

ZCE* $\neg \exists x(Fx) \rightarrow [\mathrm{JAM}_x(Fx, Gx) \equiv \neg \exists x(Gx)]$

APC* $\mathrm{JAM}_x(Fx, Gx) \wedge \neg Fa \wedge \neg Gb \rightarrow \mathrm{JAM}_x(Fx \vee x = a, Gx \vee x = b)$

RPC* $\mathrm{JAM}_x(Fx, Gx) \wedge Fa \wedge Gb \rightarrow \mathrm{JAM}_x(Fx \wedge x \neq a, Gx \wedge x \neq b)$

If it is hard to see how one could have even the most primitive grasp of the concept *sameness of number* without at least tacitly grasping the Three Principles, much the same can be said about one's grasp of *just as many* and these 'Starred Principles'. And the remarks made above, about how the Three Principles characterize sameness of number, apply here, *mutatis mutandis*: Though the Starred Principles do not fix the extension of "$\mathrm{JAM}_x(Fx, Gx)$", since they do not tell us whether there are just as many evens as odds, they do fix its extension so long as at least one of the concepts involved is finite. So the Starred Principles implicitly characterize equinumerosity, that is, *just as many*, at least for finite concepts.

Having characterized equinumerosity in terms of the Starred Principles, we are still free, of course, to characterize sameness of number, and so the concept of cardinality, in terms of equinumerosity, as follows:

$$(\mathbf{HPJ}) \quad \mathrm{N}x : Fx = \mathrm{N}x : Gx \equiv \mathrm{JAM}_x(Fx, Gx)$$

It's obvious that the original Three Principles follow immediately from this principle and the Starred Principles, so Axioms (3)–(5) follow from

HPJ, the Starred Principles, and Frege's definitions of zero and predecession. But these axioms are, remember, utterly trivial consequences of the Three Principles, given Frege's definitions: They are all but definitional transcriptions of the Principles. There is no clear sense, on this view, in which HP or HPJ underlies Axioms (3)–(5): The Starred Principles are what are fundamental.

The position just described has some attractive elements. In particular, its answer to the question what "just as many" means does have the desirable and necessary feature of being unsophisticated. But it suffers from problems not unlike those that afflict the Husserl-inspired view that gave a central place to counting. It is obvious that someone could accept the Starred Principles and yet have no idea what to make of the claim that there are just as many children as cookies.

What we need, then, is some other unsophisticated answer to the question what "just as many" and "more" mean. Suppose that there are more cookies than children. What follows? It follows that there are more than enough cookies for each child to have one. From a logical point of view, that means that there is a one-one function from the children to the cookies which omits some cookie from its range. But from a child's point of view, it need mean no more than that, if you start giving cookies to children, and you're careful not to give another cookie to anyone to whom you've already given a cookie, then you won't run out and will even have some left. Or suppose there are just as many children as cookies. Then it follows that there are just enough cookies for each child to have one. From a logical point of view, that means that there is a one-one correspondence between the cookies and the kids; but from the child's point of view, it need mean no more than that, if you start giving cookies to children, and you're careful not to give another cookie to anyone to whom you've already given one, then you won't run out, though you won't have any left. There are enough, that is to say, for everyone to have one, but not for anyone to have two.[21]

The suggestion, then, is that grasping the concepts *more, just as many*, etc., involves connecting them, in the right way, with what we might call their practical correlates: *enough, just enough*, and so on. Of the practical notions, the child has an operational understanding:[22] Her understanding of what it is for there to be just enough is bound up with the idea of giving (just) one to everyone, or putting (just) one hat on each doll's

---

[21]The notion of having should be understood abstractly, and is by children. They understand that they can have a mother, a bicycle, a cookie, a home, a teacher, a friend, and a God. A doll can have a hat, and a hat can have a doll—and in a non-metaphorical sense, too. (I am half-tempted to suggest that "have" is a free variable over relations.)

[22] My use of the term "operational" should not be taken too seriously: I do not mean to be suggesting a return to behaviorism. For all I have said here, grasp of the concept *just as many* might involve tacitly knowing that there are just as many *F*s as *G*s if, and only if, they are in one-one correspondence. My point is simply that the child's grasp of the concept involves no conscious, theoretical identification of it in other terms.

head, or what have you; to say that there are just enough is to say that one *could* give (just) one cookie to each child. Of course, to give just one cookie to each child is, in fact, to establish a one-one correspondence between the cookies and the children, and to say that one could do so is to say that there exists such a correspondence. But this sort of contrast is now familiar from the case of counting. And from the case of renates and cordates, too: The concept *just as many* has the same extension as the concept *is in one-one correspondence with*, but that simply does not imply that the concepts themselves are the same. They are not.

On this view, then, grasp of the concept *just as many* is in principle independent of the ability to count. It is, therefore, a mistake to attempt to characterize equinumerosity as Husserl did, by reference to the practice of counting. Moreover, grasp of the concept *just as many* is clearly independent even of the concept of number, that is, of any grasp of the significance of ascriptions of number: One can understand what it is for there to be just enough cookies to go around without understanding what it means to say how many cookies there are. If so, then it is wrong to attempt to explain equinumerosity in terms of sameness of number, whether characterized in terms of counting or not. On the other hand, it would seem that a grasp of the concept of cardinal number *does* require a grasp of the concept of equinumerosity: One will understand answers to how many questions *as* answers to how many questions—*as* ascriptions of number, rather than statements about the results of countings—only if one grasps the concept *just as many* and its relation to ascriptions of number. That relation, of course, is just the relation HPJ reports.

7.4   Counting and Ascriptions of Number

I have been arguing, to this point, that our concept of equinumerosity is independent of any connection with the practice of counting. As I just mentioned, however, the argument also shows that our concept of equinumerosity is independent of any grasp of the significance of ascriptions of number, that is, of the meanings of such sentences as "There are four dolls on the table". According to me, that is, our grasp of the concept of equinumerosity is independent of any grasp of the concept of cardinal number. We cannot yet conclude, however, that our grasp of the concept of cardinal number—our grasp of what specific assignments of number mean—is similarly independent of counting. Grant that when we say, "There are four dolls on the table", we don't just mean that you end with "four" if you count the dolls: If that were all we meant, then we wouldn't be ascribing number at all. But there is an obvious alternative, namely, that "There are four dolls on the table" means: There are as many dolls on the table as there are numerals between "one" and "four". If that is what we mean, then it would seem that our understanding of ascriptions of number *is* bound up with our ability to count: Establishing that there is such a relation between the dolls and the numerals between "one" and "four" would, on this conception, be just what counting the dolls did—as Frege suggested, more or less.

I am shortly going to argue that this conception of the relation between counting and ascriptions of cardinality is mistaken. I should first like to emphasize, however, that the arguments I shall be giving do not focus upon the reference to *numerals*. They apply as well, for example, to Frege's proposal that "four" means: the number of numbers between zero and three. And to the closely related view that the cardinals may be defined in terms of the ordinals, the cardinal number of a set S being the least ordinal onto whose predecessors the members of S can be mapped one-one. My specific remarks are directed against the view that ties cardinality to counting, but their extension to these similar views should be fairly obvious.

I have three objections to bring against such conceptions of cardinality. None of them, I am prepared to admit, is independently conclusive, but their combined force is substantial. Indeed, their combined force is sufficient, say I, when combined with the alternative conception I shall offer.

First of all, children, and adults, seem to be able to understand some attributions of cardinality quite independently of any connection with counting. For example, when I think that there are two blocks on the table, I do not seem to be thinking that there are as many blocks on the table as there are numerals from "one" to "two". The significance of this point is unclear, however. One might still want to hold that our conception of specific cardinalities is, in general, given by the connection with counting, even if there are specific cases in which it need not be so given.

Second, there seems to be an important distinction between the sort of knowledge one has when one knows that there are as many Fs as numerals from "one" to "nine" and when one knows that there are nine Fs. In the latter case, one actually knows how many Fs there are; or, as Kripke (to whom the observation is due) puts it, one has *de re* knowledge in the latter case but only *de dicto* knowledge in the former (Kripke, 1992). Of course, one can easily calculate how many Fs there are in the former case. But consider a different case. If I am told that there are as many Fs as there are numerals between "1" and "21" base-16, then I, anyway, do *not* thereby come to know how many Fs there are. Presumably, there are people who *do* know how many 21_{16} is, but I don't, not immediately, not without calculating how many it is. What calculating means here, for me, is finding the decimal numeral, because knowing how many Fs there are, for me, is knowing such things as that there are 33 Fs. It's not that the rule linking numerals to numbers is any more complicated in the hexadecimal case than in the decimal case: Just as there are 33 decimal numerals between "1" and "33", so there are 21_{16} hexadecimal numerals between "1_{16}" and "21_{16}". The difference, rather, is that, once this simple rule has been applied, there is more work to be done in the latter case—but not in the former, for me. That is because, in knowing that there are 33 numerals, I thereby know how many numerals there are; whereas in knowing that there are 21_{16} numerals, I do not thereby know how many.

Knowing how many is *knowing what number*. So, to reformulate the foregoing: I do not, without calculating, know what number the base-16 numeral "21" denotes, even though I do of course know that it denotes 21_{16} and know, moreover, that it denotes the number of hexadecimal numerals between "1_{16}" and "21_{16}". My knowing what number "33" denotes, therefore, can not consist in my knowing that it denotes the number of decimal numerals between "1" and "33": I have that kind of knowledge in the hexadecimal case, too.

Of course, one would like more to be said here, and I shall say none of it. But let me address one response I have frequently heard. It's easy to want to dismiss the foregoing, saying that the difference is simply that I am more familiar with decimal numerals than I am with hexadecimal ones. In some sense, that must of course be right: It's not as if there is anything intrinsically special about decimal numerals; the difference obviously has to be a matter of my relation to the decimal numerals, in the end. But it's not just my familiarity with decimal numerals that matters; it's not just a matter of my 'comfort level', as one objector once put it. It's my facility with decimal numerals that matters. The important difference is that I am able to use decimal numerals immediately in thought, whereas I almost always have to translate hexadecimal numerals before I can do much of anything with them. But that doesn't undermine the point I'm making. On the contrary, it illustrates and establishes it.[23]

Third, a point made above, in a somewhat different context, is also relevant here. The knowledge that there are five numerals between "one" and "five" is actually quite sophisticated. Isobel, who clearly had to put up with a lot, was not able to acknowledge this fact until some time after she had mastered counting and the concept *just as many*, and had developed a thorough understanding of such claims as that there are five Beanie Babies on the sofa. For a long time, she had essentially no understanding of the question how many numerals (or numbers) there are between "one" and "five". What she lacked, I think, was the concept of a numeral (or number) as an object: She had no concept of the numerals as objects that can themselves be counted. Numerals, as emphasized above, are *used* in counting; they are not mentioned. There is really no good reason that one who is able to count *with* numerals need be able to conceive of them as objects in their own right. If so, then the idea that "There are five

---

[23]Kripke (1992) emphasizes that recursion theory itself rests upon this kind of distinction. To calculate a number-theoretic function is to return its value *in a certain form*. If any expression that denoted the value would do, then every function for which we had a name would trivially be calculable.

Beanie Babies on the sofa" means that there are as many Beanie Babies as numerals between "one" and "five" confuses use and mention. In doing so, it presupposes a conception of numerals as objects that the ability to count, and even to make judgments of cardinality, does not require.

How, though, should we understand attributions of number? What *does* the statement that there are nine Fs mean, if not that there are as many Fs as numerals between "one" and "nine"? A related question is how counting establishes how many there are, if not by establishing that there are as many Fs as numerals between "one" and whatever.

The short answer to the first question is that attributions of number answer how many questions. What that means is, admittedly, less than obvious. To some extent, I think, the development of this famous Fregean doctrine awaits progress in semantics: It is not at all clear what the logical form of, say, "Two men went for a walk" should be taken to be. But Frege's central idea is really quite simple: It is that one grasps the concept *two* just in case one knows how many Fs there are when there are two Fs. In this simple case, perhaps that's just knowing that there are two Fs if there are this F and that F and no other Fs: Perhaps, that is, it's just grasping the usual first-order definition of "There are two Fs". Of course, larger numbers will pose more of a challenge: Understanding "145", say, does not consist in knowing that there are 145 Fs if, and only if—forgive me if I omit the first-order equivalent. Rather, understanding "145" has something to do with one's understanding of the decimal system. It's a nice question (an empirical one, of course) what that involves, but it is not a question that bears fundamentally upon what is at stake here. For the reasons already given above, knowing how many "145" is isn't, in any event, a matter of knowing that there are 145 Fs if, and only if, there are just as many Fs as numerals between "1" and "145".

More important is the question how we should understand counting. What alternative is there to the idea that counting establishes how many Fs there are by establishing that there are as many Fs as numerals from "1" to whatever? Some time ago, I was counting out tulip bulbs, and I did so this way: 2, 3, 5, 7, 11. That wasn't because I was using a non-standard count sequence consisting only of prime numbers. It was, rather, because I was counting the bulbs not, as one says, by twos, but by groups. We count that way all the time. Some children learn to do so at a very young age, and without ever learning to count by twos intransitively (that is, just saying the words), let alone by groups (which would just be randomly, though increasing). What are we doing when we count like that? We are, I suggest, making a series of judgments of cardinality. One is saying, "That's two bulbs so far; now three; two more, so five"; and so forth. We can think of ordinary counting in the same way: That's one; now two; now one more, so three; one more, so four; four cookies in all.

Of course, I am not denying that counting can be done mindlessly, without making judgments of cardinality along the way. It is certainly done that way by very young children, and it can even be done that way by adults. But the question here is whether counting is fundamentally a mindless exercise, and I mean to be denying that it is: I mean to be denying that we best understand what counting is, and how it functions, by conceiving it simply as a matter of tagging objects with symbols which do not, *qua* symbols used in counting, function to assign cardinal number. Conceived as my opponent wants to conceive it, counting can issue in judgments of cardinality only in one of two ways: Either "There are four *F*s" must mean: You end with "four" when you count, or it must mean: There are as many *F*s as there are numerals from "one" to "four". I have been arguing against both of these construals, but let me not simply repeat myself. What is, in the end, most worrisome, or so it seems to me, is that, if counting were mere tagging, then it wouldn't matter whether we counted with numerals or with days of the week or with letters of the alphabet. If counting were mere tagging, we should be just as happy to say "There are '*i*' dolls", meaning: There are as many dolls as there are letters from "*a*" to "*i*", as we are to say that there are nine. But we are not, and the reason is that counting is *not* mere tagging: It is the successive assignment of cardinal number to increasingly large collections of objects. And the symbols we use in counting are not mere tags: Even *qua* symbols used in counting, they function to assign cardinal number.

Note that counting, conceived as I am suggesting it should be, has nothing essential to do with equinumerosity (let alone one-one correspondence): Husserl was, we see again, absolutely right about this point. Counting 'by ones' does, of course, establish a one-one correspondence between the objects counted and an initial segment of the numerals. But that fact isn't at all fundamental to what counting is.

7.5   Closing

I have argued that something close to HP, namely, HPJ, is a conceptual truth: An appreciation of the connection between sameness of number and equinumerosity that it reports is essential to even the most primitive grasp of the concept of cardinal number. If so, then it is arguably a conceptual truth also that the (finite) cardinal numbers form (at least) an initial segment of an ω-sequence. That they do is the content of the axioms of **PA**, other than Axiom (6), existence of successor (whose inclusion would license the removal of the phrase "an initial segment of" and whose status I am setting aside). Of these, Axioms (1), (2), and (7),[24] recall, follow immediately from Frege's definitions: Acceptance of these definitions as

---

[24] As noted above, the proof of Axiom (7) requires impredicative comprehension. Someone who was bothered by that fact might, therefore, regard (7) as not being a conceptual truth—but induction has, of course, long been controversial for just that reason. See Chapter 12 for some further discussion.

capturing the content of our ordinary notions of *zero*, *one more*, and *natural number*—not uncontroversial, but hardly unreasonable—therefore gives us reason to regard these axioms as logical consequences of conceptual truths and so, I take it, as themselves being conceptual truths. The other Axioms, (3)–(5), follow, as we saw earlier, from the Three Principles and so from HPJ and the Starred Principles. But the Starred Principles themselves can arguably be established purely conceptually. Consider, for example, (APC*).

Theorem. $JAM_x(Fx, Gx) \wedge \neg Fa \wedge \neg Gb \rightarrow JAM_x(Fx \vee x = a, Gx \vee x = b)$

Proof. We shall prove a representative instance: If there are just as many children as cookies, and if one more child shows up, and we find one more cookie, then there will still be just as many children as cookies.

Suppose there are just as many children as cookies, and one more child shows up, and we find one more cookie. So there were just enough of the 'original' cookies for each of the 'original' children to have one. So, if we start giving the original cookies to the original children, and are careful not to give one to anyone to whom we've already given one, we will be able to give a cookie to each, but won't have any left. But then we can just give the new cookie to the new child, and there will have been just enough cookies for each child to have one. So there are just as many cookies as children.   □

Of course, one might question whether the sort of reasoning employed in this informal argument yields a conceptual truth—and that, of course, is a question to which Frege himself was acutely sensitive. To show that it did, we would have to formalize the argument and "follow[] it up right back to the primitive truths" from which it proceeds (Gl, §3). Doing so, however, raises difficult issues. Obviously, the informal argument relies heavily upon our informal understanding of the notion *just as many*: No formalization of the argument will be possible unless we either uncover some deductive principles—introduction and eliminations rules, say—specific to this notion or analyze it in other terms, so that other, more familiar sorts of deductive principles become applicable to it. For present purposes, we can ignore the difference between these two approaches and focus on the latter.[25]

Frege famously offers a version of this latter approach. In effect, he suggests that *just as many* should be analyzed as one-one correspondence. But even granting that Frege is right that equinumerosity *is* one-one correspondence, extensionally speaking, in what sense can we regard this as an 'analysis' of the notion of equinumerosity? I have argued already that

---

[25]In the original paper, I recorded a preference for the former approach, and the techniques used in Chapter 12 could be used to implement it. But, in that case, so far as I can see, one can do little better than to take (APC*) and the like as axioms (or transform them into rules), in which case a proof of them will be hard to come by. So I do not know if the former approach can usefully be developed.

equinumerosity is not the same concept as being in one-one correspondence with. Still, though, the connection is very tight: To say that there are just as many Fs as Gs is to imply that there are just enough Gs for each F to have one, that is, that, if we start giving Gs to Fs, and are careful not to give another F to any G to which we've already given one, then we won't run out, but won't have any left. Something like the familiar Fregean analysis of equinumerosity seems buried in there somewhere.

It is, I think, a difficult question in what sense it might be buried and what exactly one is doing when one digs it out. There is some good work by Demopoulos (1998, 2000) that seeks to articulate a notion of 'analysis' that would apply not just here but to the similar case of continuity.[26] But let me not attempt to pursue this matter here. I find this territory terribly confusing, and my views are unstable, to say the least. If, however, we assume that some such conception of analysis is available and that it is, in the relevant sense, an analytic truth that[27]

Eq There are just as many Fs as Gs if, and only if, the Fs are in one-one correspondence with the Gs,

then HP, in its predicative form,[28] ought to be an analytic truth, in the same sense, since it follows from (Eq) and HPJ. It ought similarly to be an analytic truth that the (finite) cardinals form an initial segment of an ω-sequence, since that follows from HP, in its predicative form.[29] Take that as a conjecture. Before we try to develop and defend it, however, it is worth asking what interest such a conclusion would have, even if it could be established. It would give us no reason to believe that we come by our knowledge of the relevant properties of the cardinals by deriving them from HPJ, and I see no reason to believe that we do. The suggestion that someone *could* come by knowledge of the axioms of **PA** by deriving them from HPJ suffers from problems I discussed some time ago: It is unclear what it adds to the claim that the axioms of **PA** are derivable from HPJ; if so, then someone who had sufficient powers of reasoning could of course derive them from it, but what of it? It is unclear to me, then, that there is any satisfactory way to give epistemological content to Frege's Theorem.[30]

---

[26]Another important such case, less often mentioned in this connection, but of great interest, is that of computability: Is Church's Thesis an analytic truth, in some reasonable sense?

[27]It may be that only something weaker is analytic, namely, the analogue of HPF: So long as at least one of F and G is finite, the Fs are equinumerous with the Gs if, and only if, they are in one-one correspondence. That would do for the purposes of elementary arithmetic, however.

[28]That is, in the two-sorted form that does not admit numbers into the domain of the bound first-order variables occurring in HP.

[29]Though, again, we need impredicative comprehension to get induction.

[30]One idea that might be me worth investigating is that our knowledge of the relevant axioms of PA is *based upon* our knowledge of HPJ, in the sense in which that notion is used in epistemology (Korcz, 2010).

What then does Frege's Theorem offer us? Something, I should like to suggest, like an answer to the question *why* the (finite) cardinal numbers satisfy the Dedekind-Peano axioms. I expect that question will strike some as bizarre: Certain varieties of structuralism, for example, would encourage the thought that it is ill-conceived. The finite cardinals are the objects of arithmetic, and arithmetic, the claim would be, is just the study of the properties of ω-sequences—or something like that. So, one simply can't ask why the finite cardinals form an ω-sequence: If the natural numbers just are abstract positions in such a structure, the question why they satisfy those axioms is just silly, right? But the question seems to me to be a perfectly sensible one—all the more so since it seems to have a perfectly sensible answer. The finite cardinals satisfy the axioms of **PA** because they satisfy HP, that is, because of the connection between cardinality and one-one correspondence it reports. So the interest of Frege's Theorem lies there: It offers us an explanation of the fact that the numbers satisfy the Dedekind-Peano axioms.

One might ask, though, why it should matter, if that is what Frege's Theorem offers us, whether HP or HPJ is a conceptual truth. And, in fact, I am happy to concede that, in a sense, it doesn't matter: HPJ does not need to be a conceptual truth for the Theorem to provide such an explanation; the notion of a conceptual truth doesn't even have to be coherent.[31] But HP does need to be more fundamental, in some significant way, than the axioms of **PA** if Frege's Theorem is to have any explanatory force, for not every derivation of a conclusion from premises counts as explaining why the conclusion holds in terms of the fact that the premises do. And if HP itself were less fundamental than the axioms of **PA**, the explanatory value of Frege's Theorem would be nil. What I have argued here, however, is that recognition of the truth of something very much like HP is required if one is even to have a concept of cardinality. If so, then what Frege's Theorem shows is that the fact that the finite cardinals form an initial segment of an ω-sequence is implicit in our very concept of cardinal number—'implicit' in the sense that the cardinals' forming such a sequence is logically required by the character of our concept of cardinal number.

---

[31]The account of the significance of Frege's Theorem offered here thus does not rest upon or require the elaborate framework Wright sets up and deploys in *Frege's Conception of Numbers as Objects* (Wright, 1983): It does not, in particular, need to make use of the idea that, in general, we understand a sortal concept by understanding a criterion of identity for it. Of course, for that very reason, the view I am elaborating contributes nothing to our understanding of ontological problems involving numbers. I am not, therefore, suggesting that it doesn't matter *at all* whether HP is a conceptual truth, and I stand by the argument given here that (something like) it is.

8

Syntactic Reductionism

In their "Steps Toward a Constructive Nominalism", Goodman and Quine write:

We do not believe in abstract entities. No one supposes that abstract entities—classes, relations, properties, etc.—exist in space-time; but we mean more than this. We renounce them altogether. (Goodman and Quine, 1947, p. 105)

We should distinguish three sorts of "entities" that Goodman and Quine "renounce": First, abstract objects, such as classes, numbers, and linguistic types; Second, higher-order entities, among which are relations, concepts, functions, and the like, taken in extension; Third, intensional entities, including properties and propositions. I shall not here be concerned with intensional entities, as they raise unique problems. One might want to deny that there are such things as propositions—indeed, do so for essentially Quinean reasons—without having the slightest doubt whether there are such things as classes, numbers, or relations. Nor shall I be concerned with higher-order entities. So I distinguish *higher-order Nominalism*, which specifically denies the existence of higher-order entities, from *first-order Nominalism*, which denies the existence of abstract objects. Since my attention will be focused upon first-order Nominalism, I shall mean it by "Nominalism".

Goodman and Quine's chief example of a kind of abstract object whose existence they wish to deny is that of a class. But skepticism about the existence of abstract objects is more general than skepticism about classes. One could deny the existence of classes for something like the reason given by Goodman and Quine, "that the most natural principle for abstracting classes... leads to paradoxes" (and, presumably, that there is no natural principle, or set of principles, which might serve to replace it). The natural principle in question, of course, is naïve abstraction:

$$\exists y \forall x [x \in y \equiv \phi(x)]$$

Any first-order theory that contains all such formulas is inconsistent, as Russell showed. Whether this is a good reason to "renounce" classes, or whether there is some other, 'natural' theory of sets (e.g., that provided by the so-called iterative conception), one could be skeptical about the existence of classes without being skeptical about the existence of abstract entities of all kinds. Since the more common position is a rejection, on quite general grounds, of classes together with abstract objects of other sorts, we ought not to focus too much on classes.

I have just said that abstract objects are frequently rejected on general grounds. But one of the peculiar facts about discussions of Nominalism is that few seem to think that the rejection of abstract objects needs grounds. What one frequently finds in the literature are not arguments in favor of Nominalism but attempts to defend it against certain sorts of objections, frequently against *one* sort of objection. The point of contention is whether the Nominalist can show how to 'do without' reference to abstract objects. It is in this spirit that Goodman describes "Steps" as having "accomplished" an important part of the Nominalist program, namely, "the nominalistic... treatment of most of classical mathematics" (Goodman, 1972, p. 167). And Nominalists are not the only ones who attach a great deal of importance to the question whether a theory can be 'nominalized'. For example, though Quine eventually abandoned Nominalism, his change of heart was due to the so-called Indispensability Argument, whose central premise is that science (in particular, physics) cannot do without classical mathematics which, in turn, cannot itself be nominalized.[1]

The prevalence of Indispensability Arguments as arguments against Nominalism have naturally led would-be Nominalists to view them as the most important, if not only, arguments against their position. Thus, Hartry Field's *Science Without Numbers: A Defense of Nominalism* (Field, 1980), is an attempt to answer the Indispensability Argument by showing that physics can be nominalized—i.e., that physical theories can be formulated without reference to mathematical objects and that mathematics need not be employed in reasoning within physics, e.g., for the derivation of predictions from physical laws and statements of initial conditions.[2] Similarly, in the case of linguistics, answering the Indispensability Argument would require showing that linguistic theories can be formulated without reference to linguistic types and that reference to types need not be employed in reasoning within linguistic theory.

There are two ways in which this might be done in any particular case. First, one could deny that terms purporting to refer to abstract objects are really terms at all, in much the way that Russell denied that definite descriptions—expressions of the form "the ϕ"—are singular terms. On this view, sentences such as "the type of (the word-inscription) w is F" are not of the form "$F(\phi(w))$", as they appear to be, but are instead of some more complex form.[3] Such a version of Nominalism we may call *Syntactic*

---

[1]The classic discussion of such arguments is in Putnam's "Philosophy of Logic" (Putnam, 1979). Quine is prepared to accept the existence of classes on even weaker grounds, that reference to mathematical objects simplifies scientific practice (Quine, 1960, p. 237).

[2]Field does, of course, have and offer reasons to be worried about abstract objects, namely, the epistemological ones to be mentioned shortly.

[3]One might wonder whether "the word-type of which w is a token" is itself a definite description. I am unclear what to say about this. Many functional expressions in English have the orthographic form "the ϕ of t", but this may be misleading. However that may be, my intention is that "the type of which ξ is a token" be treated as a primitive, functional

Reductionism, since the reduction is of *names* of abstract objects to more complex sorts of expressions.

On the other approach, one concedes that "the type of (the inscription) *w*" is a term, while denying that it refers to an abstract object. There are then two versions of this view. The more extreme, Fictionalism, denies that such terms refer at all; in consequence, no simple sentences containing such terms are true. The challenge to the Fictionalist is to explain how attempted reference to types can still be useful, or legitimate, in linguistic theory, even though there are no such objects. The less extreme version of the view instead denies that terms that appear, *prima facie*, to refer to abstract objects refer to *abstract* objects. Such a view, which comes in a variety of flavors, might be called *Semantic* Reductionism, since the reduction is of *reference* to abstract objects to reference to entities of other sorts.

I will not be saying anything at all about Fictionalism. What I intend to argue here is just that Syntactic Reductionism is untenable. If so, then, in so far as Nominalism can be defended without denying that a large number of the sentences we take to be true are true (that's Fictionalism), it must take the form of Semantic Reductionism. I discuss its prospects in Chapter 9.

8.1   Motivating Nominalism

As said above, the heavy focus on Indispensability Arguments that one finds in the literature might lead one to think that Nominalism is motivated by a conviction that reference to abstract objects will prove eliminable. But it is not as if the view amounts to a hunch that the Nominalist program can be successfully executed. Rather, most Nominalists believe that such reference *must* be eliminable, lest we be unable to make sense of much of what we say. Goodman, for example, writes that he "finds the notion of classes... essentially incomprehensible", describing his difficulty as one of being unable to see how more than one object can be created from, or constituted by, the very same objects (Goodman, 1977, pp. 25–6).[4] Whatever the merits of this objection, it is specific to classes and

---

expression which could be otherwise spelled. Compare Hale and Wright's discussion in "Focus Restored" (Hale and Wright, 2009a, §1).

[4] E.g., how I can have a singleton; my singleton can have a singleton; etc. That said, in *Parts of Classes*, David Lewis (1991) shows that one can treat set-theory mereologically, by taking the view that a set is the fusion of its subsets, fundamentally, of the singletons of its members. Lewis would therefore have a reasonably natural, broadly mereological theory of sets if only he could explain what the singleton of any given set is. Goodman might well say, at this point, that there is not going to be any such explanation: No sense is going to be made, in mereological terms, of the idea that there are infinitely many objects each of which arises from, or is formed from, a single object. And Lewis does finds himself utterly unable to provide any illuminating account of what singletons are; ultimately, he treats the singleton-function structurally. The problem is then, as Charles Parsons has pointed out,

cannot be used to motivate a rejection of such abstract entities as linguistic types. What really drives Goodman's Nominalism is not this particular perplexity about set-theory, but a metaphysically based skepticism about abstract objects as such. Shortly afterwards, he characterizes his opponent as filling "his world with a host of ethereal, platonistic, pseudo entities" (Goodman, 1977, p. 26). So one can understand Putnam's charge that, despite Goodman's protestations to the contrary, "Nominalists must at heart be materialists...: otherwise their scruples are unintelligible" (Putnam, 1979, pp. 328–9).[5] Though one should add: *If* their scruples are metaphysical ones.

And what motivate many Nominalists are not metaphysical but epistemological scruples. Epistemological doubts about our capacity to refer to abstract objects can be traced back to the British empiricists and beyond.[6] Now, one can imagine someone's echoing Putnam and saying, "Nominalists must at heart be empiricists: otherwise their scruples are unintelligible"—and this line has been pushed by some. But the epistemological concerns that motivate Nominalism can at least be stated without empiricist presuppositions.

The problem, expounded in especially vivid terms by Paul Benacerraf (1973), is how we can have knowledge of, say, mathematical truths if the subject-matter of mathematics is the realm of abstract objects Platonists claim it to be. Benacerraf states his argument in terms of a broadly causal theory of knowledge, according to which a belief that p will count as knowledge only if it has been caused, in some appropriate way, by the fact that p. If this is the right theory of knowledge, it is easy to see why one might be worried about our capacity to know mathematical truths. Mathematical facts are, after all, unlikely to stand in causal relationships with anything. But Benacerraf's worry should not be construed as depending upon acceptance of such a causal theory; it is more subtle.[7] The difficulty is not so much that we have a completed epistemology into which mathematical knowledge has no hope of fitting; the difficulty is that we do not seem to have much of a theory of mathematical knowledge

---

to show that *there is* a singleton function. This will not be easy, especially for one trying to do away with any prior appeal to sets: Every object has a singleton, so there are as many singletons as there are objects. In the end, Lewis is forced to rest his case upon an Indispensibility Argument.

[5]Dummett makes similar remarks in his paper "Nominalism" (Dummett, 1978a, p. 43).

[6]For example, in Book I, Part I, §vii of the *Treatise*, Hume raises the question, with a nod to Berkeley, whether "abstract or general ideas... be general or particular in the mind's conception of them" and argues that we really have no abstract ideas.

[7]This is a good thing, since, as argued in footnote 9, the argument would otherwise be fallacious. Moreover, as has often been pointed out, such an argument would overshoot its target, posing no special problem for Platonists, since it would cast doubt upon our capacity to know necessary *a priori* truths of any kind (Wright, 1983, §xi). Still, as will be argued in the next paragraph, a theory of our knowledge of the truths of logic—or, more precisely, of the capacity of valid inference to preserve knowledge—will not resolve all the problems that arise here.

at all, and it is far from obvious how any can be given which will connect that knowledge *in some way or other* with the facts of mathematics as the Platonist conceives them.

To put the point another way, we have a good enough idea how we come to know some of the facts of mathematics: We prove them. And, if we had a decent account of how logical inference leads from old knowledge to new knowledge, we could explain how proof leads from knowledge of axioms to knowledge of theorems. What we lack, however, is an explanation of how we come to know the axioms, be these the axioms of some developed mathematical theory or those propositions which are, in a less developed theory's present state, typically assumed without proof. More precisely, it is not obvious why there should be any relation at all between our belief that the axioms are true and the facts of mathematics as the Platonist conceives them, why our beliefs should reliably reflect how things stand with the sets, or the numbers, or whatever. The point of Benacerraf's mention of the causal theory of knowledge is, I think, just to suggest that, in the case of our knowledge of truths about the physical world, we are not at quite such a loss.[8]

There is a more general problem concerning how we can so much as have thoughts about abstract objects, let alone know any such thoughts to be true. How, exactly, do we succeed in making reference to such things as the number three, or the word "three", or the book *Foundations of Arithmetic*? Granted, a parallel question can be raised about such apparently concrete objects as George Clinton and the tree outside my window. But, in recent times, as causation has assumed a central role in the theory of reference to physical objects, it has come to seem that the sort of theory by means of which we might explain our capacity to refer to concrete objects will not account for our capacity to refer to abstract objects. And that leaves us not only without a theory of reference to abstract objects, but without so much as a model for one.[9]

There are thus genuine puzzles about our capacity to refer to, and to have knowledge about, abstract objects. The puzzles arise naturally when one reflects upon the fact that abstract objects are causally isolated from

---

[8]Lewis seems to me to miss this point in his discussion of Benacerraf's argument and its bearing on modal realism (Lewis, 1986, pp. 108–13). I think Burgess and Rosen (1997, §I.A.2) also underestimate the force of this worry.

[9]One should not, however, suppose that this argument shows that reference to abstract objects is impossible. One might be tempted to argue as follows: The relation of reference is a causal one, but abstract objects are causally isolated from us; hence, reference to them must be impossible. But this argument rests upon a fallacy of hasty generalization. Observations, right or wrong, about the role causality plays in reference to physical objects cannot support claims about the *general* character of the notion of reference. If one's theory of reference were informed, from the outset, by examples that included cases in which people referred to mathematical objects, a 'causal theory of reference' that implied that we never refer to mathematical objects would not be the most likely result. Similar remarks apply to arguments, based upon causal theories of knowledge, which purport to show that knowledge of truths about abstract objects is impossible. For discussion, see Lewis (1986, pp. 108–10), Burgess and Rosen (1997, pp. 35–7), and Wright (1983, §xii).

us, but they do not depend upon any specific epistemological doctrine. Moreover, the puzzles concern abstract objects of any sort, though they are especially worrying in the case of mathematical objects.

In the case of abstract objects of other kinds, it is somewhat easier to see how these concerns might be addressed. Some abstract objects, such as word- and letter-types, are instantiated by other objects, in this case word- and letter-tokens; let us call such 'instantiated' objects—however exactly this notion is to be explained—*types*. In some cases, types will be instantiated by tokens that are themselves abstract; in other cases, the tokens will not be abstract but concrete, as is arguably the case with word- and letter-tokens. Following Charles Parsons, we shall call types whose tokens are concrete *quasi-concrete*; types whose tokens are themselves quasi-concrete are also quasi-concrete. Abstract objects that are not quasi-concrete we may, following Dummett, call *purely abstract*.

Now, what does the distinction between the quasi-concrete and the purely abstract have to do with epistemological concerns about Platonism? In "Types and Tokens in Linguistic Theory", Sylvain Bromberger considers a puzzle about linguistics that is closely related to Benacerraf's puzzle about mathematics: How can linguistics both be concerned with properties of, say, sentence-types and also be an empirical discipline, one whose data are 'observations' of actual empirical events, such as utterances (Bromberger, 1992b, pp. 170–1)? His answer, very roughly, is that types are abstractions from tokens, the abstraction made in such a way that the tokens instantiating a given type form (what he calls) a quasi-natural kind: Tokens that form a quasi-natural kind not only do, but nomologically must, share certain of their properties. And properties of the types are, according to Bromberger, related in specifiable ways to these necessarily shared properties of their tokens.[10] Determining what properties a particular sentence-type has therefore requires determining (i) what properties all tokens of any sentence must share and (ii) which of these properties tokens of the given sentence actually have.

Were all of this right, we would have an outline of an answer to the questions with which Bromberger begins: What linguists observe are, indeed, actual utterances and the like; such observations yield data about the properties of sentence-types because properties of sentence-types are related, in definite ways, to properties of sentence-tokens. More generally, since quasi-concrete objects have a determinate connection with the physical world—and so are not entirely divorced from the causal stream—one might, in the case of quasi-concrete objects, not only be able to defuse the sort of epistemological worry we have been discussing, but even recon-

---

[10]Thus, for example, the answer to the question how many word-tokens a sentence contains cannot vary from one token of it to another, though the answer to the question how loudly it was uttered may. So it makes sense to think of the sentence itself as containing a certain number of words, but not as having some property corresponding to the loudness of utterances of it.

cile knowledge of facts about quasi-concrete objects with a broadly causal epistemology.

Bromberger's proposed resolution of the epistemological difficulty fits well with a proposal, originating with Frege, concerning how we are able to make reference to abstract objects. Frege's central insight was that the problem is insoluble so long as we insist upon trying to explain directly how we come into cognitive contact with abstract objects. What we ought to do instead is focus our attention on complete judgments we make about them. Rather than attempt to say directly how we are able to make reference to, e.g., *Ulysses*, we should begin by considering the meanings of such complete sentences as "*Ulysses* contains the word 'dog'". Frege's thought, embodied in his so-called context principle, is that, if we can explain the meanings of all sentences in which the word "*Ulysses*" occurs, there will be no further question what the meaning of that word is. If we can explain how we are able to understand the sentence "*Ulysses* contains the word 'dog'" as having the truth-condition that a certain object, *Ulysses*, should contain the word "dog", there will be no further question how we are able to make reference to *Ulysses*: We can refer to *Ulysses* simply by using that sentence with the meaning it has in our language.[11]

The question, then, is whether we can explain the meanings of such sentences in such a way that statements of their truth-conditions will involve reference to abstract objects. Frege's idea was that our understanding of terms referring to shapes (say) consists, at least in part, in our understanding of the conditions under which two figures have the same shape. What is needed is thus an account of the truth-conditions of sentences of the form "the shape of a is the same as the shape of b". The obvious way to give this account is to state a 'criterion of identity' for shapes, that is, to find an equivalence relation $R\xi\eta$ that satisfies:

$$\mathrm{shape}(a) = \mathrm{shape}(b) \equiv Rab$$

Predicates suitable for use with terms of the form "shape(a)" may then be explained as follows:

$$F(\mathrm{shape}(a)) \equiv f(a)$$

Here, "$f\xi$" is required to be a congruence with respect to $R\xi\eta$: That is, $fx \wedge Rxy \to fy$; or, equivalently, since $R\xi\eta$ is an equivalence relation: $Rxy \to fx \equiv fy$.

It might well appear, however, that Frege's proposal is a non-starter, for it is not at all obvious how it differs from a similar proposal that might be made by a Nominalist. The project was to explain how we are able to understand sentences containing names of shapes as having truth-conditions whose statement makes reference to shapes. But what we

---

[11] My discussion here will be framed in terms of language, but I believe that substantial parts of it could be transposed to the context of a theory of thought. As Dummett has pointed out, for example, the context principle itself can be framed at that level.

seem to have been given is a statement of the truth-conditions of such sentences which precisely do not make reference to such objects, but which make reference only to 'figures', to objects which have shapes.[12] To dramatize the problem: Wouldn't this sort of account, if developed in the case Bromberger was discussing, all but presuppose that reference to names of linguistic types could be eliminated from linguistic theory?[13] Indeed, I shall argue that it may. But this does not, by itself, imply that there are no such things as linguistic types. That conclusion will follow only if we can establish a crucial, auxiliary premise: If names of linguistic types can be eliminated from linguistic theory, there are no such objects. I shall argue, in the next section, that this premise can be resisted.

8.2 Taking Reductionism Seriously

As just suggested, the sort of elimination required by Syntactic Reductionism may be modeled on Frege's own proposal. Suppose that we have a first-order language containing terms purporting to refer to types. It is convenient to assume that these terms are constructed, one and all, from a single function-symbol "$\phi(\xi)$", the argument-place being filled by a term that purports to (and, we may assume, does) refer to an object of some other sort.[14] (The abstract terms might be of the form "the direction of (the line) l" or "the type of (the word-inscription) w". I shall use the latter as my stock example.) Thus, the language contains two sorts of terms— which, for ease of exposition, I assume to be of distinct logical types. There are 'basic' terms, "a", "b", and so on, and 'secondary' terms, "$\phi(a)$", "$\phi(b)$", etc. (and corresponding variables, which I shall write "x" and "x", respectively). The language also contains two sorts of predicate letters: Basic predicates "$f\xi$", "$g\xi$", etc., and secondary predicates "$F\xi$", "$G\xi$", etc.; there will be similar distinctions among n-place predicate-letters. We say that basic terms purport to refer to basic objects; secondary terms refer to secondary objects.

To eliminate secondary terms from this language, there are just two things we must do. First, we must find an equivalence relation that holds between two basic objects just in case the corresponding secondary objects are identical, that is, a (basic) relational expression "$R\xi\eta$" that satisfies:

---

[12]This way of conceiving the problem is central to Dummett's reflections on the context principle (Dummett, 1981a, pp. 498ff; 1978b, pp. xlii–xliv), and much of what follows can be read as a response to his worry that, if names of abstract objects are so explained, no central role is assigned to the notion of reference to abstract objects.

[13]I owe a large debt to Jim Higginbotham for helping me to see the force of this worry— and so the need, and logical space, for the arguments to be given. I should also note that Bromberger's own view has changed a bit (Bromberger and Halle, 1992).

[14]I will be making a large number of concessions here to the Syntactic Reductionist. Some of them are independently motivated, as I shall argue below; some, like this one, are simply for convenience. My purpose is to show that the sort of reduction envisaged can *never* be carried out; making these sorts of concessions thus is part of making that argument.

(1) $\phi(a) = \phi(b) \equiv Rab$

Thus, in the case of word-types, what is required is that we should find a relational expression "$R\xi\eta$" that is satisfied by word-tokens a and b just in case the type of a is the same as the type of b. We shall call such principles as (1) *abstraction principles*. Given an abstraction principle, the elimination then depends only upon our finding, for each secondary predicate, a basic predicate that is true of a given basic object x just in case the secondary predicate is true of the secondary object $\phi(x)$. What is needed is thus, for each secondary predicate "$F\xi$", a (basic) predicate "$f\xi$" that satisfies:[15]

(2) $F(\phi(a)) \equiv fa$

Thus, in the case of word-types, we must find a predicate "$f\xi$" that is true of just those tokens whose types, say, contain exactly five letters. First-order quantifiers ranging over secondary objects can then be handled by means of the following:[16]

(3) $\forall \mathbf{x} A(\mathbf{x}) \equiv \forall x A(\phi(x))$

(4) $\exists \mathbf{x} A(\mathbf{x}) \equiv \exists x A(\phi(x))$

Taken together, (1)–(4) enable us to find an equivalent of any first-order formula containing secondary terms. (The proof is by induction on the complexity of formulae and is straightforward.)

Wright objects to this attempt to eliminate names of types on a number of grounds. Why, Wright asks, should the possibility of such a reduction show that expressions of the form "$\phi(t)$" are not 'genuine singular terms'? And he goes on to argue that only an overly stringent, and questionably coherent, conception of what it is for an expression to be a 'genuine singular term' could underwrite this conclusion (Wright, 1983, pp. 29–36). But to this objection, there is a natural reply for the Syntactic Reductionist to make. If the reduction is carried out as suggested above, then terms of the form "$\phi(t)$" are not even taken seriously as semantically significant units. Perhaps there is a compelling objection, similar in spirit to that Wright presses, against that sort of treatment. But the idea that what *look* like singular terms might prove, on analysis, to be of some other semantic category, is one familiar from the theory of descriptions:

---

[15]As mentioned above, "$f\xi$" must be a congruence with respect to $R\xi\eta$. We shall also need principles like (2) for the various sorts of relational expressions in which names of types occur. But we shall ignore this fact, since no further issues arise with these, so far as I can tell.

[16]These stipulations will work, of course, only if there is, for any type one likes, a corresponding token. Such an objection can be pressed against Nominalistic construals of, e.g., linguistic types, but there are natural ways of trying to answer it, say, by treating words in general as strings of letter-tokens. But similar worries can be pressed against the Fregean, too (Peacocke, 1992, ch. 4). Hale and Wright (2001a, pp. 422–3) have some ideas about how they might be answered.

No defensible conception of what 'genuine singular terms' are should legislate against that theory, since it is true (or, at least, can be motivated quite independently of any reductionist aspirations). If, then, the reduction can be carried out in a more Russellian spirit—if there is a way of carrying out the reduction that preserves the idea that expressions of the form "$\phi(t)$" are semantically significant units, while nonetheless construing them as quantifiers rather than singular terms—then the sort of objection Wright presses will itself arguably rest upon a conception of what it is to be a 'genuine singular term' that is overly stringent.

Intuitively, one might think, a Syntactic Reductionist should construe a statement about a type as really being a statement about its tokens: Goodman suggests, for example, that "The word 'Paris' contains five letters" should be construed as meaning: Every "Paris"-inscription contains five letter-inscriptions. This proposal does indeed treat the expression "The word 'Paris'" as a quantifier, namely, as the restricted quantifier: Every "Paris"-inscription. More generally, the idea is to treat expressions of the form "$\phi(t)$" as restricted quantifiers, namely, as the restricted quantifiers: $\forall x(Rax \to \ldots x \ldots)$. On this analysis, the 'logical forms' of identity statements and simple sentences are therefore no longer given by (1) and (2), but by:

(1') $\phi(a) = \phi(b) \equiv \forall x(Rax \to \forall y(Rby \to Rxy))$

(2') $F(\phi(a)) \equiv \forall x(Rax \to fx)$

It is easy to see that these are equivalent to (1) and (2). It is hard to see what general objection along the lines suggested by Wright might be lodged against this analysis that would not equally threaten Russell's theory of descriptions. Wright's objection to Syntactic Reductionism thus will not do.

Suppose, then, that we propose to eliminate names of types in this new way. We shall still need, and it will still suffice, to find an equivalence relation that holds between tokens just in case they are of the same type and to find, for each predicate true or false of types, a corresponding predicate true of just those tokens of whose types the original predicate is true. That there is such an equivalence relation and that there are such concepts is obvious. Of course there is a relation that holds between all and only those objects which are tokens of the same type, namely: ξ is a token of the same type as η. Of course, for any given concept under which types do or do not fall, there is a concept under which fall all and only those tokens whose types fall under that concept, namely: ξ is a token whose type falls under that concept. For example, all and only those word-tokens whose types contain exactly five letters fall under the concept: ξ is a token of a type that contains exactly five letters.

It is not, however, open to a Syntactic Reductionist to refer to this relation and to these concepts in these ways: She cannot be proposing to make use of the relational expression "ξ is of the same type as η" in her 'elimination' of reference to types.[17] And the classical objection to Nominalism is that it does not seem to be possible to define this relation in terms that do not make implicit or explicit mention of types. Recall, for example, Goodman's suggestion that "The word 'Paris' consists of five letters" be recast as "Every 'Paris'-inscription consists of five letter-inscriptions". One immediately wants to know how the concept of a "Paris"-inscription is supposed to be explained without reference to types. More generally, one wants to know how Goodman will explain the relational expression "ξ is a replica of η", which he intends as a replacement for "ξ is of the same type as η". Attention will then focus on questions of circularity in his explanations, on whether some tacit reference is being made to types. The traditional objection to Goodman's proposal, and to Nominalistic treatments of types more generally, is precisely that such circularity is unavoidable.

I think this objection is at best beside the point and at worst confused. For one thing, it leaves one without any secure argument for Platonism in any case, since the charge of circularity is always open to rebuttal by a sufficiently clever would-be Nominalist. Worse, there is a straightforward answer to the objection in some cases. Most importantly for our purposes, however, the traditional objection simply cannot be pressed by anyone attracted to a neo-Fregean explanation of our capacity to refer to abstract objects. Since it is precisely such a view that I should like to defend, I in particular cannot rest with the traditional objection.

Let me explain.

As we saw above, Frege's idea was to explain our capacity to refer to abstract objects, in the first instance, by stating a 'criterion of identity' for them. Thus, in the case of shapes, Frege offers an abstraction principle of the form:

$$\text{shape}(a) = \text{shape}(b) \equiv Rab$$

If this sort of story is to be of any help at all, it had better be possible to come to understand the expression that replaces "$R\xi\eta$" here without antecedently being able to refer to shapes. And, indeed, it would certainly appear that, even if one cannot refer to shapes, one can yet understand a relational expression which will do the trick, namely, "ξ is congruent with η". Thus, if the traditional objection to Nominalism were the only one available, a Platonist view of shapes would be untenable (Dummett, 1978b, p. xliv).

Frege describes this part of his proposal like this:[18]

In our present case, we have to define the sense of the proposition "the shape of figure a is the same as that of figure b"; that is to say, we must reproduce the

---

[17]Note that the point is not that the existence of the relation and concepts depend upon their being describable in a certain way, but only that such an expression cannot be employed in the reduction.

[18]I've changed the example.

content of this proposition in other terms, avoiding the use of the expression "the shape of figure a". (Gl, §62)

In so far as this suggests that we must produce a *definition* of such identity statements in other terms, Frege has succumbed to a natural temptation to over-state the point. What we need to do in the case of linguistic types, say, is not to define "the type of token a is the same as that of token b" in terms that do not make reference to types; that may or may not be possible, depending upon the expressive resources of our language and other matters not obviously relevant, and no plausible development of the Fregean strategy should require us to do so. Our actual grasp of the criterion of identity for linguistic types is acquired, after all, not by its being defined for us in simpler terms, but by the direct method, by immersion. What we need to do is to explain in what a grasp of this criterion of identity consists, without presupposing an antecedent capacity to refer to types. Or, to put the point less technically: What we need to see is, not that it is possible to *define* Goodman's "ξ is a replica of η" without mentioning types, but that one who cannot yet refer to types can come to *understand* it.

But this means that the Fregean will have to claim either that the criterion of identity for types can be defined in simpler terms or that a grasp of that criterion can be acquired by immersion. It is at best unclear on what ground any argument the Fregean might give for one of these claims cannot just be taken over by the Syntactic Reductionist. Why, she will ask, can an understanding of Goodman's "ξ is a replica of η" not be acquired by immersion? One might perhaps want to argue that a full understanding of this expression must be coeval with a capacity to refer to types; but that argument would have to be given, and no such argument is available in the case of shapes.[19] All we can say at present is that Syntactic Reductionists and neo-Fregean Platonists have something of a common cause: Both positions require that we should explain the criteria of identity for various sorts of abstract objects without presupposing an antecedent capacity to refer to them. For a Nominalist, this is because she wishes to avoid all reference to them; for a Platonist, it is because she seeks thus to explain the genesis, and nature, of reference to them.

The conclusion for which I have just argued can be illustrated as follows. Even when Nominalism has been on the ropes, many philosophers have still paid a kind of lip-service, not just to its aspirations, but to its conception of how it might have established itself. Such philosophers reject Nominalism about, say, linguistic types because they believe there is

---

[19]On this issue, compare David Wiggins's view in *Sameness and Substance* (Wiggins, 1980, ch. 2). Wiggins holds that a grasp of the criterion of identity for, say, names of organisms is coeval with a capacity to make reference to them. I am sympathetic. Indeed, I think something similar might well be true in the case of linguistic types. But Wiggins gives no general reason to suppose that this must always be so, and, as said, it is far from obvious that it is so in all the cases with which we are concerned.

no Nominalistically acceptable definition of the relation "ξ is of the same type as η". But if this is why Nominalism has been rejected, it has been rejected for reasons that are utterly unsatisfactory, if not badly confused, and for reasons that do not apply in the case of shapes. For one thing, no Nominalist ought ever to have acceded to the demand that she provide a definition of "ξ is of the same type as η", or any other such equivalence relation, in simpler terms. And once she has declined that invitation, there is no obvious reason why she may not simply help herself to any account of our understanding of it that a Fregean Platonist might offer.

It thus seems to me that, if Syntactic Reductionists are allowed the claim that the eliminability of names of abstract objects is sufficient reason to believe that there are no such things, they are being allowed to stack the deck. The eliminability of reference to abstract objects from a first-order language is all but guaranteed by assumptions central to the sort of account suggested by Frege. I therefore assume, for methodological reasons if no others, that there are no practical obstacles to this kind of elimination. That is not to say that it is not an important question how, in particular cases, criteria of identity are to be explained. It is only to say that the possibility of such an explanation should not be at issue between Nominalists and neo-Fregean Platonists.

Actually, one should have been skeptical all along about the claim that Platonists should need to deny the eliminability of reference to abstract objects. That reference to abstract objects should be eliminable from linguistic, physical, and other theories is a necessary condition for the truth of (non-Fictionalist versions of) Nominalism; its being a sufficient condition is another matter. Somehow, Nominalists have managed to sneak a transformation of the one claim into the other past their opponents. But if the issue is whether there are such things as types, the question should not be whether one must refer to them for some purpose or other, however favored or important, but whether one *can* refer to them.

I have, in my attempt to understand Nominalism, been emphasizing the alleged peculiarity of abstract objects and the epistemological difficulties they pose. But one should not forget the utterly natural way in which we speak of them. Ordinary language does not make a distinction between names that purport to refer to abstract objects and those that refer to concrete ones, and ordinary speakers do not hesitate to describe themselves as having dealings with all sorts of abstract objects: As reading books, spelling words, understanding sentences, adding numbers, playing games, identifying shapes, and the like. No philosophical view ever has the unequivocal support of 'the common man', but I dare say that Nominalism has no more claim to his allegiance than Platonism. So if we can provide a decent resolution of the philosophical problems connected with abstract objects—if, in particular, we can explain how we are able to refer to them—I see no reason why we should be troubled if reference to them can, in the relevant sense, be eliminated.

All of these moves, however, are defensive: In effect, I have argued only that the Fregean strategy does not give the game away to the Nominalist. I shall turn to the offensive in the next section, arguing that reference to abstract objects cannot be eliminated in the way envisaged by a Syntactic Reductionist, that is, that it is not possible to eliminate reference to abstract objects by eliminating their names. This does not imply that reference to abstract objects cannot be eliminated at all: It does imply that what appear to be names purporting to refer to abstract objects really are names. If so, then, if it is not to find itself denying the truth of much of what we ordinarily believe, Nominalism must concede that those terms refer and so take the form of *Semantic* Reductionism.

8.3   The Ineliminability of Names of Abstract Objects

According to Quine (1953), the objects to which a theory is 'ontologically committed' are those over which its first-order variables range.[20] Let us consider the question, then, how the domains of theories that eliminate names of types differ from the domains of theories that do not. If there is a significant difference between the domains of these theories—a difference that can be described without begging any questions against the Syntactic Reductionist—we may be able to exploit it to show that names of types cannot be eliminated in favor of names of their tokens (and similarly for quantification). One difference, of course, is that one theory's domain consists of tokens, the other's of types—but this is hardly a difference of which we can make use in an argument against Syntactic Reductionism. But there is also a difference in the cardinality of the respective domains. The theory offered by the Syntactic Reductionist takes quantifiers that purport to range over, say, letter-types actually to range over letter-tokens. And there are more letter-tokens than there are letter-types. Indeed, quite generally, a theory that eliminates names of types in favor of names of tokens treats quantifiers purporting to range over types as ranging over tokens. And there are, in general, more tokens than there are corresponding types.[21]

---

[20]One might have been wanting to object, for some time, that the eliminability of names of abstract objects cannot be the central problem, since one can eliminate all names in favor of predicates ("Socrates" in favor of "ξ socratizes") and a description operator, itself eliminable by means of the theory of descriptions. This misses the point of the argument, which can as well be stated in terms of questions about the nature of the domain over which we must take quantifiers purporting to range over types to range. See below.

[21]One might object that, in the case of sentence-types, there are more types than tokens, since there are infinitely many English sentences but only finitely many tokens. But we have agreed to set aside the problem uninstantiated types pose for Nominalism. And what we need is not just an analysis of "Most sentences contain the word 'the'", but also of "Most six word sentences contain the word 'the'", and the like. In such cases, there are more tokens than types, and this is the situation in general.

There is thus a difference between the domains over which the competing theories take the quantifiers to range: The theory offered by the Syntactic Reductionist treats the domain as containing as many objects as there are word-tokens; the other, as containing as many objects as there are word-types. This is not a tendentious statement of the difference: The Syntactic Reductionist has to be able to make sense of such sentences as "There are more word-tokens than there are word-types", this being a perfectly good sentence of English. Now, the mere fact that there is such a difference between the competing theories is no objection to Syntactic Reductionism. But such an objection will be forthcoming if there are sentences whose truth-conditions are sensitive to the cardinality of the domain in an appropriate way. Sentences that assert something about the number of word-types having a certain property are the obvious candidates.

Sentences such as "There are at least three word-types that are Short" pose no problem for the Syntactic Reductionist, since sentences of the form "There are at least n Fs that are G" can be translated into first-order logic, whence (1'), (2'), and (4) can be used to eliminate reference to, and quantification over, word-types.[22] For this reason, quantifiers that are 'first-orderizable' (i.e., that can be defined in first-order logic) will not be problematic. But natural languages contain quantifiers that are not first-orderizable. Indeed, the word "most", used in the preceding sentence, is such a quantifier, as are "few", "just as many as", and "more" (Barwise and Cooper, 1981).[23] How, then, is a Syntactic Reductionist to represent sentences containing such expressions, such as "Most word-types are Short"? The question here is not how the quantifiers themselves are to be analyzed (i.e., how we are to analyze "Most word-types are Short" as opposed to "Few word-types are Short"), but how their argument-places are to be treated.[24] So I assume that the language into which these sentences are to be translated contains a primitive, binary quantifier "Most", using which "Most Fs are G" can be represented as: $\text{Most}_x(Fx; Gx)$.

---

[22]It is worth working through this sort of example in detail, since doing so reveals certain peculiarities about the way variables are treated by the Syntactic Reductionist.

[23]It is often possible to prove that a quantifier is non-first-orderizable by appealing to the existence of non-standard models of first-order arithmetic. Thus, in the standard model, there are more numbers less than or equal to any natural number n than there are less than n. However, if n is a non-standard number, there will be (Dedekind) infinitely many numbers less than or equal to n (in particular, all standard numbers will be less than or equal to n), so there will *not* be more numbers less than or equal to n than there are less than n. So, using "$\text{More}_x(\Phi x; \Psi x)$" to mean: There are more Φs than Ψs, the sentence

$$\forall n(\text{More}_x(x \leq n, x < n)$$

will be true only in the standard model. Hence, "More" is not first-orderizable. Similar arguments work for the other quantifiers mentioned in the text.

[24]One can cook up analyses that mix the treatment of the quantifiers with that of the argument-places, thus obscuring how the analysis does treat the argument-places. But one need only compare the analysis offered of "Most word-types that are F are G" with that offered of "Few word-types that are F are G" to disentangle its parts.

Now, it is clear that the following sort of analysis of "Most word-types are Short" will not do:

(5) $\text{Most}_x(\text{word-token}(x); \text{short}(x))$

According to this analysis, most word-types are Short if, and only if, most word-tokens are short. But it might well be that, although most word-types are not Short, there are more tokens of Short words than of Long ones, so that most word-tokens are short.

Suppose, however, that we were to choose, for each type, one and only one of its tokens to serve as a 'representative' of the type. Then, certainly, if most of these representative tokens are short, most Word-types are Short, and conversely, and we can analyze "Most word-types are Short" as "Most representative tokens are short", if we can find a way to capture the notion of a representative token in Nominalistically acceptable terms: We need to find some way of restricting the quantification on the right-hand side of (5) to a collection of representative tokens. One way of doing this is to assume the existence of a function $\rho(\xi)$ satisfying:

(6) $\forall x(Rx\rho(x) \wedge \forall y(Rxy \rightarrow \rho(x) = \rho(y))$

This should be thought of as an axiom of the theory into which the Syntactic Reductionist intends to translate talk of types: It characterizes $\rho(\xi)$ as a representation function. Any such function will have as its range a collection of representative tokens, for the range of $\rho(\xi)$ will contain one and only one token of each type.

Now, "$\exists y(x = \rho(y))$" is true of x just in case x is in the range of $\rho(\xi)$, i.e., just in case x is a representative token. So we may analyze "Most word-types are Short" as follows:

(7) $\text{Most}_x(\exists y(x = \rho(y)); \text{short}(x))$

Thus, "Most word-types are Short" means: Most representative tokens are short. Restricted quantifiers, such as appear in "Most Word-types that are F are G", can be handled as follows:

(8) $\text{Most}_x(\exists y(x = \rho(y)) \wedge fx; gx)$

A quite general method is thus available by means of which the Syntactic Reductionist can interpret sentences involving quantifiers like "Most".

Indeed, once a function-symbol such as "$\rho(\xi)$" has been admitted, there is no reason we should not make use of it in simpler contexts. Doing so will simplify the Syntactic Reductionist's theory considerably. In place of (1′) and (2′), we may use instead:

(1*) $\textbf{Type}(a) = \textbf{Type}(b) \equiv \rho(a) = \rho(b)$

(2*) $F(\textbf{Type}(a)) \equiv f(\rho(a))$

The clauses for the first-order quantifiers can then be recast so that they mirror (7):

(3*) $\forall \mathbf{x} A(\mathbf{x}) \equiv \forall x (\exists y (x = \rho(y) \rightarrow A(x)))$

(4*) $\exists \mathbf{x} A(\mathbf{x}) \equiv \exists x (\exists y (x = \rho(y) \wedge A(x)))$

Thus, "Every Type is Ψ" will be treated as meaning: Every representative is ψ.

The difficulty is not that these various analyses do not work. The difficulty is connected with the logical forms they ascribe to the statements analyzed. One cannot but notice how similar the analysis given in (7) is to this one:

(9) $\text{Most}_x (\exists y (x = \text{Type}(y); \text{Short}(x))$

This is the natural, Platonistic analysis of "Most word-types are Short", since "$\exists y (x = \text{Type}(y))$" may be read: x is a word-type. Note, too, the similarity between the logical forms a Platonist would assign to "The word-type of b is Short" and the one a Syntactic Reductionist would assign in accord with (2*):

$$\text{Short}(\text{Type}(b))$$
$$\text{short}(\rho(b))$$

To analyze simple sentences purporting to refer to word-types by means of (2*) is thus to abandon the attempt to eliminate names of word-types, expressions of the form "Type(b)". To replace "Type" by "ρ" is not to eliminate those names, but merely to re-write them. Reference to word-types therefore cannot be eliminated by eliminating names of word-types.

One might want to object that I have not shown that the Syntactic Reductionist must make use of a representation function in giving the truth-conditions of "Most types are F". And, in a sense, that is true. One could, for example, instead make use of the relational expression "η is a representative token for ξ". But functional expressions can always be eliminated in favor of relational ones: The point has nothing to do with the controversy between Nominalism and Platonism. More interestingly, one could make use of a predicate "$T\xi$", read: ξ is a representative token, and treat "Most Types that are F are G" as meaning:

$$\text{Most}_x (\text{T}(x) \wedge fx; gx)$$

And similarly, if one were prepared to accept a certain amount of set-theory, one could read "$T\xi$" as meaning: ξ is an equivalence class under $R\xi\eta$, and then analyze the sentence as meaning: Most equivalence classes under $R\xi\eta$ that contain a token that is $f\xi$ contain only tokens that are $g\xi$. Or, more in the spirit of Goodman, one could take "$T\xi$" to mean: ξ is a fusion of all and only tokens bearing $R\xi\eta$ to one another, and then analyze the crucial sentence thus: Most things satisfying Tξ all of whose token-like parts are $f\xi$ are such that all of their token-like parts are $g\xi$.

Does this show that one need not make use of a representation function? It does not, for there is a common thread to all these proposals, and that common thread is precisely the use of a represenation function. The problem is to produce an analysis of sentences of the form "Most types are *G*". To analyze this sentence, we need to specify a domain over which "Most types" is to range. The domain needs, in a strong sense, to be iso-morphic to that of the types themselves: The new domain must be of the same cardinality as the original domain,[25] and it must contain, for each type, an object that represents it, in the sense that that object's satisfy-ing some predicate appropriately related to "*Gξ*"—be this "*gξ*", or "ξ is such that its token-like parts are all *gξ*"—is equivalent to the type itself's satisfying "*Gξ*". Only if these two conditions are met will most of these objects' satisfying the corresponding predicate be equivalent to most of the types' satisfying "*Gξ*".[26]

Moreover, in order to use this new domain in an analysis of "Most types that are *F* are *G*", there needs to be a general method for going from a collection of tokens satisfying "*fξ*" to a collection of representa-tives satisfying it, so that most of those objects' satisfying the predicate corresponding to "*Gξ*"—that is, "*gξ*"—will be equivalent to most of the types' satisfying "*Gξ*". As a special case of this, there needs to be a gen-eral method for getting from any particular token to its representative (and, indeed, one would expect this to be used in getting us from the to-kens satisfying "*fξ*" to a suitable collection of representatives). But that means that any such analysis will work only if, and precisely because, it makes use of a representation function, even if the analysis does not wear the fact that it makes use of such a function on its face.[27]

Consider now the conditions the representation function is required, in the case with which the idea was introduced, to satisfy, that is, the conditions (6) imposes upon $\rho(\xi)$:

1. For each word-token, the function has as value a representative word-token of which that token is a replica.

2. The function has the same value for tokens of the same word.

3. The function has different values for tokens of different words.

---

[25]Perversely, one could allow the domain to contain exactly two tokens of each type. But this will not do as an analysis, since it will work only if we know, presumably *a posteriori*, that there are at least two tokens of every type.

[26]The point here depends upon the fact that generalized quantifiers such as "Most" are under discussion: Similar remarks do not apply to the analysis of universal and existential quantification by means of (3) and (4).

[27]The argument could be cast in terms of the question how we are to give a semantic theory that would apply to sentences containing names purporting to denote types. I have not stated the argument in this way, lest I encourage misapprehension that it depends upon claims about the correct form for a semantic theory to take.

Compare the conditions the function Type(ξ) satisfies:

(i)   For each word-token, the function has as value the word-type of which that token is a token.

(ii) The function has the same value for tokens of the same word.

(iii) The function has different values for tokens of different words.

The analyses to which these conditions correspond agree about the logical forms of sentences containing names purporting to denote types. The disagreement concerns what the range of the function is, that is, *to what* expressions such as "Type(b)" refer, not *whether* they refer. The substitution of "$\rho(\xi)$" for "Type(ξ)", pointless if one is attempting to eliminate expressions of the form "Type(b)", yet reflects a treatment of 'names of word-types' *as referring to word-tokens*, rather than to word-types.[28] That is to say: The substitution of "$\rho(\xi)$" for "Type(ξ)" suggests, in the syntax, an elimination of reference to abstract objects in the semantics. The view appropriate to an analysis in terms of (7), (2*), and the rest is that 'names of types' are referring expressions all right, but that they refer, not to types, but to representative tokens of those types. And, similarly, on other of the views we considered, names of types come out as denoting equivalence classes or certain sorts of fusions.[29]

To put the point in terms that echo Quinean views about ontological commitment: Explaining a class of terms by stating a 'criterion of identity' for them suffices to characterize a domain of quantification. It is because the 'criterion of identity' serves to introduce such a domain that the use of generalized quantifiers such as "Most" is so easily understood. I have not argued that this domain must contain the abstract objects the Platonist would like but rather that it must at least contain objects that serve to represent the types. Whatever this new domain might comprise, however, that is what quantifiers like "Most types"—and, presumably, "Some types", too—range over. To put the point in the 'material mode': Types just are what comprise that domain, and it is at best obscure why one should not just say that these objects are what terms of the form "Type(b)" denote, as well. The expressions explained by means of the 'criterion of identity' can, therefore, legitimately be treated as referring

---

[28] Such a view is suggested by Dummett (1981b, pp. 209–10).

[29] One might wonder whether a Syntactic Reductionist who was happy using higher-order logic could suggest that "Type(b)" does not refer to an object at all but instead to a second-level concept, namely, to: $\forall x(Rbx \rightarrow \ldots x \ldots)$. She could then construe "Most word-types are G" as making use of a third-order binary quantifier "Most". This position is, indeed, available: It is one of the varieties of Semantic Reductionism. And it is at this point that Wright's syntactic priority thesis enters the fray, since one way of arguing against this position would be to argue that "Type(b)" is a singular term and therefore must be taken to refer to an object, if to anything. I myself am unclear whether these arguments can work: It is not as if there aren't restricted quantifiers like the ones mentioned; such expressions in English would, like other quantifiers, have roughly the distribution of singular terms.

expressions, and any two such expressions will refer to the same object under just the conditions specified by the criterion of identity.

8.4 Where Do We Go From Here?

For a neo-Fregean, this certainly constitutes progess. But the arguments given above do not, by themselves, make a case for Platonism. Nothing in those arguments so much as suggests that names introduced by abstraction need be taken to refer to abstract objects: For all that has been said so far, they might just as well be supposed to refer to representative tokens or to certain sorts of fusions. A more sophisticated view might have it that, although "*Ulysses*" does not denote a particular one of its copies, it ambiguously denotes each of them: "*Ulysses* is Φ" would be true only if every copy of it were ϕ; false, if none were; and neither true nor false, otherwise.[30] And it would just beg the question to object, against any such view, that, according to it, names of word-types do not denote word-types. Word-types are whatever names of word-types denote. On these sorts of views, word-types just are tokens, or fusions, or what have you.

Nonetheless, I believe that the dispute between the Nominalist and the Platonist looks different now and that further progress is possible. That, however, is the topic of Chapter 9.

---

[30]This view was suggested to me by Hodes (1990).

9

The Existence (and Non-existence) of Abstract Objects

9.1 Two Problems

As George Boolos (1998f, pp. 128–9) once remarked, much of our ordinary discourse seems to involve reference to abstract objects. It's not just numbers and sets, though we do talk about them. We talk also of sentences: How many and what words they contain; how those words are spelled and pronounced; whether they were uttered on certain occasions. We talk of books, like *Die Grundlagen der Arithmetik*: We read them; talk about what sentences they contain; and argue about what is and is not said in them. One might almost be tempted to say that abstract objects are all around us, but for the fact that they aren't, since they aren't located in space.[1] It is this that gives rise to the ontological and epistemological problems that abstract objects pose. If abstract objects are not even spatial, they presumably cannot cause anything to happen. And for that reason, among others, we can have no perceptual contact with them. The notion of perception, however, seems to play a fundamental, grounding role in philosophical theories of reference to concrete, spatio-temporal objects, and in theories of knowledge about them. How, then, unless we follow Gödel (1990, p. 268) and suppose that we have "something like a perception" of sets and other abstract objects, can we explain our capacity to make reference to them or to have knowledge of them?[2] One does not have to commit a fallacy of hasty generalization,[3] supposing oneself to have proven that reference to abstracta is impossible, and that we can have no such knowledge, to find these questions pressing.

---

[1] At least, that's what one normally supposes. But, as David Lewis once warned, one should be careful not to slide from "we don't know where they are" to "we know they aren't anywhere".

[2] Charles Parsons's (1980) notion of 'quasi-concrete' objects might of course be helpful here, and it is, in many ways, I think, continuous with the Fregean ideas to be discussed below.

[3] It is a standard reply to arguments for nominalism that proceed from premises about causal conditions on knowledge that they commit this fallacy. These arguments of course derive, in the contemporary literature, from Paul Benacerraf's "Mathematical Truth" (Benacerraf, 1973). Versions of the reply can be found in Wright (1983, §xii), in Lewis (1986, §2.4), and in Burgess and Rosen (1997, pp. 35–7). For a recent defense of the claim that such concerns should still be taken seriously, see Øystein Linnebo's "Epistemological Challenges to Mathematical Platonism" (Linnebo, 2006).

I am attracted to the following sort of view, which has its origins in Frege (Gl, §§62ff). These problems will seem insoluble so long as we insist upon trying to explain directly, so to speak, how we come into cognitive contact with abstract objects. What we ought to do instead is to focus our attention on complete judgments we make about such objects. Rather than attempt to say directly how we are able to make reference to the number six, that is, we should begin by considering the meanings of complete sentences, such as "Six is not prime". Frege's thought, embodied in his famous context principle, is that, if we can explain the meanings of all sentences in which the word "six" occurs, there will be no further question what the meaning of that word is: If we can explain how we are able to understand sentences containing the word "six" as having the truth-conditions they do, there will be no further question how we are able to make reference to the number six.

Of course, simply saying that there will be no further question does not make it so. The problem facing Frege's interpreters, and those who would follow him here, is that of saying *how* explaining the meanings of sentences containing the word "six" answers the question how we are able to refer to six. This is, in part, because of the gap between meaning and reference. As Dummett (1981a, ch. 14) was the first (but hardly the last) to point out, the most natural reading of the context principle—which Frege states in terms of an undifferentiated notion of 'content'—makes it one about *sense*: The sense of a term is exhausted by the contribution it makes to determining the senses of complete sentences in which it occurs. Understood as a thesis about reference, the context principle is much less obvious. That, however, is the form in which we need it. And it is just far from obvious why the fact that "six" contributes in some regular way to the senses of sentences in which it occurs should even seem to imply that it has a reference of its own. Russell (1905) insisted, famously, that descriptions[4] have no meaning of their own, even though they do contribute regularly to the meanings of sentences in which they occur. Why should "six" be any different? Dummett pushes this sort of question quite hard.[5]

Frege's proposal is most easily developed in the sort of case with which he introduces it, the case of what we may call *types*: Abstract objects that

---

[4]And other quantificational phrases. This point seems to me often to be missed. But if one wants to understand what Russell means when he says that descriptions have no meaning of their own, then one ought first to focus on what he means when he says that "every man" has no meaning of its own, but makes a regular contribution to every sentence in which it occurs. Russell's point is the now familiar but then revolutionary one that quantificational phrases are not names, and they do not refer to strange entities called *variables*. Russell's enthusiasm for the point was no doubt connected with the fact that it cleaned up the mess that was his view of quantification in the *Principles* (Russell, 1903), on which (what we now call) quantificational phrases referred to a bewildering variety of variable entities.

[5]In some ways, the present chapter is my attempt to get at what was bothering Dummett, which is, I think, independent of the correctives issued in Wright's (1983, §x) reply.

are essentially *of* certain other objects, the types' *tokens*. Reference to such objects is to be thought of as made, most fundamentally, by means of expressions of the form "the direction of the line ℓ" or "the word-type of the inscription w". The meanings of sentences containing such expressions are then to be explained in terms of a so-called *abstraction principle*. Consider, for example, terms referring to such objects as *Die Grundlagen der Arithmetik*. I shall call these objects *editions*, since the English word "book" can refer either to tokens or to types. Thus, *Die Grundlagen* is an edition, and the physical copy of *Die Grundlagen* presently sitting on my desk is what I shall call a *book*.

Frege suggests that the first task is to explain the meanings of identity statements involving such expressions.[6] To do this, we must find an equivalence relation $R\xi\eta$ that holds between two books just in case they are copies of the same edition, whence the sentence "the edition of book a is the same as the edition of book b" may be explained as being true just in case book a bears this relation to book b. That is:

$$\mathbf{edn}(a) = \mathbf{edn}(b) \equiv Rab$$

Similarly, the meanings of sentences such as "The edition of book a is short" are to be explained by finding a predicate "$S\xi$" that is true of a book just in case its edition is short. Thus:

$$\mathbf{short}(\mathbf{edn}(a)) \equiv Sa$$

Similar clauses are needed for two- and more-place predicates.

A number of problems arise immediately. One might think, for example, that there is no guarantee that we will actually be able to find such an equivalence relation and such predicates. This is a problem, however, only if we require that they be definable in terms that do not involve the notion of an edition. But they need not be. That *there is* an equivalence relation that holds between any two books just in case they are copies of the same edition is obvious, namely: the edition of ξ is the same as the edition of η. Whether this can be defined in terms that do not involve editions is irrelevant.[7] What matters is whether one can come to *understand* an expression denoting this relation without antecedently being able to refer to editions. And one might do so by the direct method, by 'immersion', rather than by having it defined in simpler terms. That is not to

---

[6] Frege's own attitude towards the proposal is actually quite unclear, since he goes on to raise an objection to it that he seems to regard as conclusive, the Caesar objection (Gl, §66). Then again, Frege seems to emphasize the import of the basic idea in his summary of the book's results (Gl, §107).

[7] The older literature on nominalism takes this question to be central. That it isn't was made clear to me by Wright (1983, §v). In the present context, the issue is not so critical, since both the neo-Fregean and the nominalist need such an equivalence relation: the neo-Fregean, to explain names of editions; the nominalist, to eliminate them. See also Section 8.2.

say that there are no interesting problems that arise in connection with the explanation of the relevant equivalence relation, of the 'criterion of identity' for the objects in question: It is simply to say that the apparent difficulty, or even impossibility, of defining the relation in terms that do not involve editions does not itself count as an objection to the Fregean approach.

Still, there are problems of principle facing the neo-Fregean project. It is not, for one thing, obvious how it differs from certain sorts of reductionist programs. Neo-Fregeans take the availability of such abstraction principles to show us how reference to abstract objects is possible. But reductionists might equally take the availability of such principles to show that terms purporting to refer to such objects can be eliminated, since we seem to have been told how we could 'say the same things' we ordinarily say using sentences containing names of editions without using any such names. One might wonder, however, why the possibility of eliminating names of editions should even seem relevant. The most it can imply is that we 'need not' make reference to editions for some purpose or other. But the challenge, at least initially, was to explain how we *can* make reference to abstract objects, not why we must.

But, as I argue in Section 8.3, abstraction principles do not, in fact, enable us to eliminate expressions that purport to refer to editions. We have not yet considered how quantification over editions is to be handled. While first-order quantifiers do not pose a problem, there are non-first-order quantifiers—such as "most", "few", and the like—that do. Ultimately, what turns out to be requried is a functional expression "$\varphi(\xi)$" which, given a book as argument, returns as value a representative copy that bears $R\xi\eta$ to it.[8] Quantification over editions is then understood as quantification over representative copies.

So the attempt to eliminate apparent reference to editions fails. The formal properties of the functional expression "$\varphi(\xi)$" are all but indistinguishable from those of "edn(ξ)": Hence, terms of the form "$\varphi(a)$" are all but indistinguishable from those of the form "edn(a)". In fact, the only relevant difference between "$\varphi(\xi)$" and "edn(ξ)" is that the range of the former consists of representative copies; that of the latter, apparently, of editions. Similarly, the only relevant difference between "$\varphi(a)$" and "edn(a)" is that the former refers to a representative copy; the latter, apparently, to an edition. The disagreement therefore concerns not whether names of editions are referring expressions—nor, irrelevantly, how those names are to be spelled—but *to what* they refer.

The fact that it is impossible to eliminate expressions that purport to refer to editions does not, therefore, imply that it is not possible to eliminate *reference* to editions: One could hold that names of editions denote, not abstract objects, but representative books. Indeed, that may

---

[8]That is, ϕ must satisfy: $R(x, \phi(x))$, which implies: $\phi(x) = \phi(y) \equiv Rxy$, since R is an equivalence relation.

not even be the best way to put it. Surely editions are just whatever names of editions denote. If so, then, on this view, editions just *are* books. We are thus left, quite generally, with the question whether editions are books; whether word-types are their tokens; whether either is instead an equivalence class; or what have you. Following Dummett, I call this problem the *problem of trans-sortal identification.*[9]

There is another, related problem. Equivalence relations are not hard to come by. Let $Q\xi\eta$ be an equivalence relation, chosen completely at random: It might, for example, have as one of its equivalence classes the set containing each of my shoes, my daughter Isobel, Brown University, and some other things. We can now introduce names purporting to stand for objects of a certain sort, call them *duds*, just as we introduced names of editions:

$$\mathrm{dud}(a) = \mathrm{dud}(b) \equiv Qab$$

But are we really to believe that there are such objects as duds? I, at least, have a reasonably strong intuition that there are no such things. This problem I call the *proliferation problem.*[10] One might reply that the existence of duds follows trivially from the explanation given of names of duds: According to that explanation, $\mathrm{dud}(a) = \mathrm{dud}(b) \equiv Qab$; but $Q\xi\eta$ is an equivalence relation, so certainly Qaa;[11] so, trivially, $\mathrm{dud}(a) = \mathrm{dud}(a)$ and so by existential generalization, $\exists x(x = \mathrm{dud}(a))$. Moreover, the argument recently rehearsed, that expressions introduced by abstraction must be construed as referring expressions, applies here. So it is hard to see how we can avoid the claim that duds exist. And therefore, the reply might conclude, whatever intuition we might have that there are no such objects as duds ought just to be abandoned in the light of theory.

My worry, however, is that this seems to make the notion of an object, and the conception of what it is for an object to exist, so thin that the resulting view does not obviously deserve to be regarded as any form of realism. Editions, directions, word- and letter-types are all *said* to 'exist' all right, but they 'exist' in no more robust a sense than duds do, not to mention all the other objects whose names could be introduced in terms of random equivalence relations. In so far as we have an intuition that these latter objects simply *don't* exist, any view that affirms that they do will seem to *deny* that word- and letter-types exist, even while it affirms our right to *say* that they do.[12]

---

[9]Note that the problem of trans-sortal identification is *not*, as I understand it, the Caesar problem. It is one aspect of the Caesar problem, but the Caesar problem has more heads than the hydra. See Section 1.2 for a discussion of some of them.

[10]This problem is mentioned at the end of the original version of Chapter 8 (Heck, 2000b), but it was not developed there. It is mentioned by Hale and Wright in the postscript to *The Reason's Proper Study* (Hale and Wright, 2001b, pp. 423–4), where it is treated somewhat dismissively.

[11]Of course, $Q\xi\eta$ need not be a full equivalence relation. It need only be symmetric and transitive, and reflexive on its domain. But we can account for that easily enough.

[12]Compare Lewis (1999) on Meinongian views of existence.

The worry, then, is that it is becoming difficult to distinguish the neo-Fregean view from what one might call a 'permissive' nominalism. Permissive nominalists are perfectly happy to let us 'speak of abstract objects', so long as we do not take their existence seriously.[13] And what makes this all the more troublesome is that permissive nominalism is not just the most resilient form of nominalism but one of the oldest. It is, for example, what one finds in Berkeley and Hume. Their view was not so much that we should not talk in ways that seem to make reference to abstracta, but that what passes for thinking about the abstract is really just abstract thinking about the concrete. Thus, Berkeley writes:

... [S]uppose a geometrician is demonstrating the method of cutting a line in two equal parts. He draws, for instance, a black line of an inch in length: this, which in itself is a particular line, is nevertheless with regard to its signification general, since, as it is there used, it represents all particular lines whatsoever; so that what is demonstrated of it is demonstrated of all lines, or, in other words, of a line in general. And, as that *particular* line becomes general by being made a sign, so the *name* "line", which taken absolutely is particular, by being a sign is made general. And as the former owes its generality not to its being the sign of an abstract or general line, but of all particular right lines that may possibly exist, so the latter must be thought to derive its generality from the same cause, namely, the various particular lines which it indifferently denotes. (Berkeley, 1930, pp. 14–15, emphasis in original)

Berkeley would thus have been perfectly happy to let us talk of editions and even to say that there are such things as editions. The truth of such a claim, for him, is a more or less immediate consequence of facts about how 'talk about editions' is to be understood. It is, in particular, a more or less immediate consequence of the fact that there are books and a suitable equivalence relation on them.

Neo-Fregeans have tended to reply to this kind of remark by saying: Exactly! Their view is that there is no 'metaphysical distance' between, on the one hand, the fact that there are books and a certain equivalence relation on them and, on the other hand, the fact that there are editions; the former is supposed to be entirely sufficient for the latter. And if one wants to say that we now seem to be at stalemate, then neo-Fregeans have generally regarded that as good enough, insisting that the burden of proof is on their opponents. But is it really? Too much seems to depend on which case one takes to be exemplary. The neo-Fregean wants to say that we should take reference to duds as seriously as reference to editions, intuitions to the contrary be damned; the permissive nominalist wants to say that we should take reference to editions no more seriously than we take reference to duds. But the former perspective is the right one only if the neo-Fregean explanation of names of editions has accomplished what is claimed for it. Only if we assume that the explanation in terms of abstraction principles has succeeded in explaining reference to editions

---

[13] It is this sort of view that I take to be the one defended by Dummett (1981a, ch. 14).

are we under any obligation to assume that the parallel explanation must have succeeded in explaining reference to duds. And that can't just be assumed, no matter who has the burden of proof. Indeed, the objection is precisely that the parallel explanation does *not* explain reference to duds, since there are no such things, whence the original explanation can't have explained reference to editions, either. To suggest that we should simply set the intuition that there are no such things as duds aside is to suggest that we should simply assume the correctness of the neo-Fregean account.

So what might a fan of the neo-Fregean approach do? The classical form of neo-Fregean view, developed by Bob Hale and Crispin Wright, and which I shall henceforth call *Naïve Platonism*, is that *all* terms introduced by abstraction refer, and refer to abstract objects. The view that such terms refer all right, but that none refer to abstract objects, is what I shall henceforth call *Semantic Reductionism*. And the view that such terms do not refer at all is what I shall call *Fictionalism*. What I want to explore is the possibility of a more discriminating view, according to which *some though not all* expressions introduced by abstraction refer to abstract objects.[14]

In the next section, I shall begin searching for such a discriminating view by exploring the differences between the cases of duds and editions. By the end of Section 9.2, I hope to have formulated a position naturally described as between Naïve Platonism and Semantic Reductionism. In Section 9.3, however, I shall offer various reasons for dissatisfaction with this position. These criticisms will lead us to a new position, one whose statement requires only a minor reformulation of the original one—almost just a shift of emphasis. This view is better described as between Naïve Platonism and Fictionalism.

If such a view can be found and properly motivated, then whatever serves to motivate it will also serve as the raw materials for an argument against all three of the alternatives I mentioned. Each of them will stand charged of ignoring the differences we will have identified: of treating all abstraction principles the same, when they ought to be treated differently. I will not, however, be attempting to complete that argument here. We will have enough to do just to describe the middle ground.

9.2   Semantic Reductionism and Projectible Predicates

According to Semantic Reductionism, names apparently of abstract objects do not refer to abstract objects, but to objects of some concrete sort. To evaluate this position, we need an account of what determines the

---

[14]Some such view may in fact be that of Hale and Wright. It depends upon how exactly their response to the Caesar objection is ultimately developed.

sort of object to which a proper name refers. There is obviously no way that we are going to answer that question in full generality here. I shall limit attention to the case of terms purporting to denote types, expressions which are, according to the neo-Fregean approach, to be explained *via* abstraction principles.

Many who have discussed this question have attempted to characterize the sort of object to which a term refers in terms of the criterion of identity associated with that term, that is, in terms of the equivalence relation mentioned in the abstraction principle. The simplest such view is that the sort of object to which a name refers is wholly determined by the criterion of identity, whence no names with distinct criteria of identity can refer to objects of the same sort (Hale, 1988, p. 215). It is important here that the notion of a criterion of identity be intensional, in the sense that substitution of a co-extensive relation for the equivalence relation mentioned in the statement of the abstraction principle need not preserve its status as a correct specification of the criterion of identity for the objects in question. Otherwise, objects introduced by different abstraction principles might be of the same sort in one world—in which the equivalence relations were, purely accidentially, extensionally co-incident—but not in other worlds, and that seems bad.

This simple view would arguably imply that editions are neither books nor sets, but are *sui generis*. Identity statements of the form "edn(a) = edn(b)" are true if, and only if, as I shall henceforth say, a copies b: The identity of sets, however, is determined by sameness of membership; the identity of books, by something else still. But consider these abstraction principles:[15]

$$\mathbf{dir}(a) = \mathbf{dir}(b) \equiv a \parallel b$$
$$\mathbf{dor}(a) = \mathbf{dor}(b) \equiv \exists x(a \perp x \wedge b \perp x)$$

It seems to me that 'dorections' might well be our old friends directions. Perhaps they are distinct, but any principle that immediately entails that we cannot identify them is too strong.

The simple view can be weakened: Wright's generalization of his condition N^d is an example of such a weakening. Suppose that Fx is a sortal concept, and that names of Fs have been explained by means of an abstraction principle formulated in terms of some equivalence relation $R\xi\eta$, which itself holds between objects of sort S. Then, Wright's proposal is:

Gx is a sortal under which [Fs] fall if and only if there are, or could be, terms, "a" and "b", which recognisably purport to denote instances of Gx, such that the sense of the identity statement, "$a = b$", can be adequately explained by fixing its truth-conditions to be the same as those of a statement which asserts that the given equivalence relation [$R\xi\eta$] holds between a pair of objects [of sort S]. (Wright, 1983, p. 114)

---

[15]For simplicity, I assume here that we are dealing with two-dimensional Euclidean geometry.

To put it slighly differently: Fs may be identified with Gs if, but only if, identity statements concerning some Gs may be explained in the same way that identity statements concerning Fs are explained. In particular, the Fs will be identifiable with objects of sort S if, and only if, identity statements containing names of some objects of sort S may be explained by means of the abstraction principle in terms of which names of Fs are to be explained.[16]

This proposal probably resolves the problem of directions and dorections. Identity statements of the form "$dor(a) = dor(b)$" plausibly can be explained in terms of the parallelism of the two lines a and b. But, in fact, that isn't at all clear, in large part because the notion of explanation to which Wright appeals is not very clear. And that makes it hard to know how to apply Wright's proposal in general. We may presumably take the names which "recognizably purport" to refer to sets to be those of the form "the set of books that copy ξ". Is it or is it not possible to 'explain' the senses of statements of the form "the set of books that copy a = the set of books that copy b" in terms of "ξ copies η"? Or consider terms of the form "the oldest extant copy of ξ". Does

the oldest extant copy of a = the oldest extant copy of b iff
$\quad$ a copies b

count as an explanation of identity statements involving these terms? Maybe one has intuitions about the matter, but one would like more than a brute appeal to intuition. For this reason, I shall offer a different sort of solution, leaving open the question to what extent it is compatible with Wright's.[17]

Consider the expression "the father of ξ". It seems obvious enough that the father of John is a person. And one might suppose that the fact that John and Jane have the same father if, and only if, the same male begat them is what determines that "the father of John" refers to a person. It is, no doubt, of great importance that the father of a = the father of b if, and only if, the same male begat a and b. But this does not, by itself, entail that "the father of John" refers to a person. For consider the following expressions:[18]

---

[16]There seems no reason not to suppose that the converse must also be true: If some Fs are Gs, then some Gs are Fs, so we must, presumably, be able to explain identity statements concerning (some) Fs in the same way we explain identity statements concerning (some) Gs. Of course, since the condition, as formulated, applies only to sortals introduced by abstraction, we will not be able to apply it unless G is also such a sortal. But some such converse seems reasonable. And if so, then Wright's proposal implies that it will be possible to explain the truth-conditions of *mixed* identity statements—such as "$dor(a) = dir(b)$"—both in terms of the criterion of identity for directions and in terms of the criterion of identity for dorections. That seems *very* reasonable.

[17]The modifications to this view made by Hale and Wright in "To Bury Caesar..." (2001c, §6) leave untouched its near total reliance upon criteria of identity, though there are points of contact with the present view, for which see footnote 20.

[18]This kind of point has the status of what mathematicians call 'folklore'. I first heard a

- the set of persons who have the same father as John
- the oldest paternal half-sibling of John
- the singleton of the father of John
- the location of John's oldest paternal half-sibling

Each of these expressions has the same 'weak identity-conditions' as "the father of John": That is, the reference of any one of these expressions will remain unchanged when we substitute a new name for "John" if and only if the same male begat John and the person whose name is substituted for his. But not all of these expressions refer to objects of the same sort, and those which do refer to different objects.

A similar point applies to names of editions, as the following set of examples shows:

- the edition of book a
- the oldest extant copy of book a
- the set of books that copy book a

Again, substitution of the name of any book that copies a will leave the referent of each of these expressions unchanged; and *only* the substitution of such names will do so. Nonetheless, not all of these expressions refer to objects of the same sort: One refers to an edition; one, to a book; one, to a set.

The point I am illustrating with these examples can be stated precisely. Let $\varphi(\xi)$ be a function from objects of sort S to objects of sort T (not necessarily different from S). Then $\varphi(\xi)$ induces an equivalence relation $\Phi\xi\eta$ on objects of sort S, which we define as follows:

$$\Phi xy \stackrel{df}{\equiv} \phi(x) = \phi(y)$$

Distinct functions from S to T induce the same equivalence relation, as do various functions from S to sorts T' distinct from T. There are thus many distinct functions on S that have the same weak identity-conditions.

There can therefore be no objection to our introducing a functional-expression which will satisfy the abstraction principle by means of which names of editions are explained and whose range will consist of books, or sets, or objects of many other sorts. In principle, a Semantic Reductionist could hold that names of editions refer to just about anything: People, rocks, trees, books, or sets, so long as there are enough of them to go

---

version of it made by Dummett in his 1989 Hilary Term lectures on *Die Grundlagen*. Dummett remarked that if (what I am calling) weak identity-conditions determine the sort of object to which a name refers, then it is philosophically confused to think that the eccentricity of an ellipse is a real number (Dummett, 1991b, pp. 162–3). I heard Lewis make points in the same vicinity at MIT a few years later; Warren Goldfarb once mentioned a similar example to me; and it surfaces in a paper by Sullivan and Potter (1997, 139ff). It is discussed by Hale and Wright in "To Bury Caesar..." (2001c, pp. 375ff).

around. But the most principled such views are that names of editions refer to representative copies and that they refer to equivalence classes. The latter option is of course not one acceptable to a Nominalist (unless she has a Nominalistic treament of set-theory waiting in the bushes). So I shall focus attention on the former proposal, that "edn(ξ)" is a functional-expression whose range consists of actual, physical books. What we need to ask now, then, is what, if any, features of the use of this expression are sensitive to the sort of object in its range: We need to ask, that is, what difference it would make if expressions like "*Ulysses*" actually were treated as referring to books.

Suppose, then, that "*Ulysses*" really does refers to a particular, physical book. That book must have some physical location; one of its pages might be torn; someone probably owns it; perhaps someone is holding it at this very moment. If that book is the reference of "*Ulysses*", then some such sentences as "*Ulysses* is in Texas", "*Ulysses* has a torn page", and so forth, must be true. That seems very odd. Some of the oddity can be avoided if one takes a slightly different view, one that actually seems implicit in the passage from Berkeley quoted earlier: Instead of saying that "*Ulysses*" denotes some particular one of its copies, we should say that it "indifferently" denotes each of them, a view naturally explained in terms of supervaluations.[19] The sentence "*Ulysses* is in Texas" would then be true only if *every* copy of it were in Texas; false, if none were; and neither true nor false, otherwise. Still, though, "*Ulysses* is on planet Earth" will probably come out true. And it is perfectly possible, though unlikely, that "*Ulysses* has a torn page" should also be true.

One does not usually think of such sentences as having even the remotest chance of being true. There is, indeed, a temptation to deny that these sentences so much as make sense, on the ground that they involve some kind of category mistake. But, as Frege in effect remarks, there is nothing to prevent us from saying such things as that *Ulysses* has a torn page, meaning by this that every copy of *Ulysses* has a torn page, if such a way of speaking should seem useful (Gl, §69).

Still, there is a felt difference between these sorts of claims and the claim that *Ulysses* contains the word "dog". The intuition that there is such a difference is presumably what is behind the intuition that expressions like "*Ulysses*" do not refer to books. This difference has nothing to do with abstract objects as such. There is a similar difference between saying, of Frege, that he had blue eyes and saying, of Frege, that he had only blue-eyed children. The temptation here is to say that, in the former case, though not the latter, one is speaking of a *property* of Frege: It is a property of Frege that he had (or did not have) blue eyes; it is no property of him, in this strict sense, that he had (or did not have) only blue-

---

[19] A view of this sort has been elaborated and defended by Hodes (1990). It is because of the availability of this view that I suspect the sort of modal differences explored by Uzquiano (2005) will not do all the work needed here.

eyed children. This distinction—between 'real' and 'merely Cambridge' properties—is infamously difficult to explain clearly. Present purposes, however, require only that it be explained for the case of types, and that is a good deal easier.

There is much that can be said about books: That they are dirty, that they have some mass, that they contain an inscription of some word, and so forth. There are, that is, many predicates that can sensibly be attached to names of books. Of these predicates, some play a special role in our talk of editions: Those that are satisfied by a given book if, and only if, they are satisfied by every book that copies it. More important still are predicates whose satisfaction, by a given book, can always be determined even if one does not know which other books copy it. Predicates like "contains an inscription of the word 'dog'" are of this kind: One need not know which other books copy a given one to be able to determine whether that book satisfies this predicate; knowledge that a given book does satisfy it suffices for knowledge that every other copy also satisfies it. It is because there are such predicates that our ordinary practices involving editions are possible. Scholarly discussion could not be carried on as it is if one had to stop, every time one wanted to make a claim about what is said in *Die Grundlagen*, to determine what copies of it exist, what is said in them, and so forth. (For much the same reason, one can read an edition without reading any single copy of it, but rather parts of different books.) Borrowing a term, let us call such predicates as "contains an inscription of the word 'dog'" *copy-Projectible* predicates of books.

Why think that predicates of types that are introduced in terms of Projectible predicates of tokens should express properties of the types?[20] The allusion to projectibility,[21] as that notion is employed in the philosophy of science, is intentional. Projectible predicates are ones whose satisfaction by a particular sample of a substance, or by a particular member of a species, in some sense implies its satisfaction by all samples and all members. It is natural, for that reason, to think of these predicates as expressing properties of the substance or species itself. There is a clear analogy between such predicates and ones that are Projectible in my sense. Properties of the type, in this sense, do not depend upon what tokens happen to exist (just as properties of a substance should not depend upon what samples happen to exist):[22] The creation or destruc-

---

[20]Note that the idea that it matters whether there are Projectible predicates of the tokens subsumes the proposal made by Hale and Wright (2001b, p. 424, footnote 8) that it matters that the equivalence relation itself should in some sense be projectible: that we should know how to extend it to non-actual things. That is a clearly a weaker requirement than the one considered here, but it is very much in the same spirit. The idea that modal considerations might be relevant also surfaces in "To Bury Caesar..." (Hale and Wright, 2001c, pp. 357ff). Similar ideas have surfaced elsewhere, too.

[21]I shall capitalize my term "Projectible" to remind the reader that I may or may not be using it in its usual sense.

[22]It would be really nice if we could also say: Properties of the type, in this sense, do not depend upon *whether* any tokens of the type happen to exist. If so, then the view we are

tion of particular tokens will not, on this analysis, affect what properties *Ulysses* has—though it might affect whether it has only copies with torn pages.

There is another, more technical reason to think that Projectible predicates of tokens are especially important here. To over-state the point slightly: Only if a predicate of types is explained in terms of a Projectible predicate of tokens is it possible to make genuinely informative applications of Leibniz's Law involving that predicate of types. Consider sentences of the form "$F(\varphi(a))$" and "$F(\varphi(b))$", where $R\xi\eta$ is the equivalence relation figuring in the abstraction principle for "$\phi(\xi)$". If the predicate "$f(\xi)$" in terms of which "$F(\xi)$" is explained is not R-Projectible, then determining whether "$F(\varphi(a))$" is true will, in general, require one to know which other objects bear $R\xi\eta$ to a and whether those objects satisfy "$f(\xi)$". One will, in the course of determining whether "$F(\varphi(a))$" is true, therefore have to go through essentially the same procedure one would have to follow to determine whether "$F(\varphi(b))$" is true: In particular, one will (normally) have to determine whether Rab. If $\phi(a) = \phi(b)$, then one will then have all the information necessary to determine whether "$F(\varphi(b))$" is true—though, of course, one need not draw the conclusion explicitly and may not even realize as much. By contrast, if "$f(\xi)$" is R-Projectible, one can determine whether "$F(\varphi(a))$" is true simply by determining whether $f(a)$: In particular, one need not know whether Rab or whether $f(b)$. Hence, discovering that Rab—that is, that $\phi(a) = \phi(b)$—can lead to genuinely new information.

Why should that matter? The possibility of informative applications of Leibniz's Law seems to me to be bound up with the independence of an object from our ways of conceiving it. That we can know that the direction of a is such-and-such without being in a position to know whether the direction of b is such-and-such—even if, in fact, $\mathrm{dir}(a) = \mathrm{dir}(b)$—suggests not just the familiar gap between sense and reference but an even more important gap between our ability to refer to (or think about) the object and the object itself. To the extent that informative applications of Leibniz's Law are possible, to that extent the object will seem independent from our ways of conceiving it, and hence to that extent its existence will seem to be independent of the fact that we can conceive of it at all.

To sum up, then, in so far as we have an intuition that, say, "*Ulysses*" does not denote a particular copy of *Ulysses*, that intuition rests upon the thought that not everything that can be said about books can be said about *Ulysses*. But, as we saw, that can't be all there is to the intuition, since we could easily introduce conventions allowing us to say such things as "*Ulysses* has a torn page", meaning thereby that all of its copies have a torn page (or that some of its copies do, or that one of its copies does, or whatever). The intuition is thus more refined: It is that, even if we

---

discussing might also help with what Hale and Wright (2001b, pp. 422–3) call the "problem of plenitude". But more work needs to be done here.

did introduce such new predicates, they would not express *properties* of editions. So the problem became to say what a property of an edition is, and I proposed that a predicate expresses a property of editions if it is explained in terms of copy-Projectible predicates of tokens. And so, to the original question, what determines the sort of object to which expressions introduced by abstraction refer, my proposed answer is: It is determined by what predicates of its tokens are Projectible over the relevant equivalence relation.

In the case of duds, there seem to be almost no Projectible predicates of the tokens: The randomness of the equivalence relation $Q\xi\eta$ in terms of which names of duds were introduced essentially guarantees that there are no such predicates. That means that the introduction of names of duds would be largely without point. Any predicate introduced by means of abstraction must be introduced in terms of a predicate of tokens that is a congruence with respect to $Q\xi\eta$: That is, any such predicate must be true of all tokens of a given dud or else false of all of them. It is easy to produce such predicates. Most of these, however, will be similar to "ξ has only copies which have a torn page", for example: "ξ is such that all objects to which it bears $Q\xi\eta$ weigh at least ten pounds". Determining whether such a predicate is satisfied by a particular dud will, in general, require knowing to which other objects a given object bears $Q\xi\eta$. That there would be little point in talking about duds is less important, however, than *why* there would be little point: In the sense in which to speak of someone's eye-color is to speak about them, and to speak of the eye-color of their friends is not, there would be almost nothing to be said about duds; duds will have almost no 'properties' at all, other than being identical with or different from each other.

Assuming that names of duds refer, then, to what sort of object should they be taken to refer?[23] I would suggest that, in this case, we have as yet no reason to deny that they refer to equivalence classes under the relation $Q\xi\eta$: What it is true to say of an equivalence class is wholly determined by what is true of its members; what it is true to say of a given dud will, in general, be determined by what happens to be true of all, some, most, few, etc., of its tokens. Of course, that is not much of an argument for the identification. But we need not pursue the matter, as I shall shortly be rejecting the claim that names of duds refer at all.

Nonetheless, we have made progress, since the present proposal does at least illustrate how we might catch what we are chasing: a view that does not treat all abstraction principles alike. I complained earlier that the going views seem to me problematic for precisely that reason, that they treat all types the same way: editions, just like duds; duds, just like

---

[23] If we were to say, as Hale and Wright suggest (see footnote 20), that the equivalence relation used in an abstraction principle must be projectible, then that would dispense with duds altogether. But we do not yet have any motivation for that proposal, and it does not solve the problem I will shortly introduce, about day-persons.

editions. That this is true of the various versions of nominalism is clear enough; that it is true of Wright's view, mentioned earlier, is perhaps less so. But Wright's generalization of N^d, so far as I can tell, appeals only to broadly formal features of abstraction principles in attempting to specify the reference of expressions introduced by those principles. If so, then it is hard to see how duds and editions can come apart. On the other hand, in characterizing the sort of object denoted by names introduced by abstraction in terms of Projectible predicates of the tokens, the present proposal treats different abstraction principles differently and so offers us at least a hope of solving the proliferation problem.

All is not well, however. Consider an abstraction principle based upon the equivalence relation, "ξ was born on the same day as η", by which I mean: within the same twenty-four hour period, Greenwich mean time. The abstraction principle is thus:

day-person(a) = day-person(b) iff
    a was born on the same day as b

What should we say about the reference of expressions introduced in terms of this relation? One might be tempted to say that there are going to be very few Projectible predicates of persons, and so that day-persons too are at best equivalence classes. But this would be incorrect. Any predicate of days, referenced to Greenwich mean time, will obviously be Projectible. The obvious thing to say, then, would be that names of day-persons in fact denote days. This is a satisfying result, in some ways. One might even wonder what reason there could be for holding that anything more is going on here than the introduction of an abbreviation for "the day on which ξ was born". On the other hand, however, the conclusion that names of day-persons simply denote days is troubling. It isn't very hard to explain the equivalence relation *born on the same GMT day* without making a direct appeal to anything about days. If not, then it seems something of a surprise that expressions introduced by abstraction on that relation should end up referring to days.

The sort of worry this example illustrates can be generalized. Let $R\xi\eta$ be the equivalence relation for a given abstraction principle with base sort B, and let S be an arbitrary sort of object of which there are at least as many as there are equivalence classes under R. Now define a function $\phi(\xi)$ from B to S that respects R, in the sense that $\phi(a) = \phi(b) \equiv Rab$; assume further that we can do so in such a way that determining what $\phi(x)$ is, for arbitrary x, does not require knowing to which objects x bears $R\xi\eta$. Now let F be any property of Ss. Then $F(\phi(\xi))$ is R-Projectible, in our sense. But, since this depends only upon how many Ss there are (and our ability to define $\phi(\xi)$ appropriately), there are going to be lots and lots of R-projectible predicates that express all sorts of different properties, quite independent of what the base sort is and what the equivalence relation is. Surely, however, such oddly defined predicates cannot be guaranteed to express properties of the types.

As we characterized the notion above, an R-Projectible predicate was required to satisfy two conditions: First, that it be a congruence with respect to $R\xi\eta$; Second, that it should be possible to determine whether the predicate is satisfied by a given token without knowing which, if any, other tokens bear $R\xi\eta$ to it. What we have omitted from the specification of R-Projectibility is another condition that was really implicit in the spirit of the proposal. The predicate, "On the day on which ξ was born, it was cloudy in London", was not the sort of thing one had in mind as day-Projectible, the sort of thing that will express a property of the types. That, I take it, is because this predicate does not even express a property of the *tokens*. So we should add this condition to our account of R-Projectibility: An R-Projectible predicate must express a property of the tokens. Indeed, we might just speak not of R-Projectible predicates, but of R-Projectible properties. And what names of day-persons refer to will then depend upon what day-Projectible *properties* of people there are.

The resulting proposal is thus this: The sort of object to which names introduced by abstraction refer is determined by what R-Projectible properties of the tokens there are. Let us call this proposal the *Projectibility View*.[24] If the Projectibility View could be sustained, it would apparently imply that names of day-persons, like names of duds, at best denote equivalence classes, as there do not appear to be any properties of persons that are suitably Projectible. And yet, it would allow us to say that names of editions do not denote equivalence classes, since there are a large number of copy-Projectible properties of books; and it would allow us to say that names of editions do not denote books either, since not all properties of books are properties of editions. So it is, again the kind of view for which we are searching.

But, once again, all is not well. Are we really certain that there are no day-Projectible properties of persons? Whether there are is an empirical matter, not one which can be settled *a priori*. On the Projectibility View, then, it is an empirical question—and by no means a question belonging to linguistics—to what sort of object expressions of the form "day-person(t)" refer. And that is a consequence I simply find incredible. It is not that I think there are no interesting empirical questions in this area. Since it plainly is an empirical (or, more generally, a substantial) question whether any properties of persons are day-Projectible, there is certainly an emprical (or substantial) question about day-persons in the vicinity. But, on the Projectibility View, one could have as good an understanding of expressions denoting day-persons as it is possible to have and yet have *no idea* to what sorts of objects they refer, even whether those objects are abstract or concrete. That is the consequence that I find simply incredi-

---

[24]This view was inspired by Bromberger's discussion in "Types and Tokens in Linguistics" (Bromberger, 1992b), and is intended to be a natural generalization of his view there.

ble. So, while there's an empirical question around here somewhere, the Projectibility View misidentifies it.

9.3   Ideology, Existence, and Abstract Objects

Even if there are no day-Projectible predicates that express properties of persons, things might have been otherwise. Some years ago, there was a fad about what were called 'bio-rhythms'. Persons born on the same day were supposed to share certain general features of their day-to-day mental and emotional states: degrees of awareness, laxity, happiness, and so forth. Persons were, that is, supposed to have the same *bio-rhythms* if they were born on the same day. Idealizing, let us suppose that there is a detailed theory, Bio-rhythm Theory, making more precise and enlarging upon this idea. It might have been true. Had it been true, there would have been a great many properties of persons that were day-Projectible. On the Projectibility View, therefore, expressions denoting day-persons would then have denoted a certain sort of abstract[25] object: They would denote structural features of a person's mental and affective states, and these features would be shared by persons born on the same day, just as parallel lines share a direction and as books that copy one another share an edition.

According to the Projectibility View, then, as things are (or, at least, as we think they are), terms of the form "day-person(a)" denote equivalence classes; if Bio-rhythm Theory had been true, they would have denoted bio-rhythms. But, again, I just find it implausible that the sort of object such expressions denote should turn on the empirical question whether Bio-rhythm Theory is true. Indeed, it is not clear to me how a proponent of the Projectibility View could avoid saying more, namely, that, as things are, day-persons *are* equivalence classes and that, were things as just imagined, day-persons would have *been* bio-rhythms. And that, I take it, would be flatly incoherent: It would amount to supposing that what are in fact equivalence classes should have been bio-rhythms.[26] I have no fixed

---

[25] At least, these *could* have been abstract. But of course one can imagine lots of ways in which Bio-rhythm Theory might have been true, and on some of them maybe bio-rhythms would be angels. The point does not actually matter to the argument, however, so far as I can see. There are lots and lots of variations on this theme, so if this one doesn't work, another one will. And, for what it's worth, I tend to agree with Lewis (1986, §1.7) that the abstract–concrete distinction is not very clear, anyway. Indeed, I think the attitude towards abstracta expressed here serves to blur the distinction even further: If I am right, then some sorts of abstracta only exist (or fail to exist) contingently, which is a stronger claim than that individual abstracta may contingently (fail to) exist.

[26] If that doesn't seem so bad, then reflect upon the fact (to be noted below) that there are other theories, incompatible with Bio-rhythm Theory, whose truth would imply that quite different properties of persons were Projectible. Had one of those theories been true, then day-persons would have been some other sort of object. In any event, the view that all abstracta turn out to be equivalence classes is out of the spirit of Frege's original proposal: That's the view to which he retreats in the face of the Caesar problem.

view about whether the Projectibility View can avoid this consequence. But it is suggestive.

Consider the following, slightly different version of the example. Suppose there were some people who mistakenly *believed* Bio-rhythm Theory to be true (as, indeed, many people more or less did). Our description of how the world would have been if Bio-rhythm Theory had been true is *eo ipso* a description of how those who believe that Bio-rhythm Theory is true believe the world actually to be. Thus, people who believe Bio-rhythm Theory to be true take there to be a variety of day-Projectible properties of persons and so, by the reasoning of the last paragraph, take themselves to refer, by means of expressions of the form "day-person(a)", to bio-rhythms, and so not just to equivalence classes. How then should we describe to what they *do* refer? Should we say that, as a matter of fact, they refer to equivalence classes and not to bio-rhythms?

I do not think our hypothetical speakers would or should accept such a description of their linguistic practice. In speaking of bio-rhythms, they take themselves to be speaking of objects of a particular kind, ones that are shared by people born on the same day. If our hypothetical speakers were to become convinced that nothing like the properties they think are day-Projectible actually are day-Projectible—if they were to become convinced that people do not, in general, share anything like the properties they think they do—then what they would say is not that bio-rhythms have turned out to be equivalence classes, but that the objects to which they thought they were referring, bio-rhythms, *do not exist*. If this were not so, then it would be obscure why they were no longer prepared to accept the truth of a sentence such as "the day-person (that is, bio-rhythm) of a is lethargic", even if everyone born on the same day as a was, as it happened, lerthargic. It is not sufficient for the truth of that sentence, as they understand it, that everyone who happens to have been born on the same day as a just happens to be lethargic. The truth of such a sentence, as they understand it, depends upon the truth of Bio-rhythm Theory (or something like it). And that is because the existence of the objects to which they purport to be referring itself depends upon the truth of Bio-rhythm Theory (or something like it).

What I am suggesting is thus this. Instead of saying that our hypothetical speakers think they are referring to bio-rhythms but are in fact referring to equivalence classes, we should say that they think they are referring to bio-rhythms and are in fact referring to *nothing*. I will return shortly to the question why this description of the situation should be preferred. First, we need to develop the view further.

The Projectibility View, the reader will recall, is that the sort of object to which an expression introduced by abstraction *in fact* refers is determined by which properites of the tokens are *in fact* Projectible. I have been arguing for the last several paragraphs that this cannot be right. But it is almost right. The right view is: The sort of object to which

expressions introduced by abstraction *purport* to refer is determined by
what properties of the tokens users of those expressions *presume* to be
Projectible.

Presume in what sense? Contrary to what is usually supposed, an
understanding of terms introduced by abstraction is not exhausted by a
grasp of the criterion of identity for those terms. That is, abstraction
principles are not, by themselves, (always) adequate to introduce a class
of expressions. We must recognize a second component, which, borrowing
a term from Quine, we might call the *Ideology* associated with the abstraction principle. The Ideology is not a theory about the objects whose
names the abstraction principle characterizes, like Bio-rhythm Theory,
but it is closely related to such a theory. The Ideology is something like
a collection of properties of the tokens that use of the expressions introduced by the abstraction principle presupposes[27] are Projectible. So, as
said, it is not that the sort of object such expressions in fact denote is
determined by what properties are in fact Projectible, but rather that the
sort of object to which these expressions purport to refer is determined by
which properties users of these expressions presuppose are Projectible.

One reason to recognize the independence of the Ideology from the abstraction principle with which it is associated is that there is no particular
reason that the example discussed above in connection with day-persons
had to take the form it did: Bio-rhythm Theory is only one of many theories which, had they been true, would have assured us of the existence of
day-Projectible properties of persons. We might, instead, have considered
a theory asserting that all persons born on the same day have common
physiological features, be these gross anatomical ones or such properties
as heart-rate, blood-pressure, and the like. Had such a theory been true,
it seems to me that certain sorts of abstract objects would then have existed. On the other hand, I should want to deny that *bio-rhythms* should
then have existed, if for no other reason than that it seems to me that
both sorts of objects might have existed together. What *kinds* of objects
day-persons are supposed to be is thus connected with the sort of properties they are supposed to have, and so with the Ideology that is associated
with them. Only if one knows the associated Ideology can one know what
kinds of objects are supposed to be denoted by the terms in question.

I said earlier that the Projectibility View wrongly supposes that it
might be a substantial question to what sort of objects expressions of the

---

[27]I think we really do want the notion of presupposition here, but I am less sure which
notion of presupposition we want. The general idea that understanding an expression of
a certain type involves accepting certain presuppositions is nowadays fairly common. The
advantage to using it here is that we do not need to modify the abstraction principles themselves. In particular, we can still have

$$\phi(a) = \phi(b) \equiv Rab$$

and allow that this might fail to be true because use of the expressions on the left-hand
presupposes something that is not true.

form "day-person(t)" refer. I also said that there was a substantial question in the vicinity that had been misidentified. We can now see which question that is. On the view being developed, the Projectibility of the properties contained in the Ideology is a necessary condition of the truth of any sentence containing such expressions.[28] There being such objects as those the introduced terms purport to denote thus depends upon the (actual) Projectibility of the properties contained in the Ideology. So the substantial question toward which the Projectibility View was groping is just that one: Whether *there are* any such objects as the ones the expressions in question purport to denote.

It is hard to see how it could be an intelligible question whether names of abstracta denote what they purport to denote. (I speak from my own experience.) Why? It is, I think, because we have tended to suppose that the only constraint on the assignment of reference to expressions introduced by abstraction is given by the abstraction principle itself. If so, then the only constraint on the *existence* of those objects is also given by the abstraction principle, and if there is an assignment that verifies the abstraction principle, then that is enough to show that the types in question exist. If so, however, then we are stuck with the proliferation problem, because it is hard to see how any one abstraction principle could be preferred over any other. But that, in turn, is because we miss the importance of the associated Ideology. Once it is in place, there is an additional constraint on the sort of object to which terms introduced by abstraction purport to refer, and there is therefore an additional constraint on the existence of referents for those terms. In short: The objects to which the terms refer must be ones that would not exist unless the properties in the Ideology were in fact Projectible.

Another reason the presence of the Ideology is easily overlooked is that it is often obvious from context what it is supposed to be. Consider, for example, Frege's definition of names of what he calls "orientations": The orientation of a plane a is the same as the orientation of a plane b if, and only if, a is parallel to b (Gl, §64). Immediately upon encountering this definition, one immediately knows precisely what Frege means to be talking about: It is obvious what sort of object an orientation is supposed to be. By contrast, few will have discerned a conception of 'bio-rhythms' in the abstraction principle introducing names of day-persons. Whence the difference? It lies in the fact that, in the case of orientations, it is obvious what the Ideology is intended to be. As is clear from the context of Frege's discussion, the theory of orientations is to be a geometrical theory; so orientations are geometrical objects.[29]

---

[28]This is too strong, as I have occasionally indicated parenthetically. We surely want to allow for near misses and second chances, so it is really only something like the Projectibility of enough predicates sufficiently like the ones contained in the Ideology that is required. We shall return to this point.

[29]Special thanks to George Boolos here.

What, then, of duds, which haven't been mentioned for a while? Does the view outlined support the intuition that there are no such objects? From the perspective of the present view, we can only say that the question is ill-posed: No Ideology associated with names of duds has been offered, so we have no idea what kind of object it is on whose existence we are being asked to pronounce. The intuition that there are no such objects comes, I would suggest, from the thought that, whatever duds are supposed to be, they are supposed to stand to their tokens in the same sort of relation that editions, words, and the like, stand to theirs. And what I have been suggesting is simply that there is more to there being a relation "of that sort" between a type and its tokens than what is captured in an abstraction principle. Bearing *that* kind of relationship to one's tokens amounts to having properties determined by R-Projectible properties of the tokens. And, confronted with such a peculiar example as that of duds, we are essentially unable to imagine any sensible Ideology that might be associated with it. And if there is no sensible Ideology, then there are no duds, *whatever* sort of object they might be supposed to be.

So, to sum up, again, the view I'm suggesting has two parts. First, the sort of object to which names introduced by abstraction purport to refer is determined by the Ideology associated with those names; a specification of the Ideology is as fundamental a part of the explanation of those names as the specification of the abstraction principle itself. Second, there being any such objects as those to which the names so introduced purport to refer depends upon whether the properites contained in the Ideology are in fact Projectible properties of the tokens.

Why should one accept this view? We need to distinguish two sorts of issues here. The ontological issue concerns what we *ought* to say about the various cases at which we have been looking: Whether we should say there are no such objects as duds, but that there are editions; and, if so, whether we should say that editions are abstract and *sui generis* or, instead, are just books, equivalence classes, or what have you. There is also, however, a purely descriptive problem: To give some account of the felt difference between names of duds and day-persons, on the one hand, and names of editions, on the other. However seriously it should be taken, there is an intuition to say, in the former cases, either that there are no such objects or that they are 'just' something else. And one might want to understand the source of this intuition, understand to what features of the (imagined) use of these names we are responding, whatever one's view about the ontological question.

So far as the descriptive problem is concerned, it seems to me that the view I've described does get a lot of the intuitions right, in all their varied and confusing forms. Moreover, in my experience, many people's first response to the suggestion that editions might be books is that, if they were, there would be all sorts of things one could say about editions which just don't seem to make much sense. The view on offer explains the relevance of this thought. It isn't that no sense can be made of the claim that *Die Grundlagen* weighs twelve ounces, but rather that regarding such a claim as intelligible involves committing oneself to the copy-Projectibility of properties like *weighs twelve ounces*, contrary to obvious fact.[30]

As concerns the ontological problem, the view's getting these intuitions right constitutes at least some evidence in its favor. My primary goal here has been, as I said at the outset, not so much to argue for but simply to characterize a view that responds, in some principled and discriminating manner, both to the problem of trans-sortal identification and to the proliferation problem. That the view does so is another consideration in its favor.

But once one has started to take the significance of the Ideology associated with a given abstraction principle seriously, it is natural to start to wonder if it is not the Ideology that is really fundamental, and so if the concentration on abstraction principles, common to Naïve Platonism and Semantic Reductionism, is misplaced. Consider, for example, the case of linguistic types, such as sentences. Discussion of linguistic objects in the literature on ontology generally supposes that linguistic tokens can unproblematically be identified as physical entities, bits of ink or chalk, say, or disturbances in the air.[31] The problem then seems to be to identify the abstraction principle that underlies our use of names of types. In fact, however, the question what sorts of things sentences are is an empirical question. We have good reason to believe, for example, that there are *two* sentences both of which are written "Flying airplanes can be dangerous", not one sentence that sometimes means that it can be dangerous to fly airplanes and other times means that airplanes in flight may pose a hazard. If so, then it is hard to see how tokens *could* be just blotches of ink or chalk, for there is no difference between the blotches that are tokens of the one sentence and the ones that are tokens of the other. Or perhaps the right thing to say is just that the relation *is of the same type as* is not going to be explicable in terms of anything like *having the same shape*, as philosophers since Frege have often assumed, nor in terms of any other purely physical properties.[32]

The evidence that there are two sentences written "Flying airplanes can be dangerous" lies deep in linguistic theory: in phenomena whose explanation, given other principles that seem well supported, seems to

---

[30]This sort of point seems to me relevant to certain sorts of examples Chomsky often mentions when deriding semantics, examples like: John wrote a book that weighs over a pound. What is puzzling about this example is that the book John wrote is an edition; the one that weighs over a pound is a copy; and yet "that" seems to refer to the latter but to have as antecedent the former.

[31]Unhappiness with that conception surfaces from time to time, however, for example, as in Kaplan's (1990) paper "Words".

[32]There are interesting discussions of these issues, to which I am greatly indebted, in the paper by Bromberger (1992) mentioned earlier, and in another that he wrote with the phonologist Morris Halle (Bromberger and Halle, 1992).

demand such a conception of what sentences are. What informs our conception of when we have two sentences and when we have one is thus *the kinds of things we can say about sentences*. The Ideology here is thus not, as our earlier examples may have led one to suppose, simply independent of the abstraction principle. On the contrary, the Ideology *shapes* the abstraction principle, whose discovery is thus an empirical enterprise, not one of *a priori* conceptual analysis.[33]

This conclusion does not really conflict with the neo-Fregean approach to abstracta, though it does cast it in a different light. One might have thought it did conflict with it, on the ground that Wright and Hale tend to emphasize the idea that abstraction principles can be freely stipulated, in the sense that someone who makes such a stipulation incurs no obligation to guarantee that there really are objects of the sort required. Well, the view being elaborated here is, of course, in tension with part of this view, since I am claiming that we do not have any *a priori* guarantee that objects of the right sort exist. But that observation is really intended as a friendly amendment. Wright often emphasizes that, in laying down an abstraction principle, we do not thereby stipulate the existence of objects but simply the truth-conditions of certain sentences (Wright, 2001a, p. 162; 2001b, p. 311). But it is hard to take this suggestion seriously when the truth-condition of "$\phi(a) = \phi(a)$" ends up being: Raa, which is itself guaranteed to be true in virtue of the fact that $R\xi\eta$ is an equivalence relation. The view I am developing here makes space for the sort of distinction on which Wright correctly wants to insist.

Moreover, the emphasis on free stipluation of abstraction principles is dispensable. It is an artifact of idealizations Wright and Hale make, ones that allow us to abstract away from messy questions about what the abstraction principle for linguistic types (say) really is. To think that this idealization was important to Wright and Hale would be to overlook a point made earlier: The question whether the *same type* relation can be defined in other terms, so central to earlier discussions of nominalism, has turned out to be a distraction. Frege's great insight was that our understanding of what sentence-types are is intimately bound up with our appreciation of when we have one sentence-type and when we have two. It does not matter whether *same sentence* can be defined in a way that would be comprehensible to someone ignorant of sentences, or of linguistic-types generally, and it does not matter either whether our grasp of "same sentence" is or is not independent of our grasp of what can be said about sentences.

What is true, however, on my view, is that an understanding of what a sentence is does not issue simply from a grasp of when we have one

---

[33] Cognate points concerning concrete objects are made by Wiggins in *Sameness and Substance* (Wiggins, 1980, 2001). Indeed, the notion of an Ideology plays a role in the present view not unlike that played by a 'principle of activity' in Wiggins's view. I very much wish I could make the relation between these two notions precise.

sentence and when we have two but also requires an appreciation of what sorts of things can be said about sentences. But that, though certainly in conflict with Naïve Platonism, is not, so far as I can see, in any conflict at all with the spirit of Frege's position. Or so I am about to argue.

9.4   The Julius Caesar Problem

As I have just been saying, the idea of accounting for our capacity to make reference to abstract objects by means of abstraction principles has its origins in Frege. But, in the end, Frege rejects the view that names of abstracta can be explained in this way, his reason being that such a view does not resolve the Caesar problem: This explanation does not determine the truth-values of such sentences as "the edition of a is Julius Caesar" nor, for that matter, the truth-values of any sentences of the form "the edition of a is t", unless t is itself of the form "the edition of b". There has been little agreement, however, about why this is supposed to be a problem, and, in closing, I would like to explain what I take the problem to be and how the foregoing might allow us to address it.

Frege raises the Caesar objection against a proposed answer to the famous question of *Die Grundlagen* §62: "How, then, are numbers to be given to us, if we cannot have any ideas or intuitions of them?" The proposal he considers is that we may explain the senses of identity statements in which number-words occur by means of an abstraction principle, namely, HP: The number of Fs is the same as the number of Gs just in case the Fs and Gs are in one-one correspondence. So the view against which the Caesar objection is offered is this: We recognize numbers as the referents of expressions of the form "the number of Fs", and our understanding of these expressions consists (in large part) in our grasp of HP. Frege's objection to this view is that HP "will not, for instance, decide for us whether [Caesar] is the same as the [number zero]..." (Gl, §66).[34]

It is generally supposed that Frege is here raising an instance of the problem of trans-sortal identification, and in some sense that must surely be true. What is not so widely noted, however, is that Frege takes for granted we *do* recognize that Caesar is not a number. His objection is *not* that HP does not decide the truth-values of *all* 'mixed' identity statements. If that were the objection, then our intuitions about whether Caesar is a number would be irrelevant. But the problem Frege raises is not, say, that HP does not decide whether zero is the singleton of the null set—which, on Frege's explicit definition, it happens to be. Rather, the problem is that HP does not decide a question about which he takes us to have strong intuitions: Whatever numbers may be, Caesar is not among them. If so, then one might suppose that any complete account of our

---

[34]Note that the objection is not so much that the abstraction principle fails to decide the truth-value of this sentence, but that it fails to give any sense to it at all.

apprehension of numbers as objects must include an account of how we distinguish people from numbers, and Frege's objection to HP, regarded as constituting a complete explanation of how we apprehend numbers as objects, is that it alone yields no such explanation. That is why Frege writes: "Naturally, no one is going to confuse England with the direction of the earth's axis [or Caesar with the number of non-self identicals]; but that is no thanks to our definition of direction [or of Number]" (Gl, §66). So our 'definition' of direction or number must include more than just an abstraction principle.

I have suggested here that a full account of our understanding of expressions introduced by abstraction must include an account of the Ideology associated with those expressions, and that the wanted explanation of why people are not numbers can be given in terms of the Ideology. My own view, however, is that the Caesar problem is not just about transsortal identification. Rather, Frege uses the Caesar problem to raise another, more semantical sort of issue.[35] HP, and other abstraction principles, are supposed to feature identity statements on their left-hand sides, and they are supposed to explain a range of *terms*: expressions that are to be treated, semantically, as purporting to refer to objects. Only if expressions of the form "the number of Fs" are terms in this sense can our capacity to refer to numbers be explained in the way Frege considers.

Now, if expressions of the form "the number of Fs" are to be understood as referring to objects, and if a statment of the form "the number of Fs is the same as the number of Gs" really is an identity statement, then a complex predicate such as "ξ is the number of Gs" must itself be a predicate that is true or false of objects. And if HP is truly sufficient to ground an understanding of a class of terms purporting to refer to objects, then it must also issue in an understanding of such predicates as being true or false of objects. To put the point differently, if "the number of Fs" is truly a semantic constituent of "the number of Fs is the number of Gs", then it must be replaceable by a variable: It must be an intelligible question whether the open sentence "x is the number of Gs" is true or false of any particular object, independently of how that object is given to us. This is the point of Frege's remarks about sentences of the form "q is the direction of a": The definition of directions gives us no purchase whatsoever on the question whether this open sentence is true or false of England (Gl, §66). And, indeed, it gives us no more purchase on the question whether it is true or false *of* the direction of the Earth's axis, independently of how that object is given to us. It is only when we imagine the object *given as* the direction of the Earth's axis that we understand the question (Gl, §67).

The reason this matters so much to Frege is that it implies that HP does not suffice to explain the concept of cardinal number. The local point

---

[35]To the best of my knowledge, this point was first made by Parsons in "Frege's Theory of Number" (Parsons, 1995a).

is that the abstraction principle for directions does not suffice to explain the concept of direction. If it did, Frege tells us, then "q is the direction of a" could explained by distinguishing cases: If q is not a direction, then the proposition is false; if it is a direction, then it is the direction of some line, and the abstraction principle will take over from there (Gl, §66). Another route to the same point proceeds from the observation that the complex predicate "$\exists x(\xi = \mathrm{dir}(x))$" clearly defines "$\xi$ is a direction". So we understand this concept only if we understand the complex predicate in question. Indeed, one might reasonably suppose that, if the abstraction principle issues in a grasp of the concept of direction, then it does so *via* an understanding of "$\exists x(\xi = \mathrm{dir}(x))$". But this complex predicate just embeds the very context "$\xi = \mathrm{dir}(x)$" that is causing all the trouble.

So we have, on the one hand, something that the abstraction principle does not give us: an understanding of the question whether "$\xi = \mathrm{dir}(a)$" is true or false of an object, independently of how that object is given to us. And we have, on the other hand, the fact that the abstraction principle does give us an ability to understand the question whether the result of substituting a certain sort of expression—one of the form "$\mathrm{dir}(t)$"—for the placeholder is true. The contrast is thus between what we might call an 'objectual' understanding of such predicates and a merely 'substitutional' understanding. But this issue, as should now be apparent, is itself intimately connected with our understanding of quantification over directions. Consider, for example, the claim that the direction of a exists:

$$\exists y(y = \mathrm{dir}(a))$$

This too embeds the very sort of context we have been discussing. And the question whether we are equipped with an objectual understanding of the quantifier thus becomes the question whether we understand this sentence as saying that there is an *object* that is the direction of a or merely as saying that there is an expression—*not* necessarily a term, in any real sense—whose substitution for "y" would yield a truth.[36] Among the many things at issue here is thus the question whether abstraction principles can provide for an understanding of objectual quantification over the range of objects whose names they are supposed to introduce.

What I want to suggest, in closing, is that this problem—that of securing an objectual interpretation of quantification—may actually be a form of the proliferation problem. For note that the substitutional reading of the quantifier is as available in the case of day-persons or duds as it is in the case of directions, numbers, or editions. Indeed, it is often suggested that quantification over types should be explained in terms of quantification over tokens, so that "$\exists y(y = \mathrm{dir}(a))$" would just become "$\exists l(\mathrm{dir}(l) = \mathrm{dir}(a))$" and so "$\exists l(l \parallel a)$". But then, of course, by the same reasoning "$\exists y(y = \mathrm{dud}(a))$" becomes "$\exists d(\mathrm{dud}(d) = \mathrm{dud}(a))$" and so

---

[36] It's important to remember here that the substitution class for a substitutional quantifier can consist of almost anything, e.g., parentheses or suffixes (Kripke, 1976).

"$\exists d(Qda)$", and we are back in the soup. We need to keep "$\exists x(x = \mathrm{dir}(b))$" from just reducing to "$\exists l(l \parallel b)$", but it will be impossible to do so as long as the truth of "$a \parallel b$" is supposed to be wholly adequate for the truth of "$\mathrm{dir}(a) = \mathrm{dir}(b)$". Giving some teeth to the notion of existence is thus of a piece with securing an objectual interpretation of the quantifiers: In both cases, the issue is one of making "there is a direction..." have some force not had by "there is a line...".

Defenders of the neo-Fregean approach to abstracta often say that the core of their view is that, in introducing a class of expressions by abstraction, we assume no new epistemological burden as regards the existence of referents for those expressions: The epistemology of direction-talk is supposed to be wholly reduced to the epitemology of line-talk and knowledge of the abstraction principle that links the two sorts of discourse. I am, clearly, denying this. But my view is simply not vulnerable to the strongest objection typically brought against such denials, namely: The question whether there are directions makes sense only if we know what the existence of directions invovles; but the neo-Fregean account of what it is for there to be directions has been rejected, and no alternative has been offered (Wright, 2001a). I have said as precisely as I can what the existence of directions and other types involves. And, in so far as my view imposes an epistemic obligation on those who claim the existence of directions, sentences, editions, or numbers, it is not a philosophical obligation but a (broadly) scientific one, and it is one that can clearly be met, at least in some cases.

10

On the Consistency of Second-order Contextual Definitions

A great deal of attention has recently been focused upon a principle now known as HP:

$$\mathbf{N}x : Fx = \mathbf{N}x : Gx \equiv \mathbf{Eq}_x(Fx, Gx)$$

Here, "$\mathbf{N}x : Fx$" is a functional expression to be read "the number of Fs", and "$\mathbf{Eq}_x(Fx, Gx)$" abbreviates: The Fs and Gs can be correlated one-to-one.

Crispin Wright has shown that it is possible to derive the standard axioms of second-order arithmetic from HP, if that HP is taken as the sole non-logical axiom within a standard second-order logic (Wright, 1983, pp. 158–69). Our attitude towards this result is a consequence of our attitude towards HP. If we understand HP merely as stating a necessary and sufficient condition for numerical equality, the result is of much mathematical interest, but it hasn't the specifically philosophical consequences Frege would have wished to attach to it. Wright's resurrected version of Frege's logicist project, on the other hand, depends upon taking HP as a *contextual definition*: HP is to be understood as *defining* names of numbers, which are, in turn, to be thought of as (able to be) introduced into our ontology by means of this very definition (Wright, 1983, pp. 104–17, 140–2, 145–9).

HP is an instance of the general form of second-order contextual definitions:

$$Ax : Fx = Ax : Gx \equiv R_x(Fx, Gx)$$

Here, "R" is some equivalence relation on concepts. An important difficulty concerning such definitions is that not every second-order contextual definition is consistent.[1] Frege's Basic Law V is one example of an inconsistent, (second-order) definition of this sort; there are others.[2]

---

[1] I shall speak, somewhat sloppily, of consistent and inconsistent 'definitions': This should be understood as short for talk of consistent and inconsistent systems to which appropriate forms of the definitions are added as non-logical axioms. Note that HP is consistent if second-order arithmetic is (Boolos, 1998b).

[2] Boolos (1998k, pp. 214ff) mentions several examples. Boolos also mentions that the consistency of such definitions is not sufficient, if we want to construe those definitions as capable of introducing names of objects: There are individually consistent contextual

Surely, we are not to accept Frege's Basic Law V as introducing value-ranges into our ontology. Why, then, should we accept HP as so introducing numbers? It is true that the one is consistent and the other is not. But are we to accept that the mere consistency of HP is sufficient to guarantee it such powers? It is not at all obvious that the danger of its inconsistency is not sufficient to prohibit it from being acceptable as a 'definition' (Dummett, 1991b, chs. 15–17, esp. pp. 187ff). In any event, what Wright's resurrected logicism now requires is either a defense of the claim that the consistency of HP is, in fact, all that is required here, or else it requires some characterization of a subset of all second-order contextual definitions which we may rightly take as introducing a new sort of object into our ontology. That subset will include only, but need not include all, consistent definitions of this sort; moreover, the characterization itself should not amount merely to a re-statement of the condition that the definition should be consistent (else this option should collapse into the view that consistency alone is sufficient).

There is an obstacle to the provision of a characterization of this latter sort: It is recursively undecidable whether an arbitrary second-order contextual definition is consistent. For suppose our view was that the consistency of the definition is necessary and sufficient for the definition to be in good order and that, if it is in good order, then the names so introduced refer. Since consistency is here a condition on the acceptibility of a *definition*, it is natural to require that we be able, at least in principle, to tell whether the definition is acceptable. (Indeed, for this purpose, to require a recursive characterization of the condition may be too weak.)

Note, first, that the following definition is trivially consistent:

$$Tx : Fx = Tx : Gx \equiv (A \equiv A)$$

Here, A is an arbitrary second-order sentence. This definition is satisfiable with respect to every (non-empty) domain: If we fix some such domain, since $Tx : Fx = Tx : Gx$ is true for all F and G, we simply choose an element of the domain and assign it as value to "$Tx : Fx$", for each F.

Now let A be an arbitrary second-order sentence. Consider the contextual definition:

$$Cx : Fx = Cx : Gx \equiv \forall x (Fx \equiv Gx) \vee A$$

(Note that the formula on the right-hand side does indeed express an equivalence relation.) I intend to show that this definition is consistent if, and only if, A is satisfiable.

Let us first settle terminology. For fixed sentence A, call the definition in question "$K(A)$". We shall use the same name for the second-order theory whose sole non-logical axiom is $K(A)$.

---

definitions that are jointly inconsistent.

Suppose that A is satisfiable: Then there is some second-order model $\mathcal{M}$ with respect to which A is true. We may expand $\mathcal{M}$ to a model $\mathcal{M}'$ of $K(A)$ as follows. Since A is true in $\mathcal{M}$, the right-hand side of $K(A)$ will be true for all F and G. So, as above, we can make $K(A)$ true simply by choosing our favorite element of the domain and assigning it, as value, to "$Cx:Fx$", for all F.

Suppose, on the other hand, that A is not satisfiable; i.e., that $\neg A$ is valid. Then $K(A)$ is equivalent to:

$$Cx:Fx = Cx:Gx \equiv \forall x(Fx \equiv Gx)$$

since A is logically false. But that is just Frege's Basic Law V, and we know it to be inconsistent, thanks to Russell.

Hence, if we could decide the consistency of an arbitrary second-order contextual definition, we could decide the satisfiability of an arbitrary second-order sentence. But that we cannot do. Hence, the consistency of an arbitrary second-order contextual definition is at least as undecidable as the satisfiability of an arbitrary second-order sentence.

Note that the proof shows not only that A is satisfiable if, and only if, $K(A)$ is consistent, but further: If $\mathcal{M}$ is a model with respect to which A is true, then there is an expansion of $\mathcal{M}$ to a model $\mathcal{M}'$ of $K(A)$; conversely, if $K(A)$ is consistent, then it is easy to transform a model $\mathcal{M}'$ of $K(A)$ into a model $\mathcal{M}$ in which A is true. Hence, if we wish to know, say, whether there is a contextual definition that is satisfiable only in countably infinite domains, then we need only determine whether there is a second-order sentence A that is satisfiable only in countably infinite domains. (Of course, there is.)

I should emphasize that, though the undecidability of the consistency of second-order contextual definitions is an obstacle to a reconstructed logicism, it is not necessarily one that cannot be overcome.[3] First, it is of course possible recursively to characterize various subsets of the consistent second-order contextual definitions, and one might then suggest that *those* are the ones that should be accorded special status. The characterization would, however, have to reflect some plausible condition on the acceptibility of contextual definitions that can be independently motivated.

On the other hand, if we could see our way to accepting that even Frege's Basic Law V is, as a *definition*, in good order, saying that, nonetheless, there are no value-ranges, we could view the distinction between 'good' and 'bad' contextual definitions not as a distinction between definitions that are, and definitions that are not, in order; but rather, as a distinction between definitions that do (or may) and definitions that do not (or cannot) succeed in introducing names that *refer*. On such a view, the condition that distinguishes definitions that may introduce names

---

[3] I should thank one of the referees for emphasizing the need to clarify the significance of the formal proof by means of the discussion in this and the following paragraph.

that refer is a condition on the *existence* of the objects whose names the definition purports to introduce. There is then no need to require that the characterization of such a condition be recursive. The condition could be given in purely semantical terms, for there is no reason to suppose that whether there are objects of a given sort must be recursively decidable.[4]

Postscript

The topic of this chapter is what is now known as the 'bad company' objection, so-called because the charge is that HP is guilty by association. This sort of concern was expressed by several different people, including Dummett and Boolos, as mentioned above. My own interest in it was sparked, as so often, by a question from Boolos, the question being, quite simply: Is it recursively decidable whether a given 'contextual definition' is consistent? I worked on the problem for a couple weeks before stumbling upon the simple solution presented here.

The 'bad company' objection has since seen a good deal of attention. Wright discusses it from a more philosophical point of view in a few different places (Wright, 2001b, c, e), but the most important contribution, early on, was probably Kit Fine's paper, "The Limits of Abstraction" (Fine, 1998), which was presented at the conference *Philosophy of Mathematics Today* in the summer of 1993 and which later expanded to become a book by the same title (Fine, 2002). Given this opportunity, I'd like to make a few remarks about the current state of play.

First, it is worth noting that Wright's and Fine's approaches to the problem differ in an important respect. Fine's approach is, broadly speaking, semantic. His investigation is focused on models of abstraction principles and on semantic conditions on such principles. An example of such a principle is the one mentioned above in footnote 4, which has since come to be called the condition of 'stability' (Fine, 1998, p. 511).[5] Wright's approach to the problem is, on the other hand, broadly syntactic. His idea, especially as it is developed in "The Philosophical Significance of Frege's Theorem" (Wright, 2001d), is to place various sorts of restrictions upon the form of abstraction principles themselves, or on the forms of reasoning used to derive various conclusions from them. This difference is important.

---

[4] A promising necessary condition on such definitions is this one: A (second-order) contextual definition may serve to introduce a domain of objects if there is some cardinal κ, such that, for every cardinal $\lambda \geq \kappa$, the definition is satisfiable in a domain of cardinality λ. All such definitions are jointly consistent. The question whether any given such definition does introduce a domain of objects cannot be answered so simply: The condition stated is only a necessary one.

[5] Stability as a condition on the acceptability of abstraction principles has been explored by Hale and Wright (2001a, p. 427, note 14), by Weir (2003), and by Linnebo and Uzquiano (2009), as well as by Fine. I'll have more to say about Linnebo and Uzquiano below.

It will have been noticed by anyone familiar with this literature that the present chapter uses outdated terminology, namely, the phrase "contextual definition". Wright uses this sort of terminology himself in *Frege's Conception*, but he abandoned it not long after this paper was published.[6] Principles like HP are not best understood as 'definitions' of any kind, and so they have come to be known as "abstraction principles". So the idea that one ought to able to tell when one has successfully *defined* something—the idea that was behind Boolos's question to me—falls away. Still, if one thinks of 'abstraction principles', as Wright does, as constituting explanations of concepts, in some privileged sense (Boolos, 1998d, pp. 310–11), then one might still think that one ought equally to be able to tell when one has succeeded in giving such an explanation.

It does not follow that the consistency of abstraction principles must be effectively decidable. What ought to be effectively decidable is whether an abstraction principle is acceptable. If so, however, then the right condition on the acceptability of an abstraction principle seems like it ought to be syntactic rather than semantic. It's not likely that we will always be able to tell if a condition like stability is satisfied; the sorts of syntactic criteria Wright considers, on the other hand, are ones we can tell are satisfied, or not.

I myself, however, continue to be attracted by a somewhat different approach.[7] As I say at the end of the original paper, it seems worth distinguishing the question whether an abstraction principle is 'good' in the sense that it successfully explains (or characterizes) a *concept* from the question whether it is good in the sense that it successfully introduces a new range of *objects* into our ontology. Consider, for example, the much maligned Basic Law V

$$\grave{\epsilon}(F\epsilon) = \grave{\epsilon}(G\epsilon) \equiv \forall x(Fx = Gx)$$

and focus attention on the second-level functional expression that it characterizes. It seems to me that Basic Law V *does* succeed in specifying the *sense* of "`", although it does not succeed in assigning it a reference. I argue for this suggestion in Section 1.2.3, but, in large part, the argument comes to no more than that we can intelligibly say such things as: There are no such things as value-ranges, and: Frege believed that Truth is the value-range of the concept *is Truth*. We know what value-ranges are—or, rather, what they would have been if, *per impossibile*, there had been any.

Whether such a conception of abstraction principles can be defended is an issue I discuss in Chapter 9. If it can be, however, then the question posed by the bad company objection is one about reference, not about sense, and *that* sort of question is one it is natural to answer with a semantic characterization. Here it is important to remember that the

---

[6] I do not know if the paper played any role in causing this change.

[7] I seem to remember Peter Simons suggesting something like this view, as well, at the Philosophy of Mathematics Today conference. But I'm not sure if I remember correctly.

question facing a Neo-logicist is not just, "Under what conditions will an abstractor refer?"—an abstractor being a functional expression characterized by an abstraction principle—but rather: Under what conditions is it logically (or conceptually) *guaranteed* that an abstractor will refer? And to that question, I'd like again to suggest that the answer ought to be, more or less: When an abstraction principle is stable, that is, when there is a cardinal κ such that, for all $\lambda \geq \kappa$, the abstraction principle is satisfiable in a domain of size λ.

That, however, cannot be quite right. For many cardinal numbers κ, there is a sentence of pure second-order logic that says, "There are at least κ objects". There is such a sentence, for example, for κ the least (strongly) inaccessible cardinal. Let A be such a sentence, and consider the abstraction principle:

$$Cx : Fx = Cx : Gx \equiv \forall x(Fx \equiv Gx) \vee A$$

This abstraction principle is stable, for the simple reason that A itself will be true for any cardinal greater than or equal to the first inaccessible. For much the same reason, it requires the universe to contain inaccessibly many objects. But there is obviously something fishy about this principle. What's fishy about it, it seems to me, is that it *requires* there to be inaccessibly many objects, but it does not itself *provide* them. There is but one object whose existence is provided by this principle—as we saw above, in any model of this principle, "$Cx : Fx$" will have the same reference, for each assignment to "F"—so the inaccessibly many objects whose existence it demands are ones about which it tells us nothing. And so, if one asks, "How do we know that there are all those objects?" no answer is forthcoming, other than, "Well, if there weren't, you'd get a contradiction".

Contrast the situation with HP. Given choice, HP is stable: It is satisfiable in any infinite domain; and, since it is not satisfiable in any finite domain, it requires there to be infinitely many objects. What's more, however, HP itself provides those objects: The infinitely many objects whose existence it requires are not just any objects but are precisely the cardinal numbers it purports to characterize. And to the question how we know there are that many objects, we have an informative answer, one given by the proof of Frege's Theorem. The idea, then, is that HP requires the existence of certain objects of the very sort it purports to characterize—the finite cardinals and $\aleph_0$—and it is *because* it requires the existence of those objects that it requires the existence of infinitely many objects.[8]

This intuitive condition leads us naturally to a condition stronger than stability. One way to formulate it would be this. Let P be an abstraction principle, and suppose that P is stable. So there is a cardinal κ such that, for all $\lambda \geq \kappa$, P is satisfiable in a domain of size λ. Fix the least such

---

[8]This idea first appears, but is not discussed in any detail, in the paper reprinted here as Chapter 6.

κ. Then the suggestion is that an abstraction principle is acceptable if, in any model of P of size κ, there are κ abstracts of the sort characterized by P. This doesn't seem to be enough, though. The sort of problem we've been considering can arise at a cardinal below the one at which the abstraction stabilizes. The condition of stability, as we have formulated it, permits 'gaps' below κ. That is, it allows that there should be some cardinals $\mu < \kappa$ at which P is satisfiable, while there are other cardinals ν such that $\mu < \nu < \kappa$ where P is not satisfiable. It will clearly be possible to formulate an abstraction that meets the new condition just mentioned but where, for some $\mu < \kappa$, the abstraction requires the existence of μ objects but not does provide them.

This seems to me a Bad Thing. The idea behind the strengthened notion of stability was that an abstraction principle can only restrict the size of the domain if the objects whose existence it requires are ones it characterizes. This amounts to saying that, *whenever* an abstraction principle requires there to be a certain number of objects, this must be because the principle itself provides that many abstracts. Or, contrapositively, if an abstraction principle *prohibits* there from being a certain number of objects, it must be because the number of abstracts the principle itself provides is incompatible with there being that many objects.

One could fuss a fair bit, to be sure, about how precisely to formulate the condition these intuitions motivate. But here is one natural idea.[9] Say that P is *conservatively* stable if there is a cardinal κ such that (i) for all λ, P is satisfiable in a domain of size λ if and only if $\lambda \geq \kappa$ and (ii) in any model of P, there are at least κ abstracts of the sort characterized by P.[10]

There is a problem with this sort of proposal, one raised by Linnebo and Uzquiano (2009). Let X be some fixed concept and consider the abstraction principle (R_X):

$$\S_X x : Fx = \S_X x : Gx \equiv$$
$$\forall x[Xx \equiv Fx \wedge Xx \equiv Gx] \vee [\neg\forall x(Xx \equiv Fx) \wedge \neg\forall x(Xx \equiv Gx)]$$

The right-hand side, which we shall abbreviate $R_X(Fx, Gx)$, holds just in case either both F and G are co-extensive with X or neither is. So there are just two abstracts here: One for X itself, and one for everything else. It follows that R_X is satisfiable in any domain containing at least two objects, and it therefore follows as well that it is conservatively stable. So

---

[9]The condition about to be stated rules out the sorts of 'gaps' mentioned above. Maybe that seems too strong. If so, then an alternative is this. Say that μ is a *stablization point* if P is satisfiable at μ and either (i) μ is a successor cardinal, λ^+, where P is not satisfiable at λ or (ii) μ is a limit and, for all $\lambda < \mu$ there is ν between λ and μ such that P is not satisfiable at ν. Then P is acceptable if it is stable and, at every stabilization point μ, in every model of P of cardinality μ, there are μ abstracts of the sort characterized by P.

[10]An even stronger condition, and one that seems to me to have some intuitive force, is that there should be at least κ abstracts in any model of size $\geq \kappa$: How could the fact that there are more objects of *other* sorts affect how many abstracts there are?

each such principle ought to be acceptable. The worry, however, is that, if so, then *all* principles of this form taken together ought to be acceptable, and it is there that Linnebo and Uzquiano think there is a problem.

Linnebo and Uzquiano develop the problem by asking us to regard X as a parameter, a free second-order variable. This gives rise, then, to what we might call a parameterized form of the abstraction principle R_X. The next question they consider is how different abstracts $\S_X x : Fx$ and $\S_Y x : Gx$ are related. It is reasonable to suppose—and has widely been assumed, in the literature—that, in general, two abstracts will be identical only if the abstraction principles that characterize them are based upon co-extensive equivalence relations. So, in this case, $\S_X x : Fx$ and $\S_Y x : Gx$ can be identical only if X and Y are co-extensive. It follows, for reasons I shall not discuss, that:

$$\S_X x : Fx = \S_Y x : Gx \ \rightarrow \ \forall H[R_X(F, H) \equiv R_Y(G, H)]$$

which implies:

$$\S_F x : Fx = \S_G x : Gx \ \rightarrow \ \forall H[R_F(F, H) \equiv R_G(G, H)]$$

which itself implies:

$$\S_F x : Fx = \S_G x : Gx \ \rightarrow \ \forall x(Fx \equiv Gx)$$

But that is the fatal direction of Frege's Basic Law V, $\grave{\epsilon}F\epsilon$ now being definable as: $\S_F x : Fx$.

It seems to me, however, that this argument can be resisted. As well as the usual sorts of unary abstraction principles, we can also consider binary abstraction principles: principles that characterize binary functions (rather than, as is usual, unary functions). Such a principle, in the second-order case, has the form:

$$\S x : (Fx, Xx) = \S x : (Gx, Yx) \equiv R_x(Fx, Gx; Xx, Yx)$$

where the formula on the right-hand side must be an equivalence relation in an extended sense.[11] But what Linnebo and Uzquiano write as "$\S_X x : Fx$" could as well be written: $\S x : (Fx, Xx)$, and parameterized abstraction principles are really just binary abstraction principles in disguise. What they are really asking us to consider, then, is a binary abstraction principle, and their considerations about the identity of $\S_X x : Fx$ and $\S_Y x : Gx$, seen from this perspective, amount to nothing other than an intuitive motivation for the binary abstraction principle:

$$\S x : (Fx, Xx) = \S x : (Gx, Yx) \equiv$$
$$\forall H[R_X(F, H) \equiv R_Y(G, H)] \wedge R_X(Fx, Gx) \wedge R_Y(Fx, Gx)$$

---

[11]That is: For any X and Y, we must have $R_x(Fx, Fx; Xx, Yx)$; if $R_x(Fx, Gx; Xx, Yx)$, then $R_x(Gx, Fx; Xx, Yx)$; and if $R_x(Fx, Gx; Xx, Yx)$ and $R_x(Gx, Hx; Xx, Yx)$, then $R_x(Fx, Hx; Xx, Yx)$. And similarly for the second pair of argument places.

The last two conjuncts come from the abstraction principles R_X and R_Y, respectively, and the first is what emerges from the considerations about when $\S_X x : Fx$ and $\S_Y x : Gx$ can be identical. And what Linnebo and Uzquiano show, in effect, is that this principle is inconsistent.

Their argument can be reconstructed as follows. First, $\forall H[R_X(F,H) \equiv R_Y(G,H)]$ is just equivalent to $\forall x(Fx \equiv Gx)$, so our binary principle is equivalent to:

$$\S x : (Fx, Xx) = \S x : (Gx, Yx) \equiv$$
$$\forall x(Fx \equiv Gx) \wedge R_X(Fx, Gx) \wedge R_Y(Fx, Gx)$$

But then this implies:

$$\S x : (Fx, Fx) = \S x : (Gx, Gx) \equiv$$
$$\forall x(Fx \equiv Gx) \wedge R_F(Fx, Gx) \wedge R_G(Fx, Gx)$$

the right-hand side of which is just equivalent to $\forall x(Fx \equiv Gx)$. Hence:

$$\S x : (Fx, Fx) = \S x : (Gx, Gx) \equiv \forall x(Fx \equiv Gx)$$

which is Basic Law V plus a bit of harmless repetition. So the binary abstraction principle with which we've been concerned isn't conservatively stable, and we don't have to accept it.

How does that help? The worry underlying Linnebo and Uzquiano's argument[12] is that, even though all the individual principles R_X are (conservatively) stable, all of these principles taken together cannot be true. And one might think their argument still establishes that fact. Moreover, there is a different argument for that conclusion. Assume, as before, that $\S_X x : Fx$ and $\S_Y x : Gx$ can be identical only if R_X and R_Y are co-extensive as abstraction principles. Then it follows that each such abstraction principle introduces two distinct objects. If so, then all of these principles taken together lead to a violation of Cantor's Theorem: There are more such principles than there are objects, since there is such a principle for each concept X, so all of them taken together introduce more objects than there are objects, which is bad. So it would appear that at least some of the R_X have to be rejected. In virtue of symmetry, it would seem that either some of them must go, or all of them must go.

I suggest that all of them should go. Consider a particular one of the R_X, say this one:

$$\S_{\text{Man}} x : Fx = \S_{\text{Man}} x : Gx \equiv$$
$$\forall x[\textbf{Man}(x) \equiv Fx \wedge \textbf{Man}(x) \equiv Gx]\vee$$
$$[\neg\forall x(\textbf{Man}(x) \equiv Fx) \wedge \neg\forall x(\textbf{Man}(x) \equiv Gx)]$$

This is what we might call a 'non-logical' abstraction principle, on the ground that the equivalence relation mentioned on the right-hand side

---

[12]This is in fact a form of the first problem they introduce, in section 2.

makes use of non-logical vocabulary. It is indeed a natural idea that, if such a principle is acceptable, then all principles of that form ought to be acceptable. But a different way to express much the same thought is to say that the acceptability of such a principle should stand or fall with the acceptability of its parameterized form, that is, of its binary form. Or perhaps better: If such a principle is acceptable, that must be *because* its binary form is acceptable. If so, then none of the R_X are acceptable, because the binary form is, as we saw, inconsistent.

The suggestion, then, would be as follows. First, a logical abstraction principle—one based upon an equivalence relation formulable in purely logical terms—is acceptable if, and only if, it is conservatively stable. Second, a non-logical abstraction principle is acceptable if, and only if, its parameterized form (that is, its n-ary form), which will itself be a logical abstraction principle, is acceptable.

Such a view limits the scope of abstraction principles significantly—at least, of the ones acceptable on purely conceptual grounds. For example, the familiar abstraction:

$$\mathrm{dir}(a) = \mathrm{dir}(b) \equiv a \parallel b$$

is, on this view, not acceptable on purely conceptual grounds. This is because parallelism is not a logical notion, so it has to be parameterized. This leads to:

$$\mathbf{C}xy:(a, Rxy) = \mathbf{C}xy:(b, Sxy) \equiv \forall x \forall y (Rxy \equiv Sxy) \wedge Rab \wedge Sab$$

which implies:

$$\mathbf{C}xy:(a, Rxy) = \mathbf{C}xy:(a, Sxy) \equiv \forall x \forall y (Rxy \equiv Sxy)$$

which is of course inconsistent. Indeed, it seems likely that *no* non-logical abstraction principles will turn out, on this view, to be acceptable on purely conceptual grounds. But non-logical abstractions *shouldn't* be acceptable on purely conceptual grounds, it seems to me, and I've argued for a conclusion in that vicinity in Chapter 9. The good news, of course, is that HP is a conservatively stable logical abstraction, so it might still be acceptable.

Of course, the question that just begs to be asked is: Can all the purely logical, conservatively stable abstractions be true together?

11

Finitude and Hume's Principle

With an eye towards philosophical issues I shall discuss in Section 11.2, George Boolos (1998h) has investigated the relative strengths of two sorts of systems of second-order arithmetic. The more familiar of these originates with the work of Dedekind and Peano; the less familiar, with that of Frege. Dedekind-Peano systems characterize the natural numbers in terms of properties of the sequence of natural numbers; these systems may be thought of as axiomatizations of finite ordinal arithmetic. The Fregean systems, on the other hand, characterize the natural numbers as finite cardinals.[1] Fundamental to Fregean systems is an axiom specifying the condition under which two concepts[2] have the same cardinal number, together with another specifying under what conditions a cardinal number is finite.

For our purposes, we may axiomatize (second-order) Dedekind-Peano arithmetic as follows:

1. $\mathbb{N}0$

2. $\forall x \forall y (\mathbb{N}x \wedge Pxy \rightarrow \mathbb{N}y)$

3. $\forall x \forall y \forall z (\mathbb{N}x \wedge Pxy \wedge Pxz \rightarrow y = z)$

4. $\forall x \forall y \forall z (\mathbb{N}x \wedge \mathbb{N}y \wedge Pxz \wedge Pyz \rightarrow x = y)$

5. $\neg \exists x (\mathbb{N}x \wedge Px0)$

6. $\forall x (\mathbb{N}x \rightarrow \exists y (Pxy))$

7. $\forall F (F0 \wedge \forall x \forall y (\mathbb{N}x \wedge Fx \wedge Pxy \rightarrow Fy) \rightarrow \forall x (\mathbb{N}x \rightarrow Fx))$

Let us call this system **PA2** (for second-order Peano arithmetic). We use a relational expression "$P\xi\eta$", rather than the more usual functional expression "$S\xi$", to facilitate comparison with Fregean systems.

The most familiar Fregean system has but one 'non-logical' axiom, HP, which states that the number of Fs is the same as the number of

---

[1] For further discussion of this difference, see my paper "The Finite and the Infinite in Frege's *Grundgesetze der Arithmetik*" (Heck, 1998b, §5).

[2] I shall use this term to denote whatever are in the range of the second-order variables. Though my choice of terminology certainly suggests a view about what these are, my remarks here do not depend upon it. It is, of course, essential to the logicist project that second-order logic is logic, but this is not at issue among those whose positions we shall be discussing.

Gs just in case the Fs and Gs are in one-one correspondence. Taking "$\text{Eq}_x(Fx, Gx)$" to abbreviate one of the many equivalent second-order formulae that define "the Fs correspond one-one with the Gs" (or, in Frege's terminology, "the Fs are equinumerous with the Gs"), HP is then:

$$\mathbf{N}x:Fx = \mathbf{N}x:Gx \equiv \text{Eq}_x(Fx, Gx)$$

The second-order theory whose sole non-logical axiom is HP is **FA** (for "Frege arithmetic"). Note that "$\mathbf{N}x:\Phi x$" is a unary, second-level, term-forming operator: The result of substituting any formula (possibly containing further occurences of "$\mathbf{N}x:\Phi x$") for "Φx" in "$\mathbf{N}x:\Phi x$" is a term.

The definition of *finite* or *natural* number can be given in different ways. In Frege's work (Gl, §§74, 76, 83), zero and the relation of predecession are defined and, famously, the finite numbers are defined as those to which zero stands in the weak ancestral of this relation. The necessary definitions are thus:

$$0 = \mathbf{N}x:x \neq x$$
$$Pmn \equiv \exists F \exists y[Fy \wedge n = \mathbf{N}x:Fx \wedge m = \mathbf{N}x:(Fx \wedge x \neq y)]$$

Frege defines the *strong ancestral* of a relation R as follows:

$$R^*ab \equiv \forall F[\forall z(Raz \to Fz) \wedge \forall x \forall y(Fx \wedge Rxy \to Fy) \to Fb]$$

And he defines the *weak ancestral* thus:

$$R^{*=}ab \equiv R^*ab \vee a = b$$

Frege's definition of natural number is then:

$$n \text{ is a natural number just in case } P^{*=}0n$$

There are other ways to proceed, however. In sections K and Λ of Part II of *Grundgesetze der Arithmetik*, Frege formulates a purely second-order definition of finitude, to state which we need an additional definition (Gg, v. I, §158):[3]

$$\mathbf{Btw}_{xy}(Rxy, a, b)(n) \stackrel{df}{\equiv}$$
$$\forall x \forall y \forall z(Rxy \wedge Rxz \to y = z) \wedge \neg \exists x(R^*xx) \wedge R^{*=}an \wedge R^{*=}nb$$

Thus, n is between a and b in the R-series if, and only if, R is a functional relation, in whose strong ancestral no object stands to itself (i.e., whose ancestral contains no 'loops'), such that a stands in the weak ancestral

---

[3]The definition is actually given in terms of 'value-ranges'. I have silently translated it into pure second-order logic. I shall insert the bound variables, such as "x" and "y" on the left-hand side here, into the definitions, but will drop them when doing so causes no confusion.

of R to n, which in turn stands in the weak ancestral of R to b. Frege's definition of finitude is then:[4]

$$\mathrm{Fin}_x(Fx) \;\equiv\; \exists R \exists x \exists y \forall z [Fz \equiv \mathrm{Btw}(R,x,y)(z)]$$

In words: A concept is finite just in case the objects falling under it may be ordered in a certain way, namely, as the objects between x and y in the R-series, for some R, x, and y. That this definition is correct follows from Theorems 327 and 348 of *Grundgesetze*:

$$(327) \quad \mathrm{Fin}_x(Fx) \to P^{*=}(0, \mathrm{N}x:Fx)$$

$$(348) \quad P^{*=}(0, \mathrm{N}x:Fx) \to \mathrm{Fin}_x(Fx)$$

Thus, a concept is finite, in Frege's sense, just in case its number is a natural number. Frege's definition of natural number could therefore be replaced by:

$$\mathrm{N}n \equiv \exists F(\mathrm{Fin}_x(Fx) \wedge n = \mathrm{N}x:Fx)$$

Of course, this definition will be adequate only in a theory strong enough to prove (327) and (348).[5]

Analogues of these theorems are the crucial lemmas in the proofs of the main result of this chapter (see Lemmas 11.4.1, 11.4.2, and 11.4.5, below). As we shall see, given Frege's definitions of "0" and "P", theorem (327) becomes a theorem of second-order logic. The proof of theorem (348), however, must rely upon additional assumptions, for without additional assumptions, it is consistent that all concepts have the number 0 (and, of course, it is consistent that some of these are not finite).

In investigating the relative strength of Dedekind-Peano and Fregean systems, there are two sorts of questions one might raise. First, one might inquire about the relative consistency of such theories. To ask whether **FA** is consistent relative to **PA2** is to ask whether the consistency of **FA** would follow from that of **PA2**. One familiar sort of proof that it would consists in a demonstration that **FA** can be relatively interpreted in **PA2**. Roughly speaking, to relatively interpret **FA** in **PA2** is to give definitions of the primitives of **FA** in terms of the primitives of **PA2**, which definitions, when added to **PA2**, allow one to prove relativizations of the axioms of **FA** in **PA2**: By a relativization of a formula is meant, as usual, the result of restricting quantifiers occuring in the formula by means of some formula of **PA2**.[6] If **FA** can be interpreted in **PA2**, it follows immediately that, if there is a proof of a contradiction in **FA**, that proof can be

---

[4]Frege does not explicitly formulate any such definition, but it is clear from the theorems proven in sections K and Λ that this is what he intends (Heck, 1998b).

[5]Frege proves them in the system **FA+FD**. As we shall see, they are also provable in **HPF+FD** (and so in **PAF+FD**).

[6]Of course, in the context of second-order logic, one must restrict not only the first-order, but also the second-order, quantifiers. As we replace "$\forall x A(x)$" by "$\forall x(Rx \to A(x))$", so we replace "$\forall F A(F)$" by "$\forall F \{\forall x[Fx \to Rx] \to A(F)\}$".

mimicked in **PA2**, so that, if **PA2** is (syntactically) consistent, so is **FA**. As it turns out, **FA** and **PA2** are equi-interpretable—each can be relatively interpreted in the other—and so equi-consistent—an inconsistency in either would imply an inconsistency in the other.

Still, one might wonder whether **FA** is not, in some other sense, a stronger theory than **PA2**. This question is more easily understood when we have two theories formulated in the same language. Consider, for example, the following Dedekind-Peano system, which we shall call **PAS** (for 'Strong' Peano arithmetic):

1. $\mathbb{N}0$

2. $\forall x \forall y(\mathbb{N}x \wedge Pxy \rightarrow \mathbb{N}y)$

3. $\forall x \forall y \forall z(Pxy \wedge Pxz \rightarrow y = z)$

4. $\forall x \forall y \forall z(Pxz \wedge Pyz \rightarrow x = y)$

5. $\neg \exists x(Px0)$

6. $\forall x(\mathbb{N}x \rightarrow \exists y(Pxy))$

7. $\forall F[F0 \wedge \forall x \forall y(\mathbb{N}x \wedge Fx \wedge Pxy \rightarrow Fy) \rightarrow \forall x(\mathbb{N}x \rightarrow Fx)$

Clearly, every axiom of **PA2** is a theorem of **PAS**, but the converse does not hold. As far as the axioms of **PA2** are concerned, zero could have as its predecessor Julius Caesar, so long as Caesar is not a natural number. Thus, **PAS** is strictly stronger than **PA2**. This is perfectly compatible with the fact that **PA2** and **PAS** are equi-interpretable. (To interpret **PAS** in **PA2**, no 'definitions' are needed: Just restrict all the quantifiers in the axioms of **PAS** to the natural numbers, i.e., by the formula "$\mathbb{N}x$".)

The question we are discussing concerns the proof-theoretic strength of the two systems. This question is harder to raise when the theories under discussion are not formulated in the same language: Obviously, the axioms of **FA** are not going to be theorems of **PA2**, since the axioms of **FA** are not even sentences in the language of **PA2**. Nor would expanding the language of **PA2** to include such formulae help. To consider the relative strength of theories formulated in different languages, what we require is a bridge theory that relates (the referents of) the primitives of **PA2** to those of **FA**. We can then ask whether, with the aid of one or another bridge theory, the theorems of **FA** can be proven in **PA2**.[7] One might wonder what difference there is between the question whether **FA** can be

---

[7]One might well wonder what such a bridge theory must be like, if the provability of the theorems of one system from those of another is to have the kind of interest it is here taken to have. I do not know how this question should be answered. Surely, however, it is sufficient if the axioms of the bridge theory are definitions of the primitives of one of the two theories in terms of the primitives of the other. The bridge theories we shall employ below are of this sort. The question we are considering is thus one of interpretability, rather than relative interpretability.

interpreted in **PA2** and the question whether the theorems of **FA** can be proven in **PA2**, with the aid of some bridge theory. One might ask, if **FA** can be relatively interpreted in **PA2**, will that not itself guarantee that there is *some* bridge theory with the aid of which the theorems of **FA** can be proven in **PA2**? namely, that theory whose axioms are exactly the definitions used in interpreting **FA** in **PA2**? The answer to this question is "No". One must not overlook the fact that, in relatively interpreting one theory in another, it may be essential to relativize the axioms of the former theory: The usual relative interpretation of **FA** in **PA2**, for example, requires that the quantifiers occuring in HP be restricted to the natural numbers. There is no necessity that there should be a way of mimicking this restriction in *any* bridge theory, and there is certainly no need that any particular bridge theory should impose such a restriction.

Our chief interest here is in the relative strength of various Fregean systems and various Dedekind-Peano systems. We thus must make use of a bridge theory that relates (the referents of) their primitives. The bridge theory in which we shall be interested has the following three axioms:

$$0 = \mathbb{N}x : x \neq x$$
$$Pmn \equiv \exists F \exists y [Fy \wedge n = \mathbb{N}x : Fx \wedge m = \mathbb{N}x : (Fx \wedge x \neq y)]$$
$$\mathbb{N}n \equiv P^{*=}0n$$

Let us call this theory **FD**, for "Frege's definitions", since these are the definitions Frege uses in deriving axioms for arithmetic (in particular, those of **PAS**) in **FA**.

11.1 The Systems

We will here investigate the relative strengths of five different systems of arithmetic. The Dedekind-Peano systems at which we shall look are **PA2** and **PAS**, mentioned above, and a third system, **PAF**, whose axioms are those of **PA2** plus:

$$\forall x \forall y (\mathbb{N}x \wedge Pyx \to \mathbb{N}y)$$

This axiom, which we also call PAF (for "predecessors are finite"), states that any predecessor of a natural number is a natural number.[8] As we shall see, **PAF** is stronger than **PA2** and weaker than **PAS**.

Before discussing the variations on **FA** at which we shall look, let me make a remark about the background logic in which we shall be working. In the case of the Dedekind-Peano systems, the logic is usually taken to

---

[8]Boolos remarked to me that, when presenting "On the Proof of Frege's Theorem" (Boolos, 1998h) as a lecture, he heard it objected that PAF—or, more precisely, its consequence NPZ, to be mentioned below—cannot be true, since -1 surely precedes 0. But the theories in which we are interested here are theories of cardinal or ordinal numbers, and "*P*" is defined as a relation between such numbers. Negative numbers are neither ordinals nor cardinals.

be standard (axiomatic) second-order logic, with full, impredicative comprehension. In discussing **FA** and its relations to the Dedekind-Peano systems, however, it is convenient to take the logic also to contain the axiom Boolos calls FE, for "functional equivalence":

$$\forall x (Fx \equiv Gx) \rightarrow \mathrm{N}x : Fx = \mathrm{N}x : Gx$$

This axiom is clearly valid on any extensional semantics for second-order logic and so should itself be regarded as a truth of (extensional, higher-order) logic.[9] The system whose axioms are those of second-order logic, plus FE, Boolos calls **Log**. We shall henceforth suppose our background logic to be **Log**.

The Fregean systems at which we shall look are **FA** and a variation on it, in which HP has been weakened by restricting its range of application. The axiom is HPF (for *Finite* HP):

$$\mathrm{Fin}_x(Fx) \vee \mathrm{Fin}_x(Gx) \rightarrow [\mathrm{N}x : Fx = \mathrm{N}x : Gx \ \equiv \ \mathrm{Eq}_x(Fx, Gx)]$$

Here, the formula "$\mathrm{Fin}_x(Fx)$" may be defined *via* any of the equivalent second-order definitions of finitude: We shall take it to be defined as Frege defines it.[10] HPF states that finite concepts have the same number if, and only if, they are equinumerous and that no infinite concept has the same number as any finite one—making no further claim about the conditions under which infinite concepts have the same number. (For all that HPF says, all infinite concepts could have the same number, so long as no finite concept also has that number.) Call the theory whose sole non-logical axiom is HPF, **FAF** (for finite Frege arithmetic).

The results of this chapter may now be summarized in the following diagram:

$$\mathbf{FA} \Rightarrow \mathbf{PAS} \Rightarrow \{\mathbf{PAF} \Leftrightarrow \mathbf{FAF}\} \Rightarrow \mathbf{PA2}$$

Here, "$\Rightarrow$" means: Is strictly stronger than, relative to the bridge theory **FD**; that is, "**A** $\Rightarrow$ **B**" means that every theorem of **B** is a theorem of **A+FD**, but that not every theorem of **A** is a theorem of **B+FD**. That **FAF** and **PAF** occur together in the braces indicates that they are equivalent, relative to **FD**: Every theorem of **PAF** is a theorem of **HPF+FD**, and every theorem of **FAF** is a theorem of **PAF+FD**. What we need to prove are thus the following:

1.   **FA** $\Rightarrow$ **PAS** (Theorem 11.3.1)

2.   **PAS** $\Rightarrow$ **PAF** (Theorem 11.3.2)

---

[9] As should the axiom schema: $\forall x (Fx \equiv Gx) \rightarrow A(F) \equiv A(G)$.

[10] Another possibility would be to define finitude using the third-order ancestral.

$$\mathrm{Fin}_x(Fx) \ \equiv \ \forall \Phi[\Phi_x(x \neq x) \wedge \forall G(\Phi_x Gx \rightarrow \forall z(\Phi_x(Gx \vee x = z)) \rightarrow \Phi_x Fx]$$

The definition says that a concept is finite if it has every second-level property that (i) the empty concept has and (ii) is hereditary under adjunction.

3.   **PAF** $\Rightarrow$ **FAF** and **FAF** $\Rightarrow$ **PAF**, so **PAF** is equivalent (*modulo* **FD**) to **FAF** (Theorems 11.4.4, 11.4.9, respectively)

4.   **PAF** $\Rightarrow$ **PA2** (Theorem 11.3.3)

Some of the required proofs have been discussed in detail by Boolos; I shall merely indicate how those proofs go. The main work of the present chapter consists in establishing the third result mentioned.

11.2   On the Philosophical Significance of These Results

Before turning to the proofs, let me make a few remarks about the inspiration for the present investigation and about its philosophical implications.

In *Frege's Conception of Numbers as Objects* (Wright, 1983), Crispin Wright re-discovered Frege's Theorem, which states that the axioms of **PAS** are provable in **FA+FD**, proved it in some detail, and conjectured that **FA** is consistent, which it turned out to be.[11] On the basis of this result, Wright not only revived Frege's logicist project but claimed that it was substantially vindicated by the proof of Frege's Theorem. If logicism were to vindicated completely, of course, HP would have to be shown to be a logical truth—which it certainly cannot be, given our contemporary understanding of 'logical truth'. Nevertheless, Wright argued, HP is 'analytic', whence the truths of arithmetic are logical consequences of an analytic truth and so, presumably, are themselves analytic. The sense in which HP is analytic is that, "even if inadequate as a definition, it nevertheless succeeds as an *explanation*; . . . it contrives to fix the meaning of the sorts of occurence of ["Nx : Φx"] which it fails to eliminate" (Wright, 1983, p. 140; see also p. 153).

In the paper mentioned at the outset, Boolos shows that **FA** is strictly stronger than **PAS** (and so **PA2**), relative to the bridge theory **FD**. His purpose is not primarily technical: He intends this to be one consideration in favor of the view that, *contra* Wright, HP is neither 'analytic', nor a 'conceptual truth', nor any such thing (Boolos, 1998d). Boolos does not explain in detail why his result should trouble Wright, but his point seems to be that, since **FA** is significantly stronger than **PAS** (which is itself a stronger theory even that **PA2**, which is *itself* a very strong theory), it is implausible to claim that HP is a conceptual truth. It is difficult, however, to evaluate the force of this consideration: As Boolos recognizes, Wright would likely reply that since his view *is* the view that arithmetic is analytic—and, indeed, that the general theory of cardinality that **FA** embodies is analytic—he is simply being accused of holding that very view.

---

[11]See Chapter 1 for the history.

Still, there is a stronger consideration in the vicinity. For the additional proof-theoretic strength of **FA**, as compared to **PA2**, reflects a very real, and very large, conceptual gap between second-order arithmetic and the general theory of cardinality.

As several people have pointed out, the older term "Hume's Principle" was something of a misnomer. In the passage Frege cites when introducing HP, Hume is speaking not of cardinality in general but only of the cardinality of finite concepts (or sets, or whatever). As of course he was. Prior to Cantor's work on transfinite numbers, the view that all equinumerous concepts have the same number, whether they are finite or infinite, was almost universally rejected, because it gives rise to antinomies: For example, it implies that the number of natural numbers is the same as the number of even numbers, and that can seem absurd, because there are *lots* of natural numbers that are not even—indeed, according to Cantor, as many numbers as there are natural numbers! Cantor's realization that one can coherently suppose, even in the infinite case, that all and only equinumerous sets have the same cardinality constituted as enormous a *conceptual* advance as his introduction of transfinite numbers was a mathematical advance.

It is easy to forget this, so at home are we initiates with Cantor's ideas. But it is just as easy to be reminded of it: One has an opportunity every time a student wanders into one's office puzzled about these very antinomies. Indeed, my own work on this very chapter was fundamentally altered by just such an experience. A friend of mine—a professional philosopher, and so no fool—was telling me about an objection one of his students had raised in lecture. The student had insisted that there is only one 'kind' of infinity, and my friend had been tempted to reply (but wanted to check with me first) that of course there was more than one kind of infinity, since both the natural numbers and the even numbers are infinite, and the infinities in question certainly cannot be of the same kind. He was troubled by my response. Not just philosophically troubled, mind you, but really *bothered*: As I conveyed Cantor's ideas to him, he kept saying, "That's very worrying", over and over again. I had to do a lot of explaining before he was again at ease. He made the leap, but my experience served to remind me how great a conceptual leap he made at that point—and so how great a conceptual leap Cantor himself had made.

I am not going to argue that HP *isn't* a conceptual truth: On that question I regard myself, to steal a phrase of David Wiggins's, as a militant agnostic. My point is that, once one has recognized just how great a conceptual advance is required if one is acknowledge the truth of HP, one can no longer accept that Frege's Theorem has the sort of epistemological interest Wright and others have wanted it to have. What is required if logicism is to be vindicated is not just that there is *some conceptual truth or other* from which what *look like* axioms for arithmetic follow, given certain definitions: That would not show that the truths of arithmetic, *as we ordinarily understand them*, are analytic, but only that arithmetic can be interpreted in some analytically true theory.[12] To put the point differently, if we are so much as to evaluate logicism, we must first uncover the 'basic laws of arithmetic', laws that are not just sufficient to allow us to prove translations of arithmetical truths, but laws from which arithmetical truths themselves can be proven. (The distinction is not a mathematical one, but a philosophical one.) But, if these 'basic laws' are to be the basic laws of *arithmetic*, they had better be ones upon which ordinary arithmetical reasoning relies.

If Frege's Theorem is to have the kind of interest Wright suggests, it must be possible to recognize the truth of HP by reflecting on fundamental features of arithmetical reasoning—by which I mean reasoning about, and with, *finite* numbers, since the epistemological status of arithmetic is what is at issue. For what the logicist must establish is something like this: That there is, implicit in the most basic features of arithmetical thought, a commitment to certain principles, the (tacit) recognition of whose truth is a necessary precondition of arithmetical reasoning, and from which all axioms of arithmetic follow. Having identified these basic laws, we will then be in a position to discuss the question whether they are analytic, or conceptual truths, or what have you.

What used to be my favorite argument for the analyticity of HP went roughly like this: HP is a conceptual truth, because it is part of the very concept of cardinality that equinumerous concepts have the same cardinal number.[13] Perhaps, but the argument overlooks the fact that, though this maybe true of our present concept of cardinality, 'we' did not even *have* this concept of cardinality until the late nineteenth century. A recognition of the very coherence of our present concept of cardinality requires the conceptual leap I just mentioned, whence, even if HP is analytic of our present concept of cardinality, it is extremely odd to attempt to ground our knowledge of arithmetic, of all things, upon it. Moreover, there is demonstrably no way in which a recognition of the truth of HP can arise simply from reflection on the nature of ordinary arithmetical thought— not, that is, if the principles governing 'ordinary arithmetical thought' are captured by the axioms of **PA2** (or even of **PAS**) and the outcome of 're-flection' is something that could be written down as a proof. *That* is what follows from the fact that **HP** is proof-theoretically stronger than **PAS** (and so **PAF** and **PA2**). The disparity of strength parallels the conceptual

---

[12]If analysis were analytic, as Frege thought it was, then Euclidean geometry would be interpretable in an analytically true theory, *via* Cartesian co-ordinates. Are we to conclude that Frege's position was inconsistent, since he held that geometry is not analytic but synthetic *a priori*? Surely not.

[13]Wright (2001d) makes a similar but somewhat different claim, namely: Even if HP is not analytic of any pre-existing concept of cardinality, it is perfectly in order to introduce such a concept by means of HP. The difficulty is that this makes it unclear what relation might obtain between claims in which that concept occurs and our ordinary arithmetical claims.

disparity remarkably well—so well as to remind one why well-conceived technical investigations can be so philosophically fruitful.

To summarize and emphasize: HP, conceptual truth or not, cannot be what underlies our knowledge of arithmetic. For no amount of reflection on the nature of arithmetical thought could ever convince one of HP, nor even of the coherence of the concept of cardinality of which it is purportedly analytic. Granted, any rationalist project of this sort will have to invoke a distinction between the 'order of discovery' and the 'order of justification'. But the objection is not that HP is not known by ordinary speakers, nor that there was a time when the truths of arithmetic were known but HP was not. It is that, even if HP is thought of as 'defining' or 'introducing' or 'explaining' our present concept of cardinality, the conceptual resources required if one is so much as to recognize the coherence of this concept (let alone HP's truth) vastly outstrip the conceptual resources employed in arithmetical reasoning. Wright's version of logicism is therefore untenable.

Of course, this does not imply that *no* form of logicism is tenable. Careful examination of Boolos's proofs itself reveals a way forward. The important observation is that the distinction between finitude and infinitude plays a major role in these proofs. Consider, for example, the sort of model Boolos uses to show that **FA** is stronger than **PAS**. Take the domain of the model to be the natural numbers, together with Caesar and Brutus. Given any term of the form "$Nx : Fx$", assign it a value according to the following scheme:

- Caesar, if there are infinitely many Fs and infinitely many non-Fs

- Brutus, if there are infinitely many Fs but finitely many non-Fs

- n, if there are exactly n Fs, for some natural number n

Interpret the primitives of **PAS** according to the 'definitions' of the bridge theory **FD** (thus guaranteeing that all axioms of the bridge theory are true in the model): Thus, "0" denotes the number 0; "$N\xi$" is true of x iff x is a natural number; and "P" is true of the pair $<x, y>$ just in case either $x = y =$ Caesar, or $x = y =$ Brutus, or $y = x + 1$. It should be clear that the axioms of **PAS** are all true in this model. But HP is not: For example, "Even(ξ)" having been appropriately defined, "$Nx : (Nx \wedge \text{Even}(x)) = Nx : Nx$" will be false in the model—the former term denoting Caesar, the latter, Brutus—even though the evens are equinumerous with the natural numbers.

The important point is that, in this model, *HP fails only for infinite concepts*. Indeed, as Boolos essentially observes, HPF holds in every model of **PAS+FD**: So, in any model for **PAS+FD**, HP will fail to hold, if it does, only because there are some *infinite* concepts that are assigned different numbers although they are equinumerous. The natural technical question is then: Is there a reasonable Dedekind-Peano system that is equivalent to **FAF** in the presence of **FD**? The answer is that there is: Relative to **FD**, **FAF** is equivalent to **PAF**.

Now, in *Grundgesetze*, Frege actually derives the axioms of **PAS** in **FA+FD**, and these proofs do exploit the full power of HP (since the axioms of **PAS** are *not* provable in **HPF+FD**.). But Frege's proofs can easily be adapted to yield proofs, in **HPF+FD**, of the axioms of **PAF**: One need only relativize certain of the formulae appearing in those proofs to the natural numbers. Frege's development of arithmetic thus does not depend essentially upon (though it may have been psychologically impossible without) the conceptual advance of which I have been speaking. This is striking enough, but it is all the more so since my objections to Wright's attempt to ground arithmetic on HP simply cannot be raised against an attempt to ground it on HPF. For HPF's weakness, as compared to HP, reflects the conceptual distance between them, too.

There are two points to be made here: First, that recognizing the truth of HPF does not require making the conceptual advance made by Cantor; and Second, that one can be convinced of the truth of HPF merely by reflection on ordinary arithmetical thought. To take the first point: Just as HP may be thought of as the sole axiom of a general theory of cardinal numbers, HPF may be thought of as the sole axiom of a theory of finite cardinals. And since HPF makes no claims whatsoever about the conditions under which infinite concepts have the same cardinality,[14] it will not give rise to any of the antinomies generated by HP, whence one does not need to make Cantor's leap before one can accept the truth of HPF. Indeed, not only could HPF have been recognized as true prior to Cantor's work, it almost universally was. Bolzano, who was famously skeptical about HP in the infinite case, accepted HPF (Bolzano, 1950, §§21–2),[15] as did just about everyone else who considered the matter. For all that HPF says is that, in the finite case, all and only equinumerous concepts have the same number—and who knows what we should say about the infinite ones, other than that none of them have got the same number as any of the finite ones.

The second point is that this claim really is implicit in arithmetical reasoning and that one can convince oneself of its truth, and come to understand why it is true, by (and perhaps only by) reflecting on basic aspects of arithmetical thought. Now, it is not initially obvious to what notion of finitude we might appeal in reflecting on our arithmetical thought and investigating whether a commitment to HPF is implicit in it. Nor is it clear whether that notion is itself a logical one. But I submit that the intuitive notion of a finite concept is that of one the objects falling under

---

[14]This would be all the more clear were HPF formulated in a logic that allowed partial functions, so that it was defined only for finite concepts. But working in such a logic would complicate matters quite unnecessarily. Such a formulation would also answer the objection that, in its present form, HPF does not have the form of a Fregean abstraction.

[15]In §22, Bolzano gives an argument for HPF similar to the one to be given in the next paragraph.

which can be counted, i.e., enumerated by means of some process that eventually terminates. Frege's definition of finitude directly reflects this intuitive notion: What the definition says is precisely that a concept is finite if, and only if, the objects falling under it can be ordered as a discrete sequence that has a beginning and an end.[16] Our intuitive notion of finitude can thus be straightforwardly transcribed into second-order logic—thereby showing, *modulo* the status of second-order logic itself, that this intuitive notion is a logical one.

How then can one convince oneself of the truth of HPF? It suffices to realize that the process of counting, which lies at the root of our assignment of numbers to finite concepts, already involves the notion of a one-one correspondence: As Frege frequently points out, to count is to establish a one-one correspondence between certain objects and an initial segment of a sequence of numerals, starting with "1"; the process ends with a numeral that names the number of objects counted. By the transitivity of "is equinumerous with", equinumerous concepts must be equinumerous with the same initial segments;[17] conversely, any concepts the objects falling under which can be put in one-one correspondence with the same initial segment must be equinumerous. So any two concepts the objects falling under which can be counted—i.e., any two finite concepts—will be assigned the same numeral by the process of counting—i.e., will have the same number—if, and only if, they are equinumerous. And, of course, no infinite concept will get assigned any number by the process of counting. That is enough to establish HPF.[18]

Of course, one might yet have all kinds of worries about the claim that HPF is a conceptual truth. There are two broad classes of such worries: Those that arise from its impredicativity and those that rest upon the fact that it implies the existence of a lot of objects (infinitely many). I am not going to say anything here about questions of the former sort.[19] But, with regard to the latter, let me say that one needs to be very careful with such objections. Any principle sufficient to 'ground' arithmetic in the relevant sense obviously has to imply the existence of infinitely many objects. So

---

[16]That Frege intended his definition to correspond to this intuitive notion is, furthermore, clear from the way he proves Theorems (327) and (348) of *Grundgesetze* (Heck, 1998b).

[17]That there is only one such initial segment will follow from the finitude of the segments themselves, given Frege's definition, for we shall be able to show that distinct initial segments are never in one-one correspondence. The proof will depend upon certain claims about the numerals themselves, claims corresponding to the axioms of **PAF**. See here Frege's discussion of counting in *Grundgesetze* (Gg, v. I, §108). I should emphasize that the argument being given here can be formalized: Indeed, the proof of Theorem 11.3.1 can be read as a very rough formalization of it.

[18]A quite different argument can also be given for HPF, namely, the sort discussed in Section 7.3.

[19]I mean to include so-called 'bad company' objections. For discussion of these, see the exchange between Boolos (1998d) and Wright (2001b, 2001d), and the postscript to Chapter 10. Concerning the impredicativity of HP, see the exchange between Dummett (1991b, pp. 187–9 and 217–22) and Wright (2001c), as well as Chapter 6.

one cannot object to someone who is trying to establish that the truths of arithmetic are conceptual truths, or logical consequences of such, by saying that the principle on which he proposes to base arithmetic cannot be a conceptual truth, because no conceptual truth can imply that there are infinitely many objects. One might as well object that the principle yields arithmetic, i.e., that his premises imply his conclusion, i.e., accuse him of holding his view. Or, better, one should just say, flat-footedly, that arithmetic *can't* be 'analytic', in any reasonable sense, since *it* implies the existence of lots of objects. But that is not so much an objection as a refusal even to discuss the matter. No one interested in the question whether arithmetic is 'analytic' is likely to be moved by that thought.

But, having said all of that, let me emphasize that the importance of the question whether HPF is 'analytic', in the context of discussions of logicism, should not be allowed to obscure the fact that how we answer it does not affect the philosophical interest of the modification of Frege's Theorem to be presented below. If HPF really is the 'basic law of arithmetic', in the relevant sense, that is philosophically important, whatever its epistemological status might turn out to be.

11.3   The Relative Strengths of the Systems

We turn now to the proofs of the five results mentioned at the end of Section 11.2. In this section, we prove the first, second, and fourth. We will prove the other two in the following two sections.

Theorem 11.3.1. FA $\Rightarrow$ PAS

Proof. In the last section, we saw a countermodel that establishes that HP is not a theorem of **PAS+FD**. That all the axioms of **PAS** are theorems of **FA+FD** is the content of Frege's Theorem, first proven by Frege in *Grundgesetze* (though very nearly proven by him in *Die Grundlagen*). Frege's own proofs of Frege's Theorem are discussed in detail in Chapters 2 and 3, and adaptations of his arguments will be employed below, so we need not dwell on the matter here.   □

Theorem 11.3.2. PAS $\Rightarrow$ PAF

Proof. Clearly, every axiom of **PAF** other than PAF itself is a theorem of **PAS+FD** (indeed, of **PAS** by itself). That PAF is can be proven by induction. If $a = 0$, then all of its predecessors are finite, since it has none, by Axiom 5 of **PAS**. Suppose, then, that Na (that, if Na and Pya, then Ny) and that Pab. We must show that, if Pxb, then Nx. So suppose Pxb. By Axiom 4 of **PAS**, $x = a$, so Nx.

That not every theorem of **PAS** is a theorem of **PAF+FD** should be obvious: The axioms of **PAF** make no claims whatsoever about what the predecessors of objects that are not natural numbers might be, whereas the axioms of **PAS** state that predecession is one-one, not just on the natural numbers, but universally. Construction of a model is left to the reader.   □

Theorem 11.3.3. PAF ⇒ PA2

Proof. Clearly, every theorem of **PA2** is a theorem of **PAF+FD**. Again, that the converse (roughly speaking) is not true should be obvious: **PA2** is completely silent on the question whether zero, or any other natural number, has predecessors that are not natural numbers. To construct a model, let the domain consist of the natural numbers and Julius Caesar. Assign denotations to terms of the form "$Nx\!:\!Fx$" according to the following scheme:

- 0, if there are no Fs or if everything is F

- n, if there are n Fs, for some finite $n > 0$

- Caesar, if there are infinitely many Fs, but not everything is F

Interpret the primitives of **PA2** and **PAF** according to the axioms of **FD**. The axioms of **PA2** may be verified, but PAF fails: "$P(Nx\!:\!x \neq 0, 0)$" is true in the model; that is, Caesar precedes zero.   □

Since, by Theorems 11.4.4 and 11.4.9, to be proven in the next section, **PAF** is equivalent to **FAF** (in the presence of **FD**), it follows that **FAF** is strictly stronger than **PA2**. The model just given also shows this directly. For the sentence

$$\mathrm{Fin}_x(x \neq x) \wedge (Nx\!:\!x \neq x = Nx\!:\!x = x) \wedge \neg \mathrm{Eq}_x(x \neq x, x = x)$$

is true in the model, whence HPF is false in the model.

11.4   PAF is equivalent to FAF

In the proofs to be given below, we shall appeal frequently to the following easy consequence of the second axiom of **FD**:

$$Fa \to P(Nx\!:\!(Fx \wedge x \neq a), Nx\!:\!Fx)$$

This is Theorem (102) of *Grundgesetze*, and I shall cite it as such below.

The equivalence of **PAF** and **FAF** is, in essence, a consequence of the fact that Theorems (327) and (348) of *Grundgesetze*, mentioned above, can be proven both in **PAF** and in **FAF**, with the aid of the bridge theory **FD**. That is to say,

$$\mathrm{Fin}_x(Fx) \equiv P^{*=}(0, Nx\!:\!Fx)$$

can be proven in both theories. This will allow us to work back and forth between the condition of finitude, as it appears in HPF, and the claims about the natural numbers made in the axioms of **PAF**. The proof to be given here of the left-to-right direction—which is Theorem (327) of *Grundgesetze*—shows it to be a consequence simply of Frege's definitions and the axiom FE of **Log**. "The number of a finite concept is a natural number" may thus be added to the list of arithmetical facts that are, *modulo* the status of second-order logic itself,[20] undeniably logical truths. (Others are "$P(0,1)$" and "$P(1,2)$", which Boolos shows to be provable in **FD**.)

Lemma 11.4.1 (Theorem 327). **FD** $\vdash \text{Fin}_x(Fx) \rightarrow P^{*=}(0, \text{N}x : Fx)$

Proof. If $\neg\exists x(Fx)$, then $\text{N}x : Fx = 0$ (by FE and the first axiom of **FD**), whence certainly $P^{*=}(0, \text{N}x : Fx)$. So we may suppose throughout that $\exists x(Fx)$.

Suppose F is finite, i.e., that there are objects a and b and a relation R that is functional, in whose (strong) ancestral no object stands to itself, and that is such that x is F iff $R^{*=}ax \wedge R^{*=}xb$. It will involve no loss of generality to suppose that the field of R contains only Fs, i.e., that $\forall x \forall y(Rxy \rightarrow Fx \wedge Fy)$.[21] If so, then $\forall x(Fx \equiv R^{*=}xb)$—exercise in second-order logic—whence, by FE, $\text{N}x : Fx = \text{N}x : R^{*=}xb$. It will thus suffice to prove that $P^{*=}(0, \text{N}x : R^{*=}xb)$.

Since $\exists x(Fx)$, for some y, $R^{*=}ay$ and $R^{*=}yb$, so $R^{*=}ab$, by the transitivity of the ancestral. So we can prove that $P^{*=}(0, \text{N}x : R^{*=}xb)$ by (logical) induction. For whenever $R^{*=}mn$, we can prove that Fn by showing that Fm and that: $\forall x \forall y(R^{*=}mx \wedge Fx \wedge Rxy \rightarrow Fy)$.[22] By comprehension, we may take $F\xi$ to be:

$$P^{*=}(0, \text{N}x : R^{*=}x\xi)$$

It will thus suffice to prove:

(i) $P^{*=}(0, \text{N}x : R^{*=}xa)$

(ii) $\forall y \forall z[R^{*=}ay \wedge P^{*=}(0, \text{N}x : R^{*=}xy) \wedge Ryz \rightarrow P^{*=}(0, \text{N}x : R^{*=}xz)]$

We are assuming, of course, that R satisfies the conditions mentioned above.

For (i): By (102), $P[\text{N}x : (x = a \wedge x \neq a), \text{N}x : x = a]$. Since $x = a \wedge x \neq a$ iff $x \neq x$, $\text{N}x : (x = a \wedge x \neq a) = 0$, by FE. Hence, $P(0, \text{N}x : x = a)$ and so $P^{*=}(0, \text{N}x : x = a)$. So it will suffice to show that $\text{N}x : R^{*=}xa = \text{N}x : x = a$, for which, by FE, it suffices to show that $\forall x(R^{*=}xa \equiv x = a)$. From right-to-left, this is obvious. For the other direction, suppose that $R^{*=}xa$

---

[20]I believe that the proof requires only Π^1_1 comprehension, but I have not verified this is detail.

[21]Plainly, if there is such an $R\xi\eta$ as was supposed to exist, then $R\xi\eta \wedge F\xi \wedge F\eta$ will also satisfy the four conditions.

[22]This form of induction can be proven from the definition of the weak ancestral by taking $R^{*=}m\xi \wedge F\xi$ for $F\xi$. The proof is easy, but it does need Π^1_1 comprehension.

and $x \neq a$. Then $R^* xa$. But x must then be in the field of $R\xi\eta$,[23] so Fx, so $R^{*=} ax$. But then $R^{*=} ax$ and $R^* xa$, so $R^* aa$, by a form of transitivity. Contradiction.

For (ii): Suppose the antecedent. By comprehension, we may take $F\xi$ and a in (102) to be, respectively, $R^{*=}\xi y \vee \xi = z$ and z, whence:

$$P\{Nx : [(R^{*=} xy \vee x = z) \wedge x \neq z], Nx : (R^{*=} xy \vee x = z)\}$$

But $\neg R^{*=} zy$ (lest $R^{*=} zy$ and Ryz, so $R^* zz$), so

$$\forall x\{[(R^{*=} xy \vee x = z) \wedge x \neq z] \equiv R^{*=} xy\}$$

Hence, by FE,

$$Nx : [(R^{*=} xy \vee x = z) \wedge x \neq z] = Nx : R^{*=} xy$$

So, substituting:

$$P\{Nx : R^{*=} xy, Nx : (R^{*=} xy \vee x = z)\}$$

But, I claim, $R^{*=} xz$ if, and only if, $R^{*=} xy \vee x = z$.

From right-to-left: If $x = z$, then certainly $R^{*=} xz$; and, if $R^{*=} xy$, then, since Ryz, again $R^{*=} xz$.

For the other direction, suppose that $R^{*=} xz$ and $x \neq z$. Then certainly $R^* xz$. We then have the following theorem of second-order logic, the *roll-back theorem*:[24]

$$Q^* xy \rightarrow \exists z(Qzy \wedge Q^{*=} xz)$$

By the roll-back theorem, there is some w such that Rwz and $R^{*=} xw$. Since R is one-one and Ryz, $w = y$, so $R^{*=} xy$. That establishes the claim. So, by FE, $Nx : R^{*=} xz = Nx : (R^{*=} xy \vee x = z)$. Substituting, we have: $P(Nx : R^{*=} xy, Nx : R^{*=} xz)$. Since also $P^{*=}(0, Nx : R^{*=} xy)$, by transitivity, $P^{*=}(0, Nx : R^{*=} xz)$. $\qquad\square$

11.4.1   PAF+FD Contains FAF

We turn now to the proof that HPF is provable in **PAF+FD**.

Remark. The following are theorems of **PAF**:[25]

---

[23] In general: $Q^* ab \rightarrow \exists x(Qax) \wedge \exists x(Qxb)$. The latter conjunct follows from the roll-back theorem; the former, from the roll-forward theorem. The roll-back theorem is proven in the next footnote; the roll-forward theorem will be mentioned below.

[24] The roll-back theorem is proved by induction. We must show:

(i) $Qxw \rightarrow \exists z(Qzw \wedge Q^{*=} xz)$

(ii) $\exists z(Qzw \wedge Q^{*=} xz) \wedge Qwv \rightarrow \exists z(Qzv \wedge Q^{*=} xz)$

The proof of (i) is trivial: Take z to be x. For (ii), assume the antecedent. Take z in the consequent to be w. By hypothesis, Qwv. And since, by the antecedent, $Q^{*=} xz$ and Qzw, certainly $Q^{*=} xw$.

[25] In fact, **PA2+NPZ+P1MF** is deductively equivalent to **PAF**. The proof of PAF given above, in the proof of Theorem 2, depends only upon NPZ and P1MF, and not on the full force of Axioms 4 and 5 of **PAS**.

NPZ: $\neg\exists x(Px0)$

P1MF: $\forall x\forall y\forall z(P^{*=}0z \wedge Pxz \wedge Pyz \rightarrow x = y)$

ZE: $\mathbf{N}x:Fx = 0 \equiv \neg\exists x(Fx)$

Proof. Zero has no predecessor that is a natural number and, by PAF, only natural numbers precede natural numbers. So since zero is a natural number, it can have no predecessor at all. That's NPZ.

P1MF will follow immediately from axiom (4) of **PAF** if we can show that x and y are themselves natural numbers. But this follows from PAF, since z is a natural number and both x and y precede it.

For ZE: If $\neg\exists x(Fx)$, then $\forall x(Fx \equiv x \neq x)$. So, by FE, $\mathbf{N}x:Fx = \mathbf{N}x:x \neq x = 0$. Suppose, then, that $\mathbf{N}x:Fx = 0$ and that $\exists x(Fx)$, say, a. By (102), $P(\mathbf{N}x:(Fx \wedge x \neq a), \mathbf{N}x:Fx)$, so, by NPZ, $\mathbf{N}x:Fx \neq 0$. Contradiction. $\square$

Lemma 11.4.2 (*Grundgesetze*, Theorem 348).

$$\mathbf{PAF} + \mathbf{FD} \vdash P^{*=}(0, \mathbf{N}x:Fx) \rightarrow \mathbf{Fin}_x(Fx)$$

Proof. We prove the equivalent:

$$P^{*=}0n \rightarrow \forall F[n = \mathbf{N}x:Fx \rightarrow \mathbf{Fin}_x(Fx)]$$

The proof is by (logical) induction. We must thus establish:

(i) $\forall F[0 = \mathbf{N}x:Fx \rightarrow \mathbf{Fin}_x(Fx)]$

(ii) $P^{*=}0n \wedge \forall F[n = \mathbf{N}x:Fx \rightarrow \mathbf{Fin}_x(Fx)] \wedge Pnm \rightarrow$
    $\forall F[m = \mathbf{N}x:Fx \rightarrow \mathbf{Fin}_x(Fx)]$

For (i), suppose $0 = \mathbf{N}x:Fx$. By ZE, nothing is F, so F is finite.

For (ii), suppose the antecedent; suppose further that $m = \mathbf{N}x:Fx$. We must show that F is finite. Suppose $\neg\exists x(Fx)$. Then, by ZE, $m = \mathbf{N}x:Fx = 0$, so $Pn0$, contradicting NPZ. So $\exists x(Fx)$, say a, and $P[\mathbf{N}x:(Fx \wedge x \neq a), \mathbf{N}x:Fx]$. But also, by hypothesis, $P(n, \mathbf{N}x:Fx)$, and since $P^{*=}0n$, $P^{*=}(0, \mathbf{N}x:Fx)$. So, by P1MF, $n = \mathbf{N}x:(Fx \wedge x \neq a)$. Hence, by the induction hypothesis, $\mathbf{Fin}_x(Fx \wedge x \neq a)$. It is then a simple matter to show that F too must be finite. $\square$

Corollary 11.4.3. $\mathbf{PAF} + \mathbf{FD} \vdash \mathbf{Fin}_x(Fx) \equiv P^{*=}(0, \mathbf{N}x:Fx)$

Theorem 11.4.4. $\mathbf{PAF} + \mathbf{FD} \vdash \mathrm{HPF}$

Proof. By Corollary 11.4.3, it suffices to show that

$$P^{*=}(0, \mathbf{N}x:Fx) \rightarrow [\mathbf{N}x:Fx = \mathbf{N}x:Gx \equiv \mathbf{Eq}_x(Fx, Gx)]$$

We prove the equivalent:

$$P^{*=}0n \rightarrow$$
$$\forall F\{n = \mathbf{N}x:Fx \rightarrow \forall G[\mathbf{N}x:Fx = \mathbf{N}x:Gx \equiv \mathbf{Eq}_x(Fx, Gx)]\}$$

The proof is by induction. We must prove the basis case:

$$\forall F\{0 = Nx : Fx \to \forall G[Nx : Fx = Nx : Gx \equiv \mathbf{Eq}_x(Fx, Gx)]\}$$

and the induction step:

$$P^{*=}0n \to \left\{ \right.$$
$$\forall F\{n = Nx : Fx \to \forall G[Nx : Fx = Nx : Gx \equiv \mathbf{Eq}_x(Fx, Gx)]\} \wedge Pnm \to$$
$$\left. \forall F\{m = Nx : Fx \to \forall G[Nx : Fx = Nx : Gx \equiv \mathbf{Eq}_x(Fx, Gx)]\} \right\}$$

For the basis, suppose that $0 = Nx : Fx$. Suppose further that $Nx : Fx = Nx : Gx$. So $0 = Nx : Gx$, and by ZE nothing is G. So $\mathbf{Eq}_x(Fx, Gx)$. Conversely, if $\mathbf{Eq}_x(Fx, Gx)$, then nothing is G, so, by FE, $Nx : Fx = Nx : Gx$.

For the induction step: Suppose the antecedent, and suppose further that $m = Nx : Fx$. We must show that, for every G, $Nx : Fx = Nx : Gx$ iff $\mathbf{Eq}_x(Fx, Gx)$. Note that, since $P^{*=}0n$ and Pmn, $P^{*=}0m$ and so $P^{*=}(0, Nx : Fx)$.

Left-to-right: Suppose $Nx : Fx = Nx : Gx$. Since $P(n, Nx : Fx)$, we have $Nx : Fx \neq 0$, by NPZ, and so, by ZE, $\exists x(Fx)$, say a; similarly, $\exists x(Gx)$, say b. By (102):

$$P[Nx : (Fx \wedge x \neq a), Nx : Fx]$$
$$P[Nx : (Gx \wedge x \neq b), Nx : Gx]$$

Since $Nx : Fx = Nx : Gx$, $Nx : (Fx \wedge x \neq a) = Nx : (Gx \wedge x \neq b)$, by P1MF. Moreover, since $P(n, Nx : Fx)$, by P1MF, again, $n = Nx : (Fx \wedge x \neq a)$. So, by the induction hypothesis: $\mathbf{Eq}_x(Fx \wedge x \neq a, Gx \wedge x \neq b)$. But then it is easy to show that $\mathbf{Eq}_x(Fx, Gx)$, since Fa and Gb.

Right-to-left: Suppose $\mathbf{Eq}_x(Fx, Gx)$. Once again, $\exists x(Fx)$, say a, and $\exists x(Gx)$, say b, and:

$$P[Nx : (Fx \wedge x \neq a), Nx : Fx]$$
$$P[Nx : (Gx \wedge x \neq b), Nx : Gx]$$

Since $P(n, Nx : Fx)$, $n = Nx : (Fx \wedge x \neq a)$, by P1MF. But, if $\mathbf{Eq}_x(Fx, Gx)$, Fa, and Gb, certainly $\mathbf{Eq}_x(Fx \wedge x \neq a, Gx \wedge x \neq b)$. So, by the induction hypothesis, $Nx : (Fx \wedge x \neq a) = Nx : (Gx \wedge x \neq b)$. But then, since $P[Nx : (Fx \wedge x \neq a), Nx : Fx]$ and $P[Nx : (Gx \wedge x \neq b), Nx : Gx]$, $Nx : Fx = Nx : Gx$, by Axiom 3 of **PAF**. $\square$

That, then, establishes that HPF is a theorem of **PAF+FD**.

11.4.2   HPF+FD Contains PAF

We now turn to the proof that all the axioms of **PAF** are theorems of **HPF+FD**. Our plan is simply to mimic Frege's proofs of the axioms of arithmetic, relativized in the appropriate way to the natural numbers. To make these proofs work, we need to establish an analogue of Corollary 11.4.3. From this it will follow that, when talking about natural numbers, we are dealing only with finite concepts, so HPF will do the work HP does in Frege's proofs. We divide the proof into two parts: The proof that all axioms other than Axiom 6 hold is relatively easy, and we prove this first; the proof that Axiom 6 holds is of special interest and so will be considered separately. First, we establish the Corollary, by establishing an analogue of Lemma 11.4.2.

Lemma 11.4.5 (Theorem 348, again).

$$\mathbf{HPF} + \mathbf{FD} \vdash P^{*=}(0, \mathbf{N}x : Fx) \to \mathrm{Fin}_x(Fx)$$

Proof. It will suffice to show that

$$(*) \quad P^{*=}0n \to \exists F[\mathrm{Fin}_x(Fx) \wedge n = \mathbf{N}x : Fx]$$

For then suppose that $P^{*=}(0, \mathbf{N}x : Fx)$. Then, for some finite G, $\mathbf{N}x : Fx = \mathbf{N}x : Gx$. By HPF, $\mathrm{Eq}_x(Fx, Gx)$, and so F is finite.

The proof of (*) itself is by induction. We must show:

(i) $\exists F[\mathrm{Fin}_x(Fx) \wedge 0 = \mathbf{N}x : Fx]$

(ii) $P^{*=}0n \wedge \exists F[\mathrm{Fin}_x(Fx) \wedge n = \mathbf{N}x : Fx] \wedge Pnm \to$
$\qquad \exists F[\mathrm{Fin}_x(Fx) \wedge m = \mathbf{N}x : Fx]$

For (i): $0 = \mathbf{N}x : x \neq x$, and $\mathrm{Fin}_x(x \neq x)$.

For (ii), suppose the antecedent, so that $\mathrm{Fin}_x(Fx)$ and $n = \mathbf{N}x : Fx$. Since Pnm, $P(\mathbf{N}x : Fx, m)$, so by axiom 2 of **FD**, for some G and b:

$$Gb \wedge m = \mathbf{N}x : Gx \wedge \mathbf{N}x : Fx = \mathbf{N}x : (Gx \wedge x \neq b)$$

Since $\mathrm{Fin}_x(Fx)$, by HPF, $\mathrm{Eq}_x(Fx, Gx \wedge x \neq b)$, so $\mathrm{Fin}_x(Gx \wedge x \neq b)$. But then it is easy to show that G too is finite. $\qquad \square$

Corollary 11.4.6. $\mathbf{HPF} + \mathbf{FD} \vdash \mathrm{Fin}_x(Fx) \equiv P^{*=}(0, \mathbf{N}x : Fx)$

Lemma 11.4.7. $\mathbf{HPF} + \mathbf{FD} \vdash$ *All axioms of* **PAF** *other than Axiom 6*

Proof. Axioms 1, 2, and 7 do not require any special attention: In full second-order logic, each of them is a consequence of **FD** itself, indeed, of just the third axiom of **FD**. We thus need to prove Axioms 3, 4, and 5 and the Axiom PAF itself.

Axiom 5 is: $\neg \exists x(\mathbf{N}x \wedge Px0)$. In fact, we can prove the stronger NPZ: $\neg \exists x(Px0)$. Suppose that $Pn0$. By the second axiom of **FD**, there are F and y such that:

$$Fy \wedge 0 = \mathbf{N}x : Fx \wedge n = \mathbf{N}x : (Fx \wedge x \neq y)$$

But since $0 = \mathbb{N}x : x \neq x$ and $\mathrm{Fin}_x(x \neq x)$, HPF yields that $\mathrm{Eq}_x(x \neq x, Fx)$. But $\exists y(Fy)$. Contradiction.

Axiom 3 is: $\forall x \forall y \forall z (\mathbb{N}x \wedge Pxy \wedge Pxz \rightarrow y = z)$. Suppose that $\mathbb{N}a$, that is, by the third axiom of **FD**, that $P^{*=}0a$, and that Pab and Pac. By the second axiom of **FD**, there are F and G, and y and z, such that:

$$Fy \wedge b = \mathbb{N}x : Fx \wedge a = \mathbb{N}x : (Fx \wedge x \neq y)$$
$$Gz \wedge c = \mathbb{N}x : Gx \wedge a = \mathbb{N}x : (Gx \wedge x \neq z)$$

Since $P^{*=}0a$ and $a = \mathbb{N}x : (Fx \wedge x \neq y)$, by Corollary 11.4.6, $\mathrm{Fin}_x(Fx \wedge x \neq y)$. Since $\mathbb{N}x : (Fx \wedge x \neq y) = a = \mathbb{N}x : (Gx \wedge x \neq z)$, by HPF, $\mathrm{Eq}_x(Fx \wedge x \neq y, Gx \wedge x \neq z)$. But then, since Fy and Gz, $\mathrm{Eq}_x(Fx, Gx)$; and since $\mathrm{Fin}_x(Fx \wedge x \neq y)$, $\mathrm{Fin}_x(Fx)$. So, by HPF again, $\mathbb{N}x : Fx = \mathbb{N}x : Gx$, so $b = c$.

Axiom 4 is: $\forall x \forall y \forall z (\mathbb{N}x \wedge \mathbb{N}y \wedge Pxz \wedge Pyz \rightarrow x = y)$. We can actually prove the stronger P1MF: $\forall x \forall y \forall z (\mathbb{N}z \wedge Pxz \wedge Pyz \rightarrow x = y)$, from which Axiom 4 will follow, since, if $\mathbb{N}x$—that is, $P^{*=}0x$—and Pxz, then $P^{*=}0z$, and so $\mathbb{N}z$.

Suppose that $\mathbb{N}c$, that is, $P^{*=}0c$, and that Pac and Pbc. Once again, there are F and G, and y and z, such that:

$$Fy \wedge b = \mathbb{N}x : Fx \wedge a = \mathbb{N}x : (Fx \wedge x \neq y)$$
$$Gz \wedge c = \mathbb{N}x : Gx \wedge a = \mathbb{N}x : (Gx \wedge x \neq z)$$

Since $c = \mathbb{N}x : Fx$ and $P^{*=}0c$, by Corollary 11.4.6, $\mathrm{Fin}_x(Fx)$; similarly, $\mathrm{Fin}_x(Gx)$. And since $\mathbb{N}x : Fx = c = \mathbb{N}x : Gx$, by HPF, $\mathrm{Eq}_x(Fx, Gx)$. But then, since Fy and Gz, $\mathrm{Eq}_x(Fx \wedge x \neq y, Gx \wedge x \neq z)$; these are finite, since F and G are, so by HPF again, $\mathbb{N}x : (Fx \wedge x \neq y) = \mathbb{N}x : (Gx \wedge x \neq z)$ and so $a = b$.

PAF is: $\mathbb{N}x \wedge Pyx \rightarrow \mathbb{N}y$. As noted parenthetically above, the proofs of Axioms 4 and 5 in fact suffice to prove NPZ and P1MF, from which PAF follows. But it can also be proven directly. Suppose that $P^{*=}0n$ and that Pmn. Since Pmn, there are F and a such that:

$$Fa \wedge n = \mathbb{N}x : Fx \wedge m = \mathbb{N}x : (Fx \wedge x \neq a)$$

Since $P^{*=}0n$, $\mathrm{Fin}_x(Fx)$; so $\mathrm{Fin}_x(Fx \wedge x \neq a)$, and so $P^{*=}[0, \mathbb{N}x : (Fx \wedge x \neq a)]$. But then $P^{*=}0m$. $\quad\square$

In the paper mentioned earlier, Boolos proves the surprising result that, in the presence of the bridge theory **FD**, Axiom 6 of **PA2**—that is, $\forall x(\mathbb{N}x \rightarrow \exists y(Pxy))$—is redundant, since it follows from Axioms 3, 4, and 5. Later, he observed that, in the presence of **FD**, Axiom 6 in fact follows from Axiom 3 *alone*. Boolos's original proof of this extraordinary result is somewhat indirect:[26] I shall take the opportunity to give a direct

---

[26]We have that: **FD**, $3, 4, 5 \vdash 6$. Boolos (1998c) then observed that also: **FD**, $3, \neg 6 \vdash 4 \wedge 5$. But then, by truth-functional logic: **FD**, $3, \neg 6 \vdash 6$. And so: **FD**, $3 \vdash 6$.

proof here. Of course, since **FAF+FD** proves Axiom 3, it will follow that **FAF+FD** proves Axiom 6.

We begin by discussing Frege's proof of Axiom 6 in *Grundgesetze*. How can we do without the applications of Axioms 4 and 5 that he makes?

Axiom 6 (which is Theorem 157) follows directly from Theorem 155:

$$P^{*=}0n \to P[n, \mathbf{N}x : P^{*=}xn]$$

to establish which Frege needs, as a crucial lemma, Theorem 145:

$$P^{*=}0n \to \neg P^*nn$$

It is for the proof of this lemma that Axiom 5 is needed. The first way our proof will differ from Frege's is that we do not make use of this lemma, but instead pack the necessary conditions into what we want to prove

$$P^{*=}0n \land \neg P^*nn \to P[n, \mathbf{N}x : (P^{*=}0x \land P^{*=}xn)]$$

and argue by dilemma, completing the argument by also proving

$$P^{*=}0n \land P^*nn \to \exists y(Pny)$$

Frege's appeal to Axiom 4 occurs at a crucial point in the argument, but we shall see that the necessary inference does not require it. What we shall use instead is the logicized version of the Law of Trichotomy:

$$P^{*=}0b \land P^{*=}0c \to P^*bc \lor P^*cb \lor b = c$$

This Law follows from Axiom 3 and the following strengthening of the proposition 133 of *Begriffsschrift*:[27]

$$\forall x \forall y \forall z[R^{*=}ax \land Rxy \land Rxz \to y = z] \land R^{*=}ab \land R^{*=}ac \to$$
$$R^*bc \lor R^*cb \lor b = c$$

Instantiating "*R*" with "*P*", "*a*" with "0", and noting that the first conjunct then follows from Axiom 3, the Law of Trichotomy follows immediately.

---

[27]The strengthening lies in our assuming not that R is functional, but just that it is functional on the members of the R-series beginning with a.

Proof: We assume that $\forall x \forall y \forall z[R^{*=}ax \land Rxy \land Rxz \to y = z]$ and that $R^{*=}ab$ and prove

$$R^{*=}ac \to R^*bc \lor R^*cb \lor b = c$$

by induction on c. We must thus prove:

(i) $R^{*=}aa \to R^*ba \lor R^*ab \lor b = a$

(ii) $R^{*=}ax \land [R^{*=}ax \to R^*bx \lor R^*xb \lor b = x] \land Rxy \to$
       $[R^{*=}ay \to R^*by \lor R^*yb \lor b = y]$

Since $R^{*=}ab$, (i) follows from the definition of the weak ancestral. So suppose the antecedent in (ii). Note that $R^{*=}ay$. Moreover, either R^*bx or R^*xb or $b = x$. If R^*bx, then since Rxy, R^*by. Moreover, if $b = x$, then Rby, so R^*by. So suppose that R^*xb. By the roll-forward theorem, to be mentioned shortly, for some z, Rxz and $R^{*=}zb$. But then Rxz and Rxy: And since R is functional on the R-series beginning with a and $R^{*=}ax$, we have $z = y$. So $R^{*=}yb$; hence R^*yb or $b = y$.

Lemma 11.4.8. $\mathbf{FD}$ + Axiom 3 $\vdash$ Axiom 6

Proof. We must establish $\forall x(\mathbb{N}x \rightarrow \exists y(Pxy))$, for which, in light of $\mathbf{FD}$, it will suffice to establish:

$$P^{*=}0n \rightarrow \exists y(Pny)$$

We proceed by dilemma, proving each of:

$$P^{*=}0n \wedge P^{*}nn \rightarrow \exists y(Pny)$$
$$P^{*=}0n \wedge \neg P^{*}nn \rightarrow \exists y(Pny)$$

The former follows immediately from the following theorem of second-order logic, the *roll-forward theorem*:[28]

$$R^{*}ab \rightarrow \exists y(Ray \wedge R^{*=}yb)$$

For, then, if $P^{*}nn$, for some y, $Pny \wedge P^{*=}yn$, so certainly $\exists y(Pny)$.

For the latter, we prove

$$P^{*=}0n \wedge \neg P^{*}nn \rightarrow P[n, \mathbf{N}x : (P^{*=}0x \wedge P^{*=}xn)]$$

The proof is by induction. We thus need to establish:

(i) $\mathbf{FD} \vdash \neg P^{*}00 \rightarrow P[0, \mathbf{N}x : (P^{*=}0x \wedge P^{*=}x0)]$

(ii) $\mathbf{FD}$ + Axiom 3 $\vdash$
         $P^{*=}0a \wedge \{\neg P^{*}aa \rightarrow P[a, \mathbf{N}x : (P^{*=}0x \wedge P^{*=}xa)]\} \wedge Pab \rightarrow$
         $\{\neg P^{*}bb \rightarrow P[b, \mathbf{N}x : (P^{*=}0x \wedge P^{*=}xb)]\}$

For (i): Suppose $\neg P^{*}00$. Since $P^{*=}00$, by (102):

$$P[\mathbf{N}x : (P^{*=}0x \wedge P^{*=}x0 \wedge x \neq 0), \mathbf{N}x : (P^{*=}0x \wedge P^{*=}x0)]$$

Now suppose that $P^{*=}0x \wedge P^{*=}x0 \wedge x \neq 0$. Then since $P^{*=}x0$ and $x \neq 0$, $P^{*}x0$. But then $P^{*}x0$ and $P^{*=}0x$, so $P^{*}00$, contradicting our supposition. Thus $\neg \exists x(P^{*=}0x \wedge P^{*=}x0 \wedge x \neq 0)$. By FE, $\mathbf{N}x : (P^{*=}0x \wedge P^{*=}x0 \wedge x \neq 0) = 0$ and so $P[0, \mathbf{N}x : (P^{*=}0x \wedge P^{*=}x0)]$.

For (ii): Suppose the antecedent and suppose further that $\neg P^{*}bb$. Suppose, for *reductio*, that $P^{*}aa$. By the roll-forward theorem, for some y, Pay and $P^{*=}ya$. Since Pab, Axiom 3 implies that $y = b$. But then $P^{*=}ba$ and Pab, so $P^{*}bb$. Contradiction. Hence $\neg P^{*}aa$.

By the induction hypothesis, then, $P[a, \mathbf{N}x : (P^{*=}0x \wedge P^{*=}xa)]$. We need to show that $P[b, \mathbf{N}x : (P^{*=}0x \wedge P^{*=}xb)]$. Since $P^{*=}0b$ and $P^{*=}bb$, by (102):

$$P[\mathbf{N}x : (P^{*=}0x \wedge P^{*=}xb \wedge x \neq b), \mathbf{N}x : (P^{*=}0x \wedge P^{*=}xb)]$$

---

[28]The proof of this result is similar to that of the roll-back theorem, mentioned earlier.

So it is enough to show that

$$b = \mathbf{N}x : (P^{*=}0x \wedge P^{*=}xb \wedge x \neq b)$$

Now, since Pab and $P[a, \mathbf{N}x : (P^{*=}0x \wedge P^{*=}xa)]$, we have, by Axiom 3, that $b = \mathbf{N}x : (P^{*=}0x \wedge P^{*=}xa)$. So we need only show that

$$\mathbf{N}x : (P^{*=}0x \wedge P^{*=}xa) = \mathbf{N}x : (P^{*=}0x \wedge P^{*=}xb \wedge x \neq b)$$

By FE, this will follow from:

$$\forall x (P^{*=}0x \wedge P^{*=}xa \equiv P^{*=}0x \wedge P^{*=}xb \wedge x \neq b)$$

Left-to-right: Suppose $P^{*=}0x \wedge P^{*=}xa$. Then since Pab, certainly $P^{*=}xb$ and, further, $P^{*=}0b$. Suppose $x = b$. Then $P^{*=}ba$ and Pab, so P^*bb. Contradiction.

Right-to-left: Suppose $P^{*=}0x \wedge P^{*=}xb \wedge x \neq b$. Then P^*xb. By the rollback theorem, for some y, $P^{*=}xy$ and Pyb. Now, if we had Axiom 4, we could conclude that, since Pab, $a = y$, whence $P^{*=}xa$. That is how Frege ends his proof—which we have, to this point, been following closely. But we can in fact establish that $a = y$ without appealing to Axiom 4.

We have that $P^{*=}xy$ and Pyb. Since $P^{*=}0x$, certainly $P^{*=}0y$. Since, by the initial hypotheses of the inductive step, $P^{*=}0a$, the Law of Trichotomy yields that either P^*ay or P^*ya or $a = y$. Suppose that P^*ay. By the rollforward theorem, for some z, $Paz \wedge P^{*=}zy$. But since Pab, Axiom 3 implies that $z = b$. So $P^{*=}by \wedge Pyb$, so P^*bb, contradiction. Similarly, if P^*ya, then for some z, $Pyz \wedge P^{*=}za$. But since Pyb, Axiom 3 implies that $z = b$, so $P^{*=}ba \wedge Pab$, so P^*bb, once again. Hence $a = y$, and we are done. □

Theorem 11.4.9. HPF + FD ⊢ All axioms of PAF

Proof. By 11.4.7 and 11.4.8. □

11.5 Closing

We have thus seen that **FAF** is equivalent, in the presence of the bridge theory **FD**, to **PAF**. By Theorem 4, then, **FAF** is strictly stronger than **PA2**. The following two questions now raise themselves: Whether there is some further weakening of HP that is provable in **PA2+FD** and, if so, whether some such principle is equivalent, in **FD**, to the conjunction of the axioms of **PA2**. A natural axiom at which to look would be WHP (for *Weak* HP):

$$\mathrm{Fin}_x(Fx) \wedge \mathrm{Fin}_x(Gx) \to [\mathbf{N}x : Fx = \mathbf{N}x : Gx \equiv \mathrm{Eq}_x(Fx, Gx)]$$

WHP states only that finite concepts have the same number if, and only if, they are equinumerous and makes no claim whatsoever about the conditions under which infinite concepts have the same number as any concept, finite or otherwise. (As far as WHP is concerned, some infinite concepts could have the number zero, others one, and so forth.) Call the theory whose sole non-logical axiom is WHP, **WHP**. It can be shown that, though WHP is provable in **PA2+FD**, *none* of the axioms of **PA2** that are not already theorems of **FD** itself are theorems of **WHP+FD**: Not even the disjunction of these axioms—that is, of Axioms 3, 4, 5, and 6—is a theorem of **WHP+FD**.

Still, it is easy to see that there are no finite models of **WHP+FD**. For, in any model, there must be a number $Nx : x \neq x$; there must be a number $Ny : (y = Nx : x \neq x)$, which, by WHP, will differ from $Nx : x \neq x$; there will be a number $Nz : [z = Nx : x \neq x \lor z = Ny : (y = Nx : x \neq x)]$, which again must differ from the first two, and so forth. Indeed, it is not terribly difficult to prove that **PA2** can be relatively interpreted in **WHP**, and so that **PA2** and **WHP** are equi-interpretable and therefore equi-consistent.

As we saw earlier, however, the fact that a theory **A** is relatively interpretable in another **B** is no guarantee that there is any reasonable bridge theory by using which one can prove the axioms of **A** in **B**. One might therefore wonder whether, in this case, there is some bridge theory other than **FD** by appeal to which one could prove the axioms of **PA2** in **WHP**. In fact there is, the necessary modification to the axioms of **FD** not being drastic. The bridge theory **FDF** has the same first and third axioms as **FD**, but we change the second axiom to:

$$\exists F[\text{Fin}_x(Fx) \land m = Nx : Fx] \lor \exists F[\text{Fin}_x(Gx) \land n = Nx : Gx] \to$$
$$Pmn \equiv \exists F \exists y[\text{Fin}_x(Fx) \land Fy \land n = Nx : Fx \land m = Nx : (Fx \land x \neq y)]$$

This axiom now states nothing at all about when numbers that are not the numbers of finite concepts precede one another. It requires, however, that, if a number is the number of a finite concept, then it will precede or be preceded by another number only if there is some finite concept that does the trick. The axiom, though complicated as stated, seems intuitive enough and is certainly true, since it is a theorem of **HPF+FD**, as can easily be seen. It may thus be considered a partial definition of one of the primitives of **PA2** in terms of those of **WHP**. And it can be shown that, relative to **FDF**, **WHP**, **PA2**, and **PAF** are all equivalent. What philosophical interest this result might have for a logicist is a question I shall not pursue.[29]

Postscript

There were two main motivations for this work. The first, which is explicit in the text, is to develop and then respond to what Wright (2001b, p. 317) would later call "the concern about surplus content". That is the issue discussed in Section 11.2, and there is further discussion in Section 1.3. The second motivation is left implicit. It was to respond to what

---

[29]Fernando Ferreira (2005) makes interesting use of **WHP** in an argument closely related to those given here.

Wright (2001b, p. 313) labels "the concern about the universal number", the number of all objects, also known as "anti-zero", since it is $Nx:x = x$, whereas zero is $Nx:x \neq x$. Boolos had objected that, although it is an immediate consequence of HP that $Nx:x = x$ exists, it is just not obvious that there is any such number, in part, on the ground that Zermelo-Frankel set theory seems flatly to deny that there is any such number. Similar concerns apply to the number of all ordinals, the number of all sets, and the like.

Wright's discussion of this problem focuses attention on the much discussed but not terribly clear notion of an 'indefinitely extensible concept'. Wright does not pretend to know how to resolve the issue, however, and not everyone will be prepared to take indefinite extensibility seriously. Even if the notion can be given clear content, then, it would be unfortunate if Neo-logicism had to rest upon it.

The problem posed by anti-zero is also addressed by Ian Rumfitt in his paper "Hume's Principle and the Number of All Objects". Rumfitt's strategy is to argue for a condition that restricts what concepts have numbers. He first argues that there is an essential connection between cardinality and counting: If a concept is to have a cardinal number at all, it must be possible to count the objects falling under it.[30] Of course, as Rumfitt (2001, p. 528) notes, counting must here be understood in a generalized sense, so that it does not just apply to finite concepts. In the end, Rumfitt arrives at the following suggestion: "There is such a thing as the number of *F*s iff *F* is either empty or equinumerous with a bounded initial segment of some [well ordering]" (Rumfitt, 2001, p. 529). It is easy to see that this is equivalent to: There is such a thing as the number of *F*s iff the *F*s can be well-ordered with a top, that is, in such a way that the ordering has a maximal element (assuming it is non-empty).

And so, Rumfitt says, the fate of anti-zero is sealed:

... [G]iven the additional assumption that each ordinal number is an object, [our theory] entails that there is no such number as anti-zero. On that assumption, if there were such a thing as the number of all objects, then there would have to be such a thing as the number of all ordinal numbers. That, however, [our theory] excludes. The concept *ordinal number* [can be well-ordered], but it cannot be [well-ordered with a top]. For no bound can be placed on the ordinals themselves. (Rumfitt, 2001, p. 532)

---

[30]I do not find this part of the argument convincing. It is supposed to be based upon the idea, which Frege himself discusses in §108 of the first volume of *Grundgesetze* (the passage is also discussed in Chapter 3), that the possibility of using the numerals to count depends upon their being well-ordered. Frege concludes that *the numbers themselves* must have a certain sort of ordering. It is just not clear why that must be so in the infinite case. Indeed, the assumption that the cardinals are well-ordered (or even linearly ordered) is very strong, and entails the axiom of choice. But the more serious problem is that the conclusion Rumfitt wants is that *the objects to be numbered* must be orderable in a certain way, and I just do not see how that is supposed to fall out of the claim about cardinals. Counting the *F*s, in the infinite case, certainly does not involve establishing a one-one correlation between the *F*s and some cardinals.

This, however, is just wrong. It is easy to see that, if the Fs can be well-ordered at all, then they can be well-ordered with a top: If there isn't a top already, take the first element of the ordering and move it to the end.[31]

Moreover, the notion of a well-ordering is defined here, as it must be in the present case, in terms of second-order logic, not in terms of set theory.[32] That is why Rumfitt can confidently say that the concept *ordinal number* can be well-ordered: The ordinals are well-ordered by the relation $\xi < \eta$; there is, provably in ZF, no well-ordering of the ordinals, if one requires the ordering to be given by a set.[33] So Rumfitt's proposal in fact implies that *there is* a number of all ordinals, and there will be a number of all objects, too, if the universe can be well-ordered.

My way with anti-zero was different, having more in common with a proposal of Neil Tennant's that was published the same year. Tennant (1997) suggests that we should conditionalize HP, so that it says that the number of Fs is the same as the number of Gs if they are equinumerous *and these numbers exist*.[34] We then need some principles governing existence, and Tennant gives us two:

1. $Nx : Fx$ exists, if F is empty.

2.   If $Nx : Fx$ exists, and $\neg Fa$, then $Nx : (Fx \lor x = a)$ exists.

As Rumfitt (2001, p. 523–4) notes, however, it is not clear how these principles are to be justified in a way that would serve a Neo-logicist.[35]

Of course, I would have the same problem if my proposal had been to replace (1) and (2) with a single principle to the effect that $Nx : Fx$ exists if F is finite. But that was not the proposal. A clearer statement of it emerges toward the end of Chapter 7 and rests upon a distinction, on which Dummett (1998) rightly insists, between Frege's analysis of "just

---

[31]To add a little detail, suppose R well-orders the Fs; we may suppose that R has only Fs in its field. Let a be the initial element of the ordering. Consider this relation, which exists by predicative comprehension:

$$Qxy \equiv (Rxy \land x \neq a) \lor \exists z (Rzx \land y = a)$$

Then Q well-orders the Fs, and a is its maximum element. A similar proof works in the set-theoretic case.

[32]If it were defined instead in set-theoretic terms, then Rumfitt would have to appeal to axioms of set theory to prove that various concepts can be well-ordered and therefore have a number, and so his derivation of the Dedekind-Peano axioms (Rumfitt, 2001, §III) would rest upon set-theoretic assumptions.

[33]Then we would get a set of all ordinals, which leads straight to the Burali-Forti Paradox.

[34]Note that this is different from the proposal discussed in Section 1.2.3 and Chapter 9. There, the condition concerns the existence of numbers quite generally, not the existence of the particular numbers involved.

[35]Corresponding principles concerning the existence of concepts and relations have played a significant role in recent discussions of the strength of weak fragments of Frege arithmetic. See, for example, Visser's recent paper "Hume's Principle, Beginnings" (Visser, 2011). One could reframe Tennant's suggestion in these terms. That might well be an improvement.

as many" and the abstraction principle for cardinal numbers. The abstraction principle is really just:

$$\mathrm{N}x : Fx = \mathrm{N}x : Gx \equiv \text{there are just as many } Fs \text{ as } Gs$$

and HP in its usual form results from incorporating into this principle Frege's analysis of "just as many" in terms of equinumerosity. My suggestion, then, is that the ordinary notion *just as many* cannot be so analyzed, because it does not commit itself when both F and G are infinite. The only thing you can get out of the ordinary concept *just as many* is that there are just as many Fs as Gs if they are equinumerous and at least one of them is finite. Putting that together with the abstraction principle in its pure form then gives us what I called HPF, above.

It might be objected, however, that my own arguments in Chapter 7—in particular, on pages 171ff—establish the contrary. I argue there that understanding "just as many" involves connecting it with its 'practical correlate': just enough. For there to be just as many cookies as children means that, "... if you start giving cookies to children, and you're careful not to give another cookie to anyone to whom you've already given one, then you won't run out, though you won't have any left". But isn't it implicit in this that "just as many" should be analyzed in terms of one-one correspondence? Well, is it true that, if you start giving even numbers to natural numbers and are careful not to give any natural two evens, then you won't run out but won't have any left? Clearly not. It seems as if "just enough", as we ordinarily understand it, fails to get any grip here.

The arguments I give in the present chapter against the claim that HP is implicit in ordinary arithmetical thought are challenged by Fraser MacBride in his paper "On Finite Hume" (MacBride, 2000). MacBride's central claim is that HPF cannot exhaust the commitments that underlie ordinary arithmetical thought. If it did, he argues, then there would have been no air of paradox that surrounded cases like the evens and the natural numbers. HPF leaves the question whether there are as many evens as naturals completely open; only if there is some pressure towards a positive answer will there be even a hint of paradox. And so, MacBride seems to suggest, even pre-Cantorian thinkers must in some sense have endorsed HP: "If, prior to the reception of Cantor's work, ordinary arithmetical reasoning had only been informed by [HPF], then Bolzano's *Paradoxes of the Infinite* would have been a shorter book" (MacBride, 2000, p. 154). It should be clear that all of these criticisms continue to apply if we focus not on HPF but on the claims just made about how the ordinary notion *just as many* should be analyzed.

There is something right about MacBride's point, of course. For there to be an air of paradox, we must have some reason to endorse equinumerosity as a sufficient criterion for cardinal identity. But there is no reason whatsoever that this claim has to emerge from the concept of cardinal equality itself. One might have decided for all kinds of reasons that no workable theory of infinite cardinality is possible unless cardinal equality is equinumerosity. Perhaps Bolzano had come to that conclusion on the basis of mathematical reflection. Indeed, as we shall see, centuries of it had preceded him.

It is also plausible that there were, before Cantor, two competing standards for the equality of numbers: one concerning equinumerosity; the other, the relation between wholes and their proper parts. But one can acknowledge this point, too, for the same reason, while still denying that HP is implicit in ordinary arithmetical thought.

MacBride seems to want to privilege HP over the part–whole principle. He speaks of the latter as an "intuition founded only on a parochial acquaintance with finite parts and wholes" (MacBride, 2000, p. 155); HP is not derided in the same way. But I can see no basis on which to claim such an asymmetry other than a very whiggish view of history. Granted, no less a luminary than Kurt Gödel once argued that Cantor's treatment of infinite cardinality was inevitable, in the sense the concept of cardinal number can be extended from finite to infinite sets in just one way (Gödel, 1990, p. 254). But, as has been made wonderfully clear by Paolo Mancosu (2009), both the history and the mathematics turn out to be much more complex, and therefore much more interesting, than Gödel and MacBride imagine. Cantor was not playing the role of the slave boy in the *Meno*, with the mathematical muses playing the part of Socrates.

There seems to be no reason to privilege HP over the part–whole principle, so far as the commitments of ordinary arithmetical thought are concerned. What about accepting, then, that the situation really is symmetrical and that ordinary arithmetical thought is committed to both principles? It is no objection to this view that ordinary arithmetical thought then turns out to be inconsistent, as there are well-known instances of this sort of inconsistency already. As Galileo famously pointed out, for example, our ordinary thought about velocity involves two ways of measuring speed: in terms of distance covered and in terms of 'blurriness'. These correspond to our sophisticated concepts of average and instantaneous velocity, but no such distinction is made in 'folk' physics. That is why young enough children (and even adults, under the right circumstances) can be tied in congitive knots when presented with the right kinds of demonstrations, because they are relying upon inconsistent heuristics to evaluate them. So perhaps ordinary arithmetical thought too harbors this kind of inconsistency.

I have sometimes been attracted to this view, the thought being that ordinary arithmetical thought might conflate the notions of ordinal and cardinal number in much the way folk physics conflates average and instantaneous velocity. But, on reflection, that does not seem plausible, and I know of no empirical evidence for this claim. For one thing, it's simply too obvious that (say) re-arrangement of the members of some sequence can change ordinal position but leave cardinal number unaltered. For another, most of us are aware of the distinction between cardinals and ordinals from quite a young age, since most natural languages have different words for ordinals ("first", "second") and cardinals ("one", "two"). So there looks to be no basis for the view that ordinary arithmetical thought harbors inconsistency.

If that is right, then ordinary arithmetical thought is not committed to HP as well as the part–whole principle, and it is not committed to HP instead of the part–whole principle, either. So it just isn't committed to HP.

For what it is worth, however, my discussion would have benefitted enormously had I been more aware of the relevant history. Now that I am—everything I am about to say is based upon the work by Mancosu (2009) already mentioned—it seems to me that the historical facts reinforce my position. No doubt many who concerned themselves with the 'paradoxes of the infinite' prior to Cantor accepted that, if cardinal numbers can be assigned to infinite collections at all, then the assignment must respect equinumerosity. Examples are Galileo, Leibniz, and Bolzano, all of whom conclude that there are no infinite cardinal numbers, since the equinumerosity condition conflicts with the part–whole principle, which they regard as non-negotiable. As I have already emphasized, however, that does not show that a commitment to HP is implicit in ordinary arithmetical thought. And it would be wrong to think that the rejection of infinite cardinals must ultimately rest upon an appeal to equinumerosity. It might, instead, be based upon the thought that numerable collections are ones that can be counted,[36] so that infinite collections were never eligible for cardinal number in the first place. The 'paradoxes' are then the result of trying to extend to collections that are not numerable principles that apply only to ones that are.

But what is more interesting is that, while Galileo, *et al.*, may have been in the majority, there were many others who attempted, or thought we should attempt, to develop an arithmetic of the infinite that respected the part–whole principle, among them, Thabit ibn Qurra and Emmanuel Maignan. Moreover, although no fully satisfactory, non-Cantorian account of infinite cardinals has yet appeared, some interesting mathematical work has been done, and the project is not completely hopeless (Mancosu, 2009, §6).

Mancosu's announced goal in his paper is "to establish the simple point that comparing sizes of infinite sets of natural numbers is a legit-

---

[36]On a related note, I do not understand MacBride's criticism of my argument that HPF is implicit in ordinary arithmetical practice (MacBride, 2000, pp. 155–6). I think he must be presuming that, according to me, ordinary arithmetical thought is committed to the claim that *only* collections that can be counted have numbers. The claim I need, however, is just that "no infinite concept will get assigned any number by the process of counting" (this volume, p. 248), which is weaker. Moreover, although MacBride is correct that HPF, as formulated, presupposes that infinite collections have cardinal numbers, it is not difficult to formulate it in a way that would have no such presupposition.

imate conceptual possibility" (Mancosu, 2009, p. 642). I think it is clear that he succeeds. But if it is conceptually possible that infinite cardinals do not obey HP, then it is conceptually possible that HP is false, which means that HP is not a conceptual truth, so HP is not implicit in ordinary arithmetical thought.

All of that said, however, MacBride does not actually want to defend the claim that HP is implicit in ordinary arithmetical thought. What he really thinks is that the very idea that such a principle might be implicit in ordinary thought is "contentious" and "murky", for reasons associated with "Kripke's rule-following paradox" (MacBride, 2000, p. 156). My own view, however, is that cognitive psychology has long since established that theoretical principles are implicit in ordinary thought; if Kripkenstein says otherwise, then that is his problem. And it just isn't true that I made "no attempt to justify the assumption that an arithmetical epistemology must be derived from principles that we can retrieve by reflection on ordinary arithmetical reasoning" (MacBride, 2000, p. 156). While MacBride may not care for the arguments in Section 11.2, it is not as if the matter isn't discussed. That said, however, I am happy to agree that more needs to be said here, since subsequent discussion has made it clear that there are different ways of understanding the Neo-logicist project. I take the matter up in the overview, in Section 1.3.

12

A Logic for Frege's Theorem

As is now well-known, axioms for arithmetic can be interpreted in second-order logic plus HP:

$$\mathbf{N}x:Fx = \mathbf{N}x:Gx \ \equiv$$
$$\exists R[\forall x\forall y\forall z\forall w(Rxy \wedge Rzw \rightarrow x = z \equiv y = w)\wedge$$
$$\forall x(Fx \rightarrow \exists y(Rxy \wedge Gy))\wedge$$
$$\forall y(Gy \rightarrow \exists x(Rxy \wedge Fx))]$$

This result is *Frege's Theorem*. Its philosophical significance has been a matter of some controversy, most of which has concerned the status of HP itself. To use Frege's Theorem to re-instate logicism, for example, one would have to claim that HP was a logical truth. So far as I know, no-one has really been tempted by that claim. But Crispin Wright claimed, in his book *Frege's Conception of Numbers as Objects* (1983), that, even though HP is not a logical truth, it nonetheless has the epistemological virtues that were really central to Frege's logicism. Not everyone has agreed.[1] But even if Wright's view were accepted, there would be another question to be asked, namely, whether the sorts of inferences employed in deriving axioms for arithmetic from HP preserve whatever interesting epistemological property HP is supposed to have. Only then would the axioms of arithmetic have been shown to have that interesting property.

The problem is clearest for a logicist. If the axioms of arithmetic are to be shown to be logical truths, not only must HP be a logical truth, the modes of inference used in deriving axioms of arithmetic from it must preserve logical truth. They must, that is to say, be logical modes of inference. For Wright, the crucial question is less clear. It would be enough for his purposes if these modes of inference preserved whatever interesting epistemological property HP was supposed to have. But, nonetheless, Wright has typically been content to claim that second-order reasoning is logical reasoning and to suppose, reasonably enough, that, if that claim is good enough for the logicist, it is good enough for his purposes, too.

The claim that 'second-order logic is logic', as it is often put, has had both defenders and detractors.[2] I am not going to enter that debate here.

---

[1] There's a nice back-and-forth about this between Boolos (1998d) and Wright (2001d).

[2] Quine (1986) was famously skeptical. Boolos (1998g; 1998l) was an early proponent. Even still, Boolos was not at all sure that second-order reasoning would preserve analyticity (Boolos, 1998i). And at the end of his life, Boolos claimed no longer to understand the question whether second-order logic is logic, stated so baldly.

What I want to argue here is that a Neo-logicist does not need to commit herself to any claims about second-order logic.

In a typical proof of Frege's Theorem, axioms for arithmetic are derived from HP in second-order logic, but not all of the power of second-order logic is needed for these proofs. The power of second-order logic derives from the so-called comprehension axioms, each of which states, in effect, that a given formula defines a 'concept' or 'class'—something in the domain of the second-order variables. These axioms take the form:[3]

$$\exists F \forall x [Fx \equiv A(x)].$$

In full second-order logic, one has such an axiom for every formula $A(x)$ in which "F" does not occur free. So-called 'predicative' second-order logic has comprehension only for formulae containing no bound second-order variables.[4] Predicative second-order logic is weak in a well-defined sense: Given any first-order theory Θ, adding predicative second-order logic to Θ yields a conservative extension of it.[5] Full second-order logic, on the other hand, is extremely powerful, and it is that power that underlies much of the skepticism about the appropriateness of the term 'second-order *logic*'.[6]

Between predicative second-order logic and full second-order logic are systems of intermediate strength, each admitting a different set of comprehension axioms. In principle, any set of comprehension axioms will do, and there are many that have been considered.[7] What are perhaps the most natural intermediate systems arise from syntactic restrictions on the formulae appearing in the comprehension axioms. Say that a formula containing no bound second-order variables is Π^0_∞. Then where ϕ

---

[3]There are similar axioms for many-place predicates, of course.

[4]One could also consider a system that had no comprehension axioms at all, but this adds essentially nothing to first-order logic.

[5]A theory Γ' in a language L' conservatively extends a theory Γ in $L \subseteq L'$ if every theorem of Γ' stable in L is already a theorem of Γ. We can show that Γ' does conservatively extend Γ by showing that every model of Γ can be extended to a model of Γ' by adding interpretations of the primitives in $L' \setminus L$. For suppose so and suppose A is not a theorem of Γ. By completeness, let $\mathcal{M}$ be a model of Γ in which A is false. We can extend $\mathcal{M}$ to a model $\mathcal{M}'$ of Γ', and then A will also be false in $\mathcal{M}'$ and so not a theorem of Γ'.

Now, given any model for Θ, let the second-order domain contain exactly the subsets of the first-order domain definable (with first-order parameters) in the language of Θ so interpreted. The result is a model of Θ plus predicative comprehension, which is thus a conservative extension of Θ.

There is some need for care here when Θ contains axiom schemata: The schemata must not come to have new instances as a result of the addition of second-order vocabulary.

[6]Boolos (1998d) expresses this worry. It also plays a role in some of Etchemendy's discussions (1990). Koellner (2010) has developed an extremely sophisticated version of this objection.

[7]The standard reference on second-order logic is Shapiro's *Foundations without Foundationalism* (Shapiro, 1991), especially chapters 3–4. For a shorter, more accessible overview, see his piece in the *Blackwell Guide to Philosophical Logic* (Shapiro, 2001). There are many intermediate systems other than those we shall consider here, some of which are not based upon comprehension principles but on various sorts of choice principles, and so forth.

is Π^0_∞, formulae of the form $\forall F_1 \ldots \forall F_n \phi$ and $\exists F_1 \ldots \exists F_n \phi$ are Π^1_1 and Σ^1_1, respectively. If ϕ is Σ^1_n, then $\forall F_1 \ldots \forall F_n \phi$ is Π^1_{n+1}, and if ϕ is Π^1_n, then $\exists F_1 \ldots \exists F_n \phi$ is Σ^1_{n+1}. Second-order logic with Π^1_n comprehension has only those comprehension axioms in which $A(x)$ is Π^1_n (or simpler).

It is important to note that, as I have formulated the Π^1_n comprehension scheme, free second-order variables are allowed to occur in the comprehension axioms. As a result, there is no significant difference between Π^1_n comprehension and Σ^1_n comprehension. If $A(x)$ is a Σ^1_n formula, then its negation is (trivially equivalent to) a Π^1_n formula. Hence, Π^1_n comprehension delivers a concept F such that:

$$\forall x[Fx \equiv \neg A(x)]$$

But then predicative comprehension delivers a concept G such that:

$$\forall x[Gx \equiv \neg Fx]$$

and so we have:

$$\exists G \forall x[Gx \equiv A(x)]$$

We might as well therefore regard Π^1_n comprehension as Π^1_n-or-Σ^1_n comprehension.

More significantly, consider the formula:

$$\forall F[Fa \wedge \forall x \forall y(Fx \wedge Pxy \to Fy) \to Fb]$$

This formula defines the so-called 'weak ancestral' of the relation P. It is obviously Π^1_1, so Π^1_1 comprehension delivers a concept $\mathbb{N}$ such that:

$$\mathbb{N}n \equiv \forall F[Fa \wedge \forall x \forall y(Fx \wedge Pxy \to Fy) \to Fn]$$

If we take a to be 0 and P to be the relation of predecession—read Pxy as: x is the number immediately preceding y—then that is Frege's definition of the concept of a natural number. And Π^1_1 comprehension delivers the existence of this concept even if P itself has been defined by a Σ^1_1 formula, as it normally is in Fregean arithmetics:

$$Pab \overset{df}{\equiv} \exists G \exists y[b = Nx : Gx \wedge Gy \wedge a = Nx : (Gx \wedge x \neq y)]$$

The existence of the relation P is guaranteed by Σ^1_1—equivalently, Π^1_1—comprehension.

We may seem to be cheating here: Won't such a method end up reducing *all* comprehension to Π^1_1 comprehension? That would indeed be disastrous, but no such result is forthcoming. Chaining instances of comprehension together works in this case only because the variable F does not occur within the scope of the quantifier $\exists G$ that appears in the definition of P. The method will allow us to apply Π^1_1 comprehension twice to a formula of the form:

$$\forall F[\ldots F \cdots \to \exists G(\ldots G \ldots)]$$

but *not* to one of the form:

$$\forall F[\ldots F \cdots \to \exists G(\ldots G \ldots F \ldots)]$$

But one might still think such 'chaining' impermissible, even if coherent. Comprehension, so formulated, collapses Π_1^1 and Σ_1^1 comprehension and, moreover, fails to distinguish Π_1^1 sets from sets that are Π_1^1 in Π_1^1 sets. Is that really wise? Obviously, I am not suggesting that these distinctions do not matter, and if one wishes to use second-order logic to investigate problems to which these distinctions are relevant, then comprehension should be formulated so as to prohibit such 'chaining': One need only prohibit free second-order variables from appearing in the comprehension scheme. But it is not clear that these distinctions matter in the present context. I shall discuss the matter further below (see page 285). For the moment, I appeal to authority: In "Systems of Predicative Analysis", Feferman (1964) formulates the comprehension axioms this way.

Both the concept of predecession and the concept of natural number are thus delivered by Π_1^1 comprehension: That should make it plausible that the standard proof of Frege's Theorem requires only Π_1^1 comprehension, a conjecture that can be verified by working through the proof in detail while paying careful attention to what comprehension axioms are used.[8] There is a sense in which this result is best possible. I have mentioned several times that axioms for arithmetic can be derived from HP in second-order logic, but I have not yet said which such axioms I have in mind. There are, of course, many equivalent axiomatizations—I shall present one such axiomatization below—but what is important at the moment is that standard presentations of Frege's Theorem do not include a derivation of the usual first-order axioms for addition and multiplication. The reason is that, in a standard second-order language, the recursive defintions of addition and multiplication can be converted into explicit defintions in a way due (independently) to Dedekind and to Frege. The recursion equations themselves—and these just are the first-order axioms—can then be recovered from the definition. Unsurprisingly, however, the derivation of the recursion equations from the explicit definition needs more than predicative comprehension. The proof that addition and multiplication are well-defined and satisfy the recursion equations is by induction, and the induction is on a predicate containing the definition of addition or multiplication. The legitimacy of the induction thus presumes that the predicate in question defines a relation. Since the formula that defines addition is Π_1^1, we will need at least that much comprehension even to interpret first-order PA.[9]

---

[8]The mentioned fact was first noted in the original version of Chapter 7 (Heck, 2000a). Linnebo (2004) gives a detailed proof, and he also proves the converse: PA with Π_1^1-comprehension is interpretable in FA with Π_1^1-comprehension. See also Burgess's discussion in *Fixing Frege* (Burgess, 2005, pp. 151ff).

[9]Assuming we add no other axioms. It turns out that there is a way to interpret PA in ramified predicative Frege arithmetic, if we add a limited form of reducibility (Heck, 2011).

One can at least imagine a view that would regard Π_1^1 comprehension axioms as logical truths but deny that status to any that are more complex—a view that would, in particular, deny that full second-order logic deserves the name. In light of what has been said, such a view would serve the purposes of a Neo-logicist such as Wright. I do not expect it to be obvious at this point how such a view might be motivated, and it is in fact no part of the view I want to defend here that, say, Δ_3^1 comprehension axioms are *not* logical truths. What I am going to suggest, however, is that there is a special case to be made on behalf of Π_1^1 comprehension. Or something like it.

12.1 Predecession

As it happens, the only comprehension axioms one actually needs for the proof of Frege's Theorem—besides a handful of instances of predicative comprehension—are these:

$$\exists P\{Pab \equiv \exists G \exists y[b = \mathbf{N}x:Gx \wedge Gy \wedge a = \mathbf{N}x:(Gx \wedge x \neq y)]\}$$
$$\exists R\{Ran \equiv \forall F[\forall x(Pax \to Fx) \wedge \forall x \forall y(Fx \wedge Pxy \to Fy) \to Fn]\}$$

The latter, of course, defines the relation that would usually be written: $P^*\xi\eta$. That is, it defines the relation that is the (strong) ancestral of predecession.

Øystein Linnebo (2004, pp. 172–3) suggests that there is something seriously wrong with Frege's defintion of predecession. It simply does not seem reasonable to suppose that a notion as simple as that of predecession should logically be so complex. Consider, for example, the proof of the familiar fact that every number other than zero has a predecessor. This proposition, $x \neq 0 \to \exists y(x = Sy)$, is one of the axioms of Robinson arithmetic, Q, but it is redundant in PA, since it is provable from the other axioms by induction not just in PA but in the much weaker theory known as $I\Delta_0$, which has induction only for bounded formulae.[10] In Frege arithmetic, however, a formalization of that proof would require Σ_1^1 comprehension, since the induction must now be on $x \neq 0 \to \exists y(Pyx)$.[11] Are we really to believe that such strong logical resources are needed for the proof of such a simple statement? The more plausible view is the one enshrined in the usual treatment of arithmetic: Predecession is a *primitive* notion.

Linnebo's concern is a sensible one, but I think it can be answered. Although the definition of predecession is undeniably Σ_1^1 in form, it is not, I want to suggest, Σ_1^1 in spirit. The definition one would really like to give is this one:

---

[10]The induction can be carried out on $x \neq 0 \to \exists y < x(x = Sy)$.

[11]A different proof can be given that would not require comprehension at all, but there are other examples of this same form.

(*P*-lite) $P(\mathbf{N}x : Gx, \mathbf{N}x : Fx) \equiv \exists y(Fy \wedge \mathbf{N}x : Gx = \mathbf{N}x : (Fx \wedge x \neq y))$

To be sure, (*P*-lite) is not a proper definition. It does not tell us when *Pab* but only when $P(\mathbf{N}x : Gx, \mathbf{N}x : Fx)$: Nothing in (*P*-lite) tells us whether Julius Caesar, that familiar conqueror of Gaul, precedes 0 or not. But the obvious reply is that it was supposed to be implicit in (*P*-lite) that *only numbers have or are predecessors*. If Caesar is a number, then he is the number of *F*s, for some *F*, in which case (*P*-lite) will determine which numbers he precedes and succeeds. If he is not a number, then he neither precedes nor succeeds any number. Hence, the question which numbers Caesar precedes and succeeds is equivalent to the question whether Caesar is a number and, if so, which one he is. Well, if that isn't a familiar problem! Maybe it is even a serious problem. But it is a problem the Neo-logicist had anyway.[12]

Suppose that this problem of 'trans-sortal identification' has either been solved or been justifiably ignored. (Maybe it isn't a serious problem, just an amusing one.) Then (*P*-lite) tells one everything one needs to know about predecession. How would that allow the Neo-logicist to avoid appealing to Σ_1^1 comprehension? Well, the Neo-logicist might regard predecession as primtive and regard (*P*-lite) as analytic of that notion. Moreover,

(*P*-imp) $Pab \rightarrow \exists F(a = \mathbf{N}x : Fx) \wedge \exists G(b = \mathbf{N}x : Gx)$

simply makes explicit the implicit requirement that only numbers can be or have predecessors, so it too is analytic of predecession. No appeal to comprehension is then needed to guarantee that the relation of predecession exists, any more than in the usual formulation of second-order arithmetic.

One might worry that this strategy makes everything too easy. Why can't the Neo-logicist just regard the ancestral as primitive and take the usual definition of the concept of natural number to be analytic of it? Then no appeal to comprehension would be needed! It will become clear that I am in a way sympathetic with this suggestion, but the arguments just offered on behalf of the claim that (*P*-lite) is analytic do not generalize to the case of the ancestral. Those arguments apply only to certain sorts of explicit definitions, namely, those that can be resolved into something of the form:

$$Ra_1 \ldots a_n \overset{df}{\equiv}$$
$$\exists F_1 \ldots \exists F_n[a_1 = \Phi x : F_1 x \wedge \cdots \wedge a_n = \Phi x : F_n x \wedge \mathcal{R}_x(F_1 x, \ldots, F_n x)]$$

which is equivalent to the conjunction of

(*R*-lite) $R(\Phi x : F_1 x, \ldots, \Phi x : F_n x) \equiv \mathcal{R}_x(F_1 x, \ldots, F_n x)$

_____
[12]I explore this problem in Chapters 5, 6, and 9.

and

(R-imp) $Ra_1 \ldots a_n \rightarrow \exists F_1(a_1 = \Phi x : F_1 x) \wedge \cdots \wedge \exists F_n(a_n = \Phi x : F_n x)$

The arguments presented above purport to show that, in this case, (R-lite) is already an adequate definition of R, *modulo* an instance of the problem of trans-sortal identification. But they apply only to this sort of case.

The case of the ancestral is not such a case,[13] but there is a different such case that is important. Consider the so-called predicative fragment of *Grundgesetze*, which consists of predicative second-order logic plus a schematic form of Frege's Basic Law V:

$$\hat{x}(A(x)) = \hat{x}(B(x)) \equiv \forall x (A(x) \equiv B(x))$$

This theory is known to be consistent (Heck, 1996). What saves the system from inconsistency is the fact that membership is defined in terms of a Σ_1^1 formula:

$$a \in b \equiv \exists F(b = \ddot{x}(F x) \wedge F a)$$

and we do not have comprehension for such formulae in the predicative fragment. So, crucially, we cannot prove the following instance of naïve comprehension:

$$a \in \hat{x}(x \notin x) \equiv a \notin a$$

whence the paradox that threatens to arise when we take a to be $\hat{x}(x \notin x)$ is averted. But it is averted only at the cost of our inability to prove the formula just displayed, and that has always seemed to me to be deeply counterintuitive. I can now give some content to the intuition thus countered.

The definition of membership is of precisely the form we have been discussing. The definition of membership one would really like to give is this one:

(∈-lite) $a \in \hat{x}(B(x)) \equiv B(a)$

That is not a proper definition. It does not tell us when $a \in b$ but only when $a \in \hat{x}(B(x))$, and so on and so forth. But *modulo* the problem of trans-sortal identification, or so I would argue, (∈-lite) is a perfectly good definition. Any neo-Fregean who is prepared to countenance Basic Law

---

[13] It is not such a case because we know that the ancestral cannot be defined by a Σ_1^1 formula. *Proof*: Suppose it can. Then there is a formula $\exists F_1 \ldots \exists F_n \phi(a, x, R, F_1, \ldots, F_n)$ with ϕ being Π_∞^0 such that $\exists F_1 \ldots \exists F_n \phi(a, x, , R, F_1, \ldots, F_n)$ holds if, and only if, there is a finite sequence $a = a_0, a_1, \ldots, a_n = x$ where $Ra_i a_{i+1}$. Now consider the first-order formula $\phi(a, x, R, F_1, \ldots, F_n)$ and consider the formulae $\neg Rax$, $\neg \exists x_0(Rax_0 \wedge Rx_0 x)$, $\neg \exists x_0 \exists x_1(Rax_0 \wedge Rx_0 x_1 \wedge Rx_1 x)$, etc. Obviously, there are models of $\phi(a, x, R, F_1, \ldots, F_n)$ and any finite subset of these. By compactness, there is thus a model in which all of them hold. But that is not a model in which there is a finite sequence connecting a to x, yet $\phi(a, x, R, F_1, \ldots, F_n)$ holds in the model and so *a fortiori* $\exists F_1 \ldots \exists F_n \phi(a, x, , R, F_1, \ldots, F_n)$ holds as well.

V ought to regard membership as primitive and characterized by ($\in$-lite) and

($\in$-imp) $a \in b \to \exists F(b = \hat{x}(Fx))$

But then Russell's paradox reappears, since ($\in$-lite) immediately implies:

$$\hat{x}(x \notin x) \in \hat{x}(x \notin x) \equiv \hat{x}(x \notin x) \notin \hat{x}(x \notin x)$$

And that seems to me an intuitively satisfying result. There is nothing *truly* impredicative about the definition of membership.

The predicative fragment of *Grundgesetze* may be consistent, then, but it is not really *coherent*.[14]

12.2   Ancestral Logic

As noted above, the only impredicative instances of comprehension that are needed for the proof of Frege's Theorem are these:

$$\exists P\{Pab \equiv \exists G \exists y[b = \mathbf{N}x : Gx \wedge Gy \wedge a = \mathbf{N}x : (Gx \wedge x \neq y)]\}$$
$$\exists R\{Ran \equiv \forall F[\forall x(Pax \to Fx) \wedge \forall x \forall y(Fx \wedge Pxy \to Fy) \to Fn]\}$$

The arguments of the last section purported to establish that the former has no significant epistemological cost. If that is accepted, then we may draw the following intermediate conclusion: The question whether the logic used in the proof of Frege's Theorem is epistemologically innocent reduces to the question what our attitude should be to Frege's definition of the ancestral.

The assumption that the ancestral of an arbitrary relation exists is much weaker than full Π_1^1 comprehension. In fact, there is a logic known as *ancestral logic* that characterizes the logic of the ancestral in an otherwise first-order language (Shapiro, 1991, p. 227).[15]

We begin with an ordinary first-order language $\mathcal{L}$ and form a new language $\mathcal{L}_*$ by adding an operator $*_{xy}$ that forms a binary relational expression from a formula with two free variables, these variables being bound by the operator. So we have formulae of the form: $*_{xy}(\phi xy)(a, b)$, where ϕxy is a formula. Let an interpretation of $\mathcal{L}$ be given. We expand it to an

---

[14]Note that this argument does not even purport to show that predicativity restrictions are not otherwise justified. It is entirely specfic to the case of Basic Law V. Note further that it does not really depend upon the assumption that Basic Law V and the definition of membership are given in schematic form, rather than in the forms:

$$\forall F \forall G[\hat{x}(Fx) = \hat{x}(Gx) \equiv \forall x(Fx \equiv Gx)]$$
$$\forall F[a \in \hat{x}(Fx) \equiv Fa]$$

If $\in$ is taken as primitive, then $x \notin x$ is predicative, and it can then be used to instantiate Fx.

[15]There is also some recent work by Avron (2003) on such logics.

interpretation of $\mathcal{L}_*$ as follows:[16] Suppose that ϕxy is satisfied by exactly the ordered pairs in some set Φ; then $*_{xy}(\phi xy)(a, b)$ is true if, and only if, there is a finite sequence $a = a_0, \ldots, a_n = b$ such that each $<a_i, a_{i+1}> \in \Phi$. Less formally: $*_{xy}(\phi xy)(a, b)$ is true just in case a can be linked to b by a finite sequence of ϕ-steps. We require that there should be at least one such step: $*_{xy}(\phi xy)$ is therefore the *strong* ancestral of ϕ, so-called because we do not always have: $*_{xy}(\phi xy)(a, a)$. The weak ancestral of ϕ, denoted $*_{xy}^=(\phi xy)$, may be defined in the usual way: $*_{xy}(\phi xy)(a, b) \vee a = b$. We shall use the more familiar notation $\phi^* ab$ and $\phi^{*=} ab$, omitting the bound variables when there is no danger of confusion.

It is easy to see that ancestral logic is not completely axiomatizable: It permits the formulation of a categorical theory of arithmetic (Shapiro, 1991, p. 228).[17] But, of course, that need not prevent us from partially axiomatizing the logic. One way to proceed would be to take as introduction rules:

$$\phi ab \vdash \phi^* ab$$
$$\phi^* ab, \phi bc \vdash \phi^* ac$$

and as an elimination rule:

$$\phi^* ab \vdash \forall x(\phi ax \to A(x)) \wedge \forall x \forall y(A(x) \wedge \phi xy \to A(y)) \to A(b)$$

Call this system *weak* ancestral logic. Its introduction rules reflect the 'inductive' character of the ancestral: Taking ϕab to mean: b is a's parent, they tell us that one's parents are one's ancestors and that any parent of an ancestor is an ancestor. The elimination rule is a principle of induction, in schematic form.

Weak ancestral logic incorporates, in its elimination rule, one half of Frege's definition of the ancestral. But it does not incorporate the other half of Frege's definition:

$$\forall F[\forall x(\phi ax \to Fx) \wedge \forall x \forall y(Fx \wedge \phi xy \to Fy) \to Fb] \to \phi^* ab$$

To my mind, that is an important weakness. Consider the following argument.[18]

Suppose that b is a's ancestor and c is b's ancestor. Suppose further (i) that all of a's parents are, say, blurg and (ii) that

---

[16]This specification is less precise that it would really need to be, since it does not allow for additional free variables in ϕ. But let us not be too pedantic.

[17]Add to a first-order formulation of *PA* the axiom: $\forall n[*_{xy}^=(y = Sx)(0, n)]$.

[18]The argument is even more natural if we use plurals: Suppose that whenever a's parents are among some things and any parent of one of those things is one of those things, ... It would be a simple matter to reformulate everything we are doing here in such terms. Indeed, no reformulation is needed: One may, as Boolos (1998l) noted, simply regard variables like F as being plural variables and take Fx to mean: x is among the Fs. The point would then be that we can prove Frege's Theorem in a language that allows for plural reference, even if it does not allow for plural *quantification*.

blurghood is hereditary—that is, that any parent of someone who is blurg is also blurg. Since b is a's ancestor, b is blurg, by the elimination rule. But then, by (ii), all of b's parents are blurg and so, since c is b's ancestor, c is blurg, again by the elimination rule. That is, if (i) and (ii), then c is blurg. And so, by Frege's definition of the ancestral, c is a's ancestor.

That, obviously, is an argument for the transitivity of the ancestral and, so far as I can see, nothing like it can be formalized within weak ancestral logic. That is not to say that the transitivity of the ancestral cannot be proved in weak ancestral logic. It can be, though in a different way, namely, in roughly the way Frege proves it in *Begriffsschrift*.[19] But that is a different argument, one whose formalization in standard second-order logic requires the use of Π_1^1 comprehension. No comprehension at all is needed for the formalization of the argument just given (Boolos, 1998i, pp. 158–9). That we can formalize the more complicated argument in weak ancestral logic but not the less complicated one suggests that weak ancestral logic has at least one thing upside down.

One may have been wanting to ask what the nonsense term "blurg" is doing in the above argument, and that is a perfectly reasonable question. But such reasoning is very common. At least it is very common for me to engage in such reasoning, especially when I am teaching logic to undergraduates. Perhaps that would be more obvious if I were to replace "blurg" with "F", but one does not have to use letters to engage in such reasoning. And there is no reason to dismiss it out of hand. In many cases, such reasoning can be understood as tacitly semantic. We may take "blurg" to be a variable that ranges over expressions and construe the argument as tacitly invoking semantic notions, such as satisfaction. A related proposal would construe the argument substitutionally. On either construal, however, this particular argument would only establish something about concepts we can *name*, whence it is surely invalid. But it seems to me a perfectly good argument, so some other way of understanding such reasoning is needed.

A fan of second-order logic might suggest that "blurg" is a second-order variable and that the argument as a whole tacitly involves second-order quantification, its intuitive force revealing the extent to which second-order reasoning is intuitively compelling (Boolos, 1998l, pp. 59ff). But there are two aspects to this suggestion, and they can be disentangled:

---

[19]Frege's proof is by induction on $\phi^* a\xi$. The elimination rule yields

$$\phi^* bc \to [\forall x(\phi bx \to \phi^* ax) \land \forall x \forall y(\phi^* ax \land \phi xy \to \phi^* ay) \to \phi^* ac]$$

The second conjunct follows immediately from the second introduction rule; the first follows from $\phi^* ab$ and the second introduction rule. Hence, $\phi^* bc \land \phi^* ab \to \phi^* ac$.

It would be cleaner if I had a nice example of a theorem whose proof is easily formalized using Frege's definition of the ancestral but which cannot be formalized in weak ancestral logic. There must be one.

We can interpret "blurg" as the natural language correlate of a *free* second-order variable and simultaneously deny that second-order *quantification* is involved in the argument at all.

Given a first-order language, add to it a stock of second-order variables. We do not permit these variables to be bound by quantifiers: They occur only free. Thus, there are formulae in the language such as:

$$\forall x(\phi ax \to Fx) \wedge \forall x \forall y(Fx \wedge \phi xy \to Fy) \to Fb$$

where F is a free second-order variable. An interpretation of such a formula is simply a first-order interpretation. Free second-order variables, that is to say, are to be treated just as predicate-letters are in first-order logic: They are to be assigned subsets of the domain. Implication is then defined as usual: A set of formulae Γ implies a formula A if, and only if, every interpretation that makes all formulae in Γ true also makes A true. A formula is valid if it is implied by the empty set of formulae.[20]

The proof-theory is also straightforward. I shall take us to be working in a system of natural deduction. Such a system will have some mechanism or other for keeping track of the premises used in the derivation of a given formula. I assume that we have some natural set of rules for first-order logic already in place. No special rules that govern free second-order variables are being introduced at this point. Call the resulting system *minimal schematic logic* (minimal SL). It should be clear that minimal SL is sound with respect to the semantics mentioned above. It is also complete, since, at this point, the free second-order variables differ from predicate-letters only in what they are called.[21]

We can now reformulate ancestral logic. The elimination rule, which we may call $(*-)$, remains one direction of Frege's explicit definition of the ancestral, though it now need not be formluated as a schema but can be formulated using a free second-order variable:

$$\phi^* ab \vdash \forall x(\phi ax \to Fx) \wedge \forall x \forall y(Fx \wedge \phi xy \to Fy) \to Fb$$

The introduction rule $(*+)$ is the other direction of Frege's definition of the ancestral:

$$\forall x(\phi ax \to Fx) \wedge \forall x \forall y(Fx \wedge \phi xy \to Fy) \to Fb \vdash \phi^* ab$$

where F may not be free in any premises on which the premise of this inference itself depends.[22] Call the resulting system *minimal schematic*

---

[20]A formula of schematic logic is therefore valid only if its universal closure is a valid second-order formula.

[21]The point of calling them "variables" would then be only to distinguish them from predicate constants: A schematic language would be regarded as interpreted even if it assigns no interpretations to free second-order variables.

[22]It is here that the additional expressive power provided by the presence of free second-order variables makes itself felt: No such rule could possibly be formulated in a purely first-order language.

278   *A Logic for Frege's Theorem*

ancestral logic (minimal SAL). It should again be clear that this logic is sound if the ancestral is interpreted as indicated above. It is not, of course, complete with respect to that semantics, since no recursive axiomatization can be.

What we have done is transcribe Frege's explicit definition of the ancestral into the framework of schematic logic. Why does the transcription work? Consider the introduction rule, (∗+). If we can derive

$$(\dagger) \quad \forall x(\phi ax \to Fx) \wedge \forall x \forall y(Fx \wedge \phi xy \to Fy) \to Fb$$

from premises in which F does not occur free, then, in standard second-order logic, we can use universal generalization to conclude that

$$(\dagger\dagger) \quad \forall F[\forall x(\phi ax \to Fx) \wedge \forall x \forall y(Fx \wedge \phi xy \to Fy) \to Fb]$$

and then use Frege's definition of the ancestral to conclude that ϕ^*ab. Similarly in the case of the elimination rule: ϕ^*ab and Frege's definition together imply $(\dagger\dagger)$ which in turn implies $(\dagger)$. But then $(\dagger\dagger)$ is just a layover and Frege's explicit definition is a ladder we can kick away.[23] Goodbye, ladder.

The transitivity of the ancestral can be proven in minimal SAL:[24]

[1]	(1)	ϕ^*ab	Premise
[2]	(2)	ϕ^*bc	Premise
[3]	(3)	$\forall x(\phi ax \to Fx) \wedge \forall x \forall y(Fx \wedge \phi xy \to Fy)$	Premise
[1]	(4)	$\forall x(\phi ax \to Fx) \wedge \forall x \forall y(Fx \wedge \phi xy \to Fy)$ $\to Fb$	$(1, *-)$
[1,3]	(5)	Fb	$(3,4)$
[1,3]	(6)	$\forall x(\phi bx \to Fx)$	$(3,5)$
[2]	(7)	$\forall x(\phi bx \to Fx) \wedge \forall x \forall y(Fx \wedge \phi xy \to Fy)$ $\to Fc$	$(2, *-)$
[1,2,3]	(8)	Fc	$(3,6,7)$
[1,2]	(9)	$\forall x(\phi ax \to Fx) \wedge \forall x \forall y(Fx \wedge \phi xy \to Fy)$ $\to Fc$	$(3,8, \to +)$
[1,2]	(10)	ϕ^*ac	$(9, *+)$

So minimal SAL does not suffer from the problem that plagues weak ancestral logic.

That said, minimal SAL is still a very weak logic. The transitivity of the ancestral can be proven in minimal SAL because it can be proven in second-order logic without any appeal to comprehension. But the proof of theorem (124) of *Begriffsschrift*:

$$\phi^*ab \wedge \forall x \forall y \forall z(\phi xy \wedge \phi xz \to y = z) \wedge \phi ac \to \phi^{*=}ac$$

---

[23]Thanks to Stewart Shapiro for the allusion.
[24]I have not included all the steps that would be required to make this argument formally precise, only enough to make it clear that it could be.

breaks down, as a little experimentation will show. The reason is that Frege's proof uses Π_1^1 comprehension, and there is nothing in minimal SAL that gives us the power of Π_1^1 comprehension.[25]

How are we to get that power without second-order quantifiers? Easily. There are no explicit comprehension axioms in the formal systems of *Begriffsschrift* and *Grundgesetze*, either. Rather, Frege has a rule of substitution: Given a theorem of the form $\ldots F \ldots$, infer $\ldots \phi \ldots$, for any formula ϕ (subject to the usual sorts of restrictions). The substitution rule is, as is well-known, equivalent to comprehension.[26] What we need here is thus a rule of substitution:

> Suppose we have derived A from the premises in Γ, and let $A_{F/\phi}$ be the result of replacing all occurences of F in A by the formula ϕ (subject to the usual sorts of restrictions, again). Then, if F is not free in Γ, we may infer $A_{F/\phi}$.

As a special case, of course, if A is provable (from no assumptions), then we may infer any substitution instance of it. And, given this rule, theorem (124) of *Begriffsschrift* can now be proven. (See the appendix for the proof.)

The substitution rule is clearly sound given the semantics sketched above, but it is another question whether we should regard it as *justified* and, if so, on what sort of ground. This question will be considered below, in Section 12.6.

We thus have two kinds of systems: There are the systems without the substitution rule—minimal schematic logic and minimal schematic ancestral logic—and there are the systems with the substitution rule— what we may call full schematic logic and full schematic ancestral logic. In fact, there are further distinctions to be drawn, since the substitution rule can be restricted in various ways. We might, for example, require the formula that replaces F not to contain *. Adding this rule to minimal schematic ancestral logic would give us the effect of predicative comprehension, so we might call the resulting system *predicative* schematic ancestral logic. Such subsystems will not be of much interest here, however, though they certainly would be of interest in explorations of predicative versions of Fregean systems. In particular, the known results about such systems can easily be adapted to the present framework.

---

[25]Zoltan Gendler Szabó has proved that comprehension is, in fact, required. It is also required for the proof of theorem (133):

$$\phi^{*=}ab \wedge \phi^{*=}ac \wedge \forall x\forall y\forall z(\phi xy \wedge \phi xz \to y = z) \to \phi^* bc \vee b = c \wedge \phi^* cb$$

which thus cannot be proven in minimal SAL, either.

[26]Substitution implies comprehension: Trivially, we have $\forall x(Fx \equiv Fx)$, so existential generalization yields: $\exists G\forall x[Gx \equiv Fx]$; by substitution: $\exists G\forall x[Gx \equiv \phi x]$, for each formula ϕ. The proof of the converse is messier but not difficult: It is by induction on the complexity of the formula $\ldots F \ldots$.

12.3   Schemata in Schematic Logic: A Digression

One reason second-order languages are so appealing is that principles that have to be formulated, in first-order languages, as axiom-schemata can be formulated in second-order languages as single axioms. It would be nice if schematic languages had a similar appeal—if, for example, the axiom of separation could be expressed by the single axiom:

(Sep) $\forall z \exists y \forall x (x \in y \equiv x \in z \land Fx)$

from which its various instances could then be inferred by substitution. But it can't be, not if we characterize the rule of substitution as we did above. A *theory*, after all, is a set of formulae, and its theorems are the formulae that are deducible from the sentences in that set.[27] A deduction must thus assume some of the sentences of the theory as premises and then derive a theorem from them. If (Sep) is taken as a premise in a deduction, however, the variable F will obviously be free in that premise, whence it cannot be substituted for. Indeed, it is easy to see that (Sep) does not imply all (or even most) other instances of separation, not if "implies" is defined as it was above. So it is a very good thing that such instances cannot all be deduced from it.

There is really a more basic problem here: I've yet to say what it might mean to assert something like (Sep); I've yet, that is, to say what the truth-conditions of (Sep) are. One might reasonably want to deny that (Sep) *has* truth-conditions. Since it contains a free second-order variable, one cannot speak of it as being true or false absolutely but only as being true or false under this or that assignment of a value to F. But there is an alternative. One can give free second-order variables the so-called 'closure interpretation', effectively taking (Sep) to be true just in case its universal closure is true. We do not actually need to consider the universal closure, of course, for we can define truth for formulae of schematic languages directly: A formula is true if, and only if, it is true under all assignments to its free variables.[28]

This definition of truth would solve our problem concerning separation. Unfortunately, it brings a whole host of other problems with it.[29] To

---

[27]Sometimes the term "theory" is used in a different sense—a theory is a deductively closed set of sentences, and its theorems are just the members of that set—but that is not the sense that is relevant here.

[28]Thanks, long after the fact, to Boolos for impressing this point upon me. A similar point arises in Frege. He too gives formulae with free variables the closure interpretation, but he does *not* regard them as simply abbreviations of their closures (Heck, 2010).

[29]Shapiro considers a logic $L2K-$, which is similar to the systems we have been discussing: Free second-order variables are given the closure interpretation. It turns out, however, to be surprisingly difficult to define a notion of implication with respect to which any reasonable set of deductive principles is sound (Shapiro, 1991, p. 81). In the end, Shapiro does define such a notion, but it is not really consistent with the closure interpretation. On Shapiro's definition, Fx does not imply Gx. But if the former really means $\forall F(Fx)$ and the latter really means $\forall G(Gx)$, it should. If one wanted to say that it is perhaps best if

make further progress, we need to distinguish two sorts of assumptions that are made in arguments. Sometimes, one makes an assumption 'for the sake of argument'. For example, one might assume the antecedent of a conditional and try to prove its consequent in order to prove that conditional. So, for example, if one were trying to prove $F0 \rightarrow F0 \vee G0$, one might begin by assuming $F0$: "Suppose 0 is blurg", one might say. One is not to continue with, "So 0 is odd. And prime. And, for that matter, even." One does not, that is to say, expect to be understood as having assumed that zero has every property there is. If we call an assumption made 'for the sake of argument' a *supposition*, then what has been shown is that suppositions are not to be understood in terms of their universal closures, the reason being that a supposition is relevantly like the antecedent of a conditional: No-one would suppose that $F0 \rightarrow G0$ should be understood as $\forall F(F0) \rightarrow \forall G(G0)$, however tempted she was by the thought that it should be understood as $\forall F \forall G(F0 \rightarrow G0)$.

But this observation does not make the closure interpretation of (Sep) any less available. The reason is that suppositions are mere tools of argument: They are not put forward as true in their own right. In a sense, that is obvious, since one sometimes makes a supposition only for *reductio*, but I am suggesting something stronger: that suppositions are not even *assumed* to be true; that is not the role they play in argument. Still, it ought nonetheless to be possible to assume that something is true, if only to investigate its consequences. Let us reserve the term *hypothesis* for an assumption of this kind. Then there is no bar to our understanding hypotheses in terms of their universal closures.

Formally, then, we distinguish between premises that are suppositions and premises that are hypotheses. An inference is valid if every interpretation that makes all of its suppositions and hypotheses true— where truth for hypotheses is understood in terms of the closure interpretation—also makes its conclusion true. The distinction must be tracked in the proof-theory as well, and rules of inference that discharge premises, such as *reductio* and conditional proof, will need some modification: The premise discharged must be a supposition, not an hypothesis. But the crucial observation, for our purposes, is that the rule of substitution can now be relaxed: One can infer $A_{F/\phi}$ from A so long as F is not free in any *supposition* on which A depends; it may be free in *hypotheses* on which A depends. This rule is clearly sound: Since, semantically speaking, hypotheses are treated as if they were universally closed, it is as if there aren't any free variables in the hypotheses, at least as far as the definition of implication is concerned.

There are other issues concerning the formulation of ancestral logic that we could discuss, but let me set them aside: How they are resolved does not really bear upon the philosophical issues that are motivating

---

Fx doesn't imply Gx, I'd happily agree, but that intuition isn't consistent with the closure interpretation.

this discussion.[30] The question with which we started was whether we can regard (Sep) as a formulation of separation. The answer is that we can if we regard a theory as a set of *hypotheses* from which theorems are to be deduced. As an *hypothesis*, (Sep) does imply all other instances of separation, including those containing other free variables.

One might wonder why I chose separation as my example rather than induction, since the same issues will, of course, arise with respect to induction in the context of schematic logic. They do not, however, arise in the context of *ancestral* logic, since we can formulate induction in ancestral logic as the single sentence:

$$\forall z *_{xy}^{=} (y = Sx)(0, z)$$

As we shall see, this point can be generalized: Any principle expressed in first-order logic by an axiom scheme can be expressed in Arché logic,[31] to be introduced next, by a single sentence.

12.4   Arché Logic

The methods used in Section 12.2 allowed us to transcribe Frege's explicit definition of the ancestral into schematic logic. A brief review will reveal, however, that the methods used presume only that the formula defining the ancestral is Π_1^1: They presume nothing about what formula it is. We can thus generalize that construction.

Consider first the simplest case. Let $\phi_x(Fx, y)$ be an arbitrary formula containing no free variables other than those displayed. In standard second-order logic, we can explicitly define a new predicate $\mathcal{A}_\phi$ as follows:

$$\mathcal{A}_\phi(y) \overset{df}{\equiv} \forall F \phi_x(Fx, y)$$

The definition is licensed, in effect, by Π_1^1 comprehension, which guarantees that $\mathcal{A}_\phi$ exists. But the trick used with the ancestral can also be used here. We have an introduction rule, $(\mathcal{A}_\phi+)$:

$$\phi_x(Fx, y) \vdash \mathcal{A}_\phi(y)$$

---

[30] Of course it matters that the technical issues *can* be resolved. I'll leave that as an exercise. One nice thing to do is to add structural rules that allow an hypothesis freely to be converted to a supposition and a supposition to be converted to a hypothesis so long as none of its free variables are free in any of the suppositions on which it depends.

The distinction I am drawing here is very close to a distinction I drew elsewhere, for an ostensibly quite different purpose, between what I called "rules of inference" and "rules of deduction" (Heck, 1998c). In fact, however, the formal situation is almost identical, since modal formulae are there interpreted as if they were always preceded by a universal quantifier over accessible worlds. The techniques developed there can therefore be used here.

[31] I hope it's not too presumptuous to take this name for the logic. Obviously, it is intended in tribute.

where, as usual, F may not be free in any premises on which $\phi_x(Fx, y)$ depends, and an elimination rule, $(\mathcal{A}_\phi -)$:

$$\mathcal{A}_\phi(y) \vdash \phi_x(Fx, y)$$

In effect, these rules define $\mathcal{A}_\phi(y)$ as equivalent to $\forall F \phi_x(Fx, y)$ without using an explicit universal quantifier to do so. The extension of the new predicate $\mathcal{A}_\phi$ is then the set of all those y such that $\phi_x(Fx, y)$ is true for every assignment to F.

Generalizing, we may allow more than one predicate variable to occur in ϕ;[32] we allow the predicate variables to be of various adicities; and we allow additional free first-order variables, in which case what is defined is a relation rather than just a predicate. So, in general, ϕ may be of the form

$$\phi_{x_1 \ldots x_{max(k_i)}}(F_1(x_1, \ldots, x_{k_1}), \ldots, F_n(x_1, \ldots, x_{k_n}), y_1, \ldots, y_m)$$

and, simplifying notation, we introduce a new predicate $\mathcal{A}_\phi(\overline{y})$ subject to the rules:

$$\phi_{\overline{x}}(\overline{F}, \overline{y}) \vdash \mathcal{A}_\phi(\overline{y})$$

$$\mathcal{A}_\phi(\overline{y}) \vdash \phi_{\overline{x}}(\overline{F}, \overline{y})$$

It is a more serious question whether we wish to allow additional free *second*-order variables to occur in ϕ. I'll return to this issue below (see page 292). For now, we do *not* allow additional free second-order variables.

Call what was just described the *scheme of schematic definition*. It allows us to transcribe what we would normally regard as an explicit definition of a new predicate or relation in terms of a Π_1^1 formula into schematic logic: If we can prove $\psi_x(Fx, a)$, then we can, in standard second-order logic, use universal generalization to conclude that $\forall F \psi_x(Fx, a)$ and so that $\mathcal{A}_\psi(a)$; if we have $\mathcal{A}_\psi(a)$, then by definition, $\forall F \psi_x(Fx, a)$ and so, by universal instantiation, $\psi_x(Fx, a)$. But the explicit definition of $\mathcal{A}_\psi(a)$ in terms of $\forall F \psi_x(Fx, a)$ simply mediates the transitions between $\psi_x(Fx, a)$ and $\mathcal{A}_\psi(a)$. It can be eliminated in favor of a schematic definition of $\mathcal{A}_\psi(a)$ in terms of those same transitions. Goodbye, ladder.

If we add the scheme of schematic definition to (minimal) SL, we thus get a system in which new predicates co-extensional with Π_1^1 formulae can be introduced by schematic definition. Here again, however, the scheme of schematic definition is, by itself, deductively very weak: To exploit its power, we need a rule of substitution. We thus have three sorts of systems, depending upon the strength of the substitution principle we assume. Call the system without substitution *minimal Arché logic* (minimal AL). *Predicative Arché logic* contains a restricted substitution principle:

---

[32]If we do not, we will get a logic in which only Π_1^1 formulae with a single universal quantifier can be defined. This is, I believe, enough for the proof of Frege's Theorem.

The formula replacing F may not contain any of the new predicates $\mathcal{A}_\phi$. *Full Arché logic* (full AL) allows unrestricted substitution.[33] These systems obviously include minimal, predicative, and full schematic ancestral logic, respectively, and the logical strength of minimal, predicative, and full AL are, I suspect, that of second-order logic with no comprehension, predicative comprehension, and Π^1_1 comprehension, respectively. But it should be equally clear that the language in which these systems are formulated has nothing like the expressive power of a second-order language.

The axiom scheme of separation can be expressed in full AL by a single axiom. Let $\sigma_x(Fx, w)$ be the formula:

$$\forall z \exists y \forall x (x \in y \equiv x \in z \land Fx) \land w = w$$

Then the scheme of schematic definition gives us a new predicate $\mathcal{A}_\sigma(w)$ subject to the rules:

$$\forall z \exists y \forall x (x \in y \equiv x \in z \land Fx) \land w = w \vdash \mathcal{A}_\sigma(w)$$
$$\mathcal{A}_\sigma(w) \vdash \forall z \exists y \forall x (x \in y \equiv x \in z \land Fx) \land w = w$$

or equivalently:

$$\forall z \exists y \forall x (x \in y \equiv x \in z \land Fx) \vdash \forall w \mathcal{A}_\sigma(w)$$
$$\forall w \mathcal{A}_\sigma(w) \vdash \forall z \exists y \forall x (x \in y \equiv x \in z \land Fx)$$

So separation is expressed by the single sentence: $\forall w \mathcal{A}_\sigma(w)$. A similar technique plainly applies to any axiom scheme.[34]

The scheme of schematic definition also allows us to define new predicates that are co-extensional with Σ^1_1 formulae. Let $\phi_x(Fx, a)$ be a formula. We want to define a new predicate equivalent to $\exists F \phi_x(Fx, a)$. Use the scheme of schematic definition to introduce a new predicate $\mathcal{A}_{\neg\phi}$, subject to the rules:

$$\neg\phi_x(Fx, a) \vdash \mathcal{A}_{\neg\phi}(a)$$
$$\mathcal{A}_{\neg\phi}(a) \vdash \neg\phi_x(Fx, a)$$

---

[33] In full AL, the elimination rule $(\mathcal{A}_\phi -)$ effectively takes the form

$$\mathcal{A}_\phi(a) \vdash \phi_x(B(x), a)$$

where $B(x)$ is an arbitrary formula (subject to the usual restrictions). In predicative AL, $B(x)$ is not permitted to contain new predicates of the form $\mathcal{A}_\phi$.

[34] A more elegant way to proceed is to extend the scheme of schematic definition to allow a new zero-place predicate (i.e., a sentential variable) to be defined in terms of a formula $\phi_x(F_1 x, \ldots, F_n x)$, in which case we have:

$$\phi_x(F_1 x, \ldots, F_n x) \vdash \mathcal{A}_\phi$$
$$\mathcal{A}_\phi \vdash \phi_x(F_1 x, \ldots, F_n x)$$

Then separation is expressed by the zero-place predicate thus defined when we take $\phi_x(Fx)$ to be (Sep).

So $\mathcal{A}_{\neg\phi}(a)$ is equivalent to $\forall F \neg\phi_x(Fx, a)$. We now regard $\neg\mathcal{A}_{\neg\phi}(a)$ as degeneratively of the form $\psi_x(Fx, a)$ and introduce a new predicate $\mathcal{A}_{\neg\mathcal{A}_{\neg\phi}}$, which I shall write: $\mathcal{A}^\phi$, subject to the rules:

$$\neg\mathcal{A}_{\neg\phi}(a) \vdash \mathcal{A}^\phi(a)$$

$$\mathcal{A}^\phi(a) \vdash \neg\mathcal{A}_{\neg\phi}(a)$$

So $\mathcal{A}^\phi(a)$ is equivalent to $\forall F(\neg\mathcal{A}_{\neg\phi}(a))$, that is, to $\neg\mathcal{A}_{\neg\phi}(a)$, that is, to $\neg\forall F \neg\phi_x(Fx, a)$ and so to $\exists F \phi_x(Fx, a)$, as wanted.

This argument obviously depends upon our allowing predicates defined using the scheme of schematic definition to appear in formulae used to define yet further new predicates using that same scheme. As already noted, we can, formally speaking, restrict the scheme so as not to allow such iteration, the result being predicative AL. But this restriction has no motivation in the context of this investigation. The scheme of schematic definition formalizes a certain mode of *concept-formation*. Once one has used it to form a certain concept, one has that concept, and there is simply no reason one cannot iterate the process of concept-formation in the way we have allowed.

That said, there is a more elegant way to define predicates that are equivalent to Σ^1_1 formulae. As we have seen, the scheme of schematic definition, as currently formulated, in effect characterizes $\mathcal{A}_\phi$ in terms of introduction and elimination rules that mirror those for the universal quantifier that appears in its explicit second-order definition. That suggests that we should characterize $\mathcal{A}^\phi$ in terms of introduction and elimination rules that mirror those for the existential quantifier that appears in *its* explicit second-order definition. So we may take the introduction rule for $\mathcal{A}^\phi$, $(\mathcal{A}^\phi+)$, to be:

$$\phi_x(A(x), a) \vdash \mathcal{A}^\phi(a),$$

where, in this case, variables free in $A(x)$ may be free in premises on which $\phi_x(A(x), a)$ depends. In full AL, $A(x)$ may be any formula; in predicative AL, it may not contain schematically defined predicates; in minimal AL, it must not contain schematic variables, either.

The elimination rule, $(\mathcal{A}^\phi-)$, is more complex, but only because the elimination rule for the existential quantifier is itself more complex, involving as it does the discharge of an assumption. Suppose we have derived a formula B from formulae in some set Δ together with $\phi_x(A(x), a)$, where none of the free variables occurring in $A(x)$, other than x itself—which is actually bound in $\phi_x(A(x), a)$—occur free in B or in Δ. Suppose further that we have derived $\mathcal{A}^\phi(a)$ from the formulae in some set Γ. Then we may infer B, discharging $\phi_x(A(x), a)$, so that B depends only upon Γ and Δ.[35]

---

[35]If one wants to make the distinction between suppositions and hypotheses here, then $\phi_x(A(x), a)$ must be a supposition.

We may state the rule symbolically as follows:

$$\phi_x(A(x), a), \Delta \vdash B$$
$$\Gamma \vdash \mathcal{A}^\phi(a)$$

$$\Delta, \Gamma \vdash B$$

This rule simply parallels the relevant instance of the usual elimination rule for the second-order existential quantifier, except that we have replaced $\exists F \phi_x(Fx, a)$ with $\mathcal{A}_\phi(a)$. It is convenient to expand the scheme of schematic definition to allow schematic definitions of this form, too, since doing so adds no additional strength to the logic.

I intend the term "scheme of schematic definition" to be taken seriously: I regard the introduction and elimination rules $(\mathcal{A}_\phi+)$ and $(\mathcal{A}_\phi-)$ as *defining* the new predicate $\mathcal{A}_\phi$ and therefore regard the rules themselves as effectively self-justifying, since they are consequences of (indeed, components of) a definition. In particular, then, I am proposing that we should regard $(*+)$ and $(*-)$[36] as defining the ancestral. Perhaps that would be a reason to regard the ancestral as a logical notion and to regard these rules are logical rules. I am not sure, because I am not sure what the word "logical" is supposed to mean here. But the crucial issue for the Neo-logicist is epistemological. The proof of Frege's Theorem makes heavy use of the ancestral and of inferences of the sort described by $(*+)$ and $(*-)$. A Neo-logicist must therefore show that she is entitled both to a grasp of the concept of the ancestral and to an appreciation of the validity of $(*+)$ and $(*-)$, and this entitlement must be epistemologically innocent in the sense that it does not itself import epistemological presuppositions that undermine the Neo-logicist project: It must not, for example, presuppose a grasp of the concept of finitude, and it is a common complaint that our grasp of the concept of the ancestral presupposes precisely that. But if we regard the ancestral as schematically defined by $(*+)$ and $(*-)$, we may dismiss this complaint.

To be sure, one cannot simply introduce a new expression and stipulate that it should be subject to whatever introduction and elimination rules one wishes: Inconsistency threatens, as Prior (1960) famously showed. A complete defense of the position I am developing here would thus have to contain an answer to the question when such stipulations are legitimate, and to many others besides. But my purpose here is more modest. I am trying to argue that a certain position is available and worth considering. Whether it is true is a question for another day.

---

[36] In schematic ancestral logic, $*$ is an operator, and so its logic can be characterized by the pair of rules $(*+)$ and $(*-)$. In Arché logic, there is no such operator. Rather, we have to define the ancestral of each relation separately, using the scheme of schematic definition. But since this point does not affect the present discussion, I shall ignore it.

12.5   Frege's Theorem

If we are to prove Frege's Theorem in some form of schematic logic, we must be able to formalize HP in schematic logic. HP is, of course, neither a definition nor an instance of comprehension, but the techniques developed above may nonetheless be applied to it. We may represent HP as a pair of rules. Let $Fx \approx^{Ryz}_{xyz} Gx$ abbreviate:

$$\forall x \forall y \forall z \forall w (Rxy \wedge Rzw \rightarrow x = z \equiv y = w) \wedge$$
$$\forall x (Fx \rightarrow \exists y (Rxy \wedge Gy)) \wedge \forall y (Gy \rightarrow \exists x (Rxy \wedge Fx))$$

so that $Fx \approx^{Ryz}_{xyz} Gx$ says: R correlates the Fs one-one with the Gs. Then the introduction rule (N+) is easy enough to state:

$$Fx \approx^{Rxy}_{xyz} Gx \vdash \mathbf{N}x : Fx = \mathbf{N}x : Gx$$

The elimination rule (N−) is more complicated, but it simply parallels $(\mathcal{A}^{\phi}-)$:

$$\Delta, Fx \approx^{Rxy}_{xyz} Gx \vdash B$$
$$\Gamma \vdash \mathbf{N}x : Fx = \mathbf{N}x : Gx$$

$$\overline{\qquad\qquad \Delta, \Gamma \vdash B \qquad\qquad}$$

That is: If we have derived a formula B from assumptions in some set Δ together with the assumption that $Fx \approx^{Rxy}_{xyz} Gx$ and we have derived $\mathbf{N}x : Fx = \mathbf{N}x : Gx$ from the assumptions in some set Γ, then we may infer B, discharging $Fx \approx^{Rxy}_{xyz} Gx$. *Arché arithmetic* is full Arché logic plus these two rules.

I should emphasize before continuing that I am *not* claiming that (N+) and (N−) schematically define the cardinality operator. I am not even claiming that the possibility of formulating HP in schematic logic should do anything to ease any concerns one might have had about its epistemological status. My point here concerns only the logical resources needed for the proof of Frege's Theorem.

Frege's definitions of arithmetical notions can all be formalized in Arché arithmetic. Zero may be defined in the usual way:

$$0 = \mathbf{N}x : x \neq x$$

Frege's definition of predecession becomes a schematic definition of a new relation-symbol P subject to the rules $(P+)$:

$$\exists y [\mathbf{N}x : A(x) = b \wedge A(y) \wedge \mathbf{N}x : (A(x) \wedge x \neq y) = a] \vdash Pab$$

and $(P-)$:

$$\exists y[\mathbb{N}x : A(x) = b \wedge A(y) \wedge \mathbb{N}x : (A(x) \wedge x \neq y) = a], \Delta \vdash B$$
$$\Gamma \vdash Pab$$

$$\Delta, \Gamma \vdash B$$

where $A(x)$ is an arbitrary formula. The strong ancestral of this relation is schematically defined as subject to the two rules:

$(P*+)$ $\quad \forall x(Pax \rightarrow Fx) \wedge \forall x \forall y(Fx \wedge Pxy \rightarrow Fy) \rightarrow Fb \vdash P^*ab$

$(P*-)$ $\quad P^*ab \vdash \forall x(Pax \rightarrow Fx) \wedge \forall x \forall y(Fx \wedge Pxy \rightarrow Fy) \rightarrow Fb$

subject to the usual restrictions. The weak ancestral is defined as usual, and the definition of the concept of natural number is then:

$$\mathbb{N}n \equiv P^{*=}0n$$

Frege's proofs of axioms of arithmetic can then be formalized straightforwardly.

We may take the axioms of arithmetic to be as follows:[37]

1. $\quad \mathbb{N}0$

2. $\quad \forall x \forall y(\mathbb{N}x \wedge Pxy \rightarrow \mathbb{N}y)$

3. $\quad \forall x \forall y \forall z(\mathbb{N}x \wedge Pxy \wedge Pxz \rightarrow y = z)$

4. $\quad \forall x \forall y \forall z(\mathbb{N}x \wedge \mathbb{N}y \wedge Pxz \wedge Pyz \rightarrow x = y)$

5. $\quad \neg \exists x(\mathbb{N}x \wedge Px0)$

6. $\quad \forall x(\mathbb{N}x \rightarrow \exists y(Pxy))$

7. $\quad A(0) \wedge \forall x \forall y(\mathbb{N}x \wedge A(x) \wedge Pxy \rightarrow A(y)) \rightarrow \forall x(\mathbb{N}x \rightarrow A(x))$

For convenience, induction has been formulated as a schema. As we saw above, this can be avoided, but let us work with the schema, as it is a bit simpler.[38]

---

[37] Addition and multiplication can be defined using the scheme of schematic definition, so we need no special axioms governing them.

[38] Consider the formula: $F0 \wedge \forall x \forall y(\mathbb{N}x \wedge Fx \wedge Pxy \rightarrow Fy) \rightarrow Fa$. The scheme of schematic definition yields a new predicate $\mathcal{I}a$ subject to the two rules:

$$F0 \wedge \forall x \forall y(\mathbb{N}x \wedge Fx \wedge Pxy \rightarrow Fy) \rightarrow Fa \vdash \mathcal{I}a$$
$$\mathcal{I}a \vdash F0 \wedge \forall x \forall y(\mathbb{N}x \wedge Fx \wedge Pxy \rightarrow Fy) \rightarrow Fa$$

As before, $\mathcal{I}a$ is then equivalent to $\forall F[F0 \wedge \forall x \forall y(\mathbb{N}x \wedge Fx \wedge Pxy \rightarrow Fy) \rightarrow Fa]$, and induction is: $\forall x(\mathbb{N}x \rightarrow \mathcal{I}x)$.

In Arché arithmetic, just as in Frege arithmetic (second-order logic plus HP), Axioms (1) and (2) follow easily from the definition of $\mathbb{N}$: (1) is immediate, and (2) follows from the transitivity of the ancestral. The proofs of Axioms (3), (4), and (5) in Frege arithmetic all appeal to HP, but they use only predicative comprehension and so are easily formalized in Arché arithmetic.

Axiom (7) is stronger than what the definition of the ancestral by itself delivers. Simple manipulations give us:

$$A(0) \wedge \forall x \forall y (A(x) \wedge Pxy \rightarrow A(y)) \rightarrow \forall x (\mathbb{N}x \rightarrow A(x))$$

But (7) is stronger than this. The second conjunct of its antecedent is $\forall x \forall y (\mathbb{N}x \wedge A(x) \wedge Pxy \rightarrow A(y))$, which is weaker than $\forall x \forall y (A(x) \wedge Pxy \rightarrow A(y))$, since it contains the additional conjunct $\mathbb{N}x$ in its antecedent. But the instances of (7) can be proven, as usual, by induction on $\mathbb{N}\xi \wedge A(\xi)$.[39]

And finally, as was first noticed by Boolos (1998h), axiom (6) can be derived from axioms (3), (4), and (5), the definitions, and the following very weak consequence of HP :

(Log) $\forall x (Fx \equiv Gx) \vdash \mathbb{N}x : Fx = \mathbb{N}x : Gx$

The proof uses only Π_1^1 comprehension, so it, too, can be formalized in full Arché logic.[40]

A form of Frege's Theorem can thus be proven in Arché arithmetic. Exactly how strong the resulting fragment of second-order arithmetic is, I do not know for sure. It seems exceedingly likely that it is equivalent to second-order arithmetic with Π_1^1 comprehension. It is certainly stronger than first-order PA, since it contains ancestral logic, and ancestral logic plus PA is, as noted earlier, categorical. For another illustration of the strength of Arché arithmetic, note that the explicit definition of satisfaction for the language of first-order arithmetic is Π_1^1, so that definition can be converted into a schematic definition in Arché arithmetic. The usual induction will then establish the consistency of first-order PA, so Arché arithmetic is a non-conservative extension of first-order PA.

---

[39]It is not often noted that, for this reason, the proof of induction in Frege arithmetic requires Π_1^1 comprehension. That is yet another reason one should not expect to get by with much less if one is trying to derive the axioms of PA from HP. As it happens, however, if one is willing to forego induction and interpret a weaker theory, such as Robinson arithmetic, then one can do so in predicative second-order logic (Heck, 1996; Burgess, 2005; Heck, 2011).

[40]Boolos's proof uses Axioms (1), (2), and (7), as well, but these can be derived from Frege's definitions, as we have seen. As we have also seen, however, the proof of Axiom (7) uses Π_1^1 comprehension, which is also used in the proofs of certain other, very general facts about the ancestral that are used in the proof of Axiom (6). All such uses can, however, be avoided in ramified Frege arithmetic (Heck, 2011). The proof of Axiom (6) therefore does not *require* impredicative reasoning. The lesson is that the place impredicative reasoning is required is in the proof of induction, just as Poincaré famously suspected.

12.6   Philosophical Considerations

A close examination of the proofs just sketched will show that, if one regards predecession as primitive and subject to the rules $(P+)$ and $(P-)$, as suggested in Section 12.1, then Frege's Theorem can be proven in full schematic ancestral logic. If one regards the ancestral too as primitive and subject to the rules $(*+)$ and $(*-)$, then Frege's Theorem can be proven in *predicative* schematic ancestral logic. Predicative systems are generally regarded as epistemologically innocent. So if both predecession and the ancestral could be regarded as primitive, the mentioned rules being analytic of these notions, the logic needed for the proof of Frege's Theorem would be epistemologically innocent, and uncontroversially so.

But the question which notions are primitive does not seem to me to be the right question to ask here: It is far too slippery. It is better, I think, to regard both predecession and its ancestral as defined by means of the scheme of schematic definition and to regard $(P+)$, $(P-)$, $(P*+)$, and $(P*-)$ as analytic on the ground that they are consequences of, because components of, those definitions.[41] The proof of Frege's Theorem, in that case, needs full Arché logic. In particular, the proof needs the unrestricted rule of substitution. A philosopher with principled concerns about impredicativity might therefore be tempted to say—and, indeed, might long have been wanting to say—that the foregoing is largely beside the point if the question at issue is whether the logic required for the proof of Frege's Theorem is epistemologically innocent. The unrestricted substitution rule is impredicative, and the only question, really, was where the impredicativity would ultimately surface. The bump has been pushed around a fair bit, to be sure, but Frege's carpet is no flatter now than it was before.

I disagree. The scheme of schematic definition allows us to introduce a new predicate $\mathcal{A}_\phi$ subject to the rules:

$(\mathcal{A}_\phi+)$ $\phi_x(Fx, a) \vdash \mathcal{A}_\phi(a)$

$(\mathcal{A}_\phi-)$ $\mathcal{A}_\phi(a) \vdash \phi_x(Fx, a)$

In the presence of the unrestricted substitution rule, the elimination rule is equivalent to:

$(\mathcal{A}_\phi\text{sub})$ $\mathcal{A}_\phi(a) \vdash \phi_x(B(x), a)$

where $B(x)$ is now a formula rather than a variable. In particular, in the case of the ancestral, the elimination rule is equivalent, in the presence of unrestricted substitution, to:

$(\phi^*\text{sub})$ $\phi^* ab \vdash \forall x(\phi ax \rightarrow B(x)) \wedge \forall x \forall y(B(x) \wedge \phi xy \rightarrow B(y)) \rightarrow B(b)$

---

[41]As mentioned in footnote 37, addition and multiplication can be similarly defined.

Such a rule can be understood in two ways.[42] One way takes the set of formulae $B(x)$ that can appear in the rule to be determined by reference to
some fixed language. That is how axiom schemata, such as the induction
scheme in PA:

$$A(0) \wedge \forall x(A(x) \to A(Sx)) \to \forall x A(x)$$

are usually understood. The induction scheme is usually regarded as ab-
breviating an infinite list of axioms, one for each formula of the language
of arithmetic. When we consider expansions of the language of PA, then—
say, the result of adding a truth-predicate T—that expansion does not, in
itself, result in any new axioms' being added to the original theory. In
particular, sentences containing T, such as:[43]

$$T0 \wedge \forall x(Tx \to T(Sx)) \to \forall x(Tx)$$

do not automatically become axioms of the new theory, though such sentences do have the form of induction axioms. That is why adding a truth-
predicate to PA, and even adding the Tarskian clauses for it, yields a
conservative extension: One can't do much with the truth-predicate if it
doesn't occur in the induction axioms.

Formally, of course, one can proceed how one likes, but this way of
thinking of the induction scheme is not obviously best. Even if our theory of arithmetic is formulated in a first-order language, one would have
thought that the induction scheme should be regarded as one that *does*,
as it were, automatically import new instances of that scheme into our
theory as our language expands. In everyday mathematics, we do not
so much as ask, as our language expands, whether the new instances of
induction that become available should be accepted as true.[44] As it is
ordinarily understood, that is to say, the induction scheme expresses an
open-ended commitment to the truth of all sentences of a certain form,
both those we can presently formulate and those we cannot. The rule
(ϕ^*sub) is to be understood in this way, too: It expresses an open-ended
commitment to the validity of all inferences of a certain form.[45]

The unrestricted substitution rule thus expresses the open-ended nature of the commitments we undertake when we define a new predicate
via an instance of the scheme of schematic definition. And that, simply,

---

[42]Feferman (1991) discusses this distinction at some length. Regarding his historical remarks in §1.5, it is perhaps worth noting that the first accurate formulation of the sort of
substitution rule that is needed here is due to Frege: See Rule 9 in section 48 of *Grundgesetze*. As Feferman notes, citing Church, such a rule is missing from *Begriffsschrift*.

[43]Or, much more importantly, the instance of induction you get if you take $A(x)$ to be:
$\forall y < x(\text{Bew}_{\text{PA}}(y) \to T(y))$.

[44]One might suggest that vagueness and, in particular, the Sorites paradox refute this
suggestion, since accepting induction for predicates like "heap" leads to problems. But the
common wisdom nowadays is that induction is irrelevant, since the paradox can be reformulated without it.

[45]I borrow the term "open-ended" from Vann McGee (1997).

is how the scheme is intended to be understood. Someone whose under-standing of the ancestral was completely constituted by her grasp of the rules (ϕ^*+) and (ϕ^*-) would, it seems to me, be quite surprised to hear that these rules do not license the sort of inference required for the proof of Theorem (124) of *Begriffsschrift*. I am not saying that it would be *incoherent* to refuse to accept that inference. I am simply saying that it would not be a natural reaction.

One might object that this justification of the substitution rule, if it is defensible at all, ought to apply just as well in the context of full second-order logic. Perhaps it does so extend. But in the case of second-order logic, there is another and more fundamental problem with which we must contend: We must explain the second-order quantifiers. Absent such an explanation, we do not so much as understand second-order languages, and the question how the substitution rule should be justified doesn't arise. Now, to understand the second-order universal quantifier, one must understand what it means to say that all concepts are thus-and-so. But to understand that sort of claim, or so it is often argued, one must have a conception of what the second-order domain comprises. One must, in particular, have a conception of (something essentially equivalent to) the full power-set of the first-order domain, and many arguments have been offered that purport to show that we simply do not have a definite conception of $\wp(\omega)$. It is not my purpose here to evaluate such arguments. Maybe they work, and maybe they do not. My purpose here is to identify an epistemologically relevant difference between second-order logic and Arché logic. Here it is: Since there are no second-order quantifiers in schematic languages, the problem of explaining the second-order quantifier simply does not arise.

A similar point arises in connection with a question set aside earlier: whether second-order parameters should be permitted in schematic definitions, that is, whether one should be allowed to define what Frege would have called a "relation of mixed level" in the following sort of way:

$$A_{yz}(x, Fy, Gz) \vdash \Phi_y(x, Fy)$$
$$\Phi_y(x, Fy) \vdash A_{yz}(x, Fy, B(z))$$

The first of these is the introduction-rule, of course, and $A_{yz}(x, Fy, Gz)$ is formula with the free variables shown; what is being defined here is the 'mixed level' relational expression $\Phi_y(x, Fy)$.

Permitting this sort of definition leads to a system with the expressive power of full second-order logic.[46] A friend of second-order logic might welcome this news, of course: If Arché logic really is epistemologically in-

---

[46]The proof is by induction. Suppose that, for each Σ_n^1 formula $\phi(x_1, \ldots, x_n, F_1, \ldots, F_m)$, where the F_i are second-order parameters, we can schematicaly define a new formula $\Phi(x_1, \ldots, x_n, F_1, \ldots, F_m)$ that is equivalent to it. (The other case is similar.) Now consider a Π_{n+1}^1 formula $\forall F_1 \phi(x_1, \ldots, x_n, F_1, F_2, \ldots, F_m)$. Since $\phi(x_1, \ldots, x_n, F_1, F_2, \ldots, F_m)$ is Σ_n^1, we can schematically define a formula $\Phi(x_1, \ldots, x_n, F_1, F_2, \ldots, F_m)$ that is equivalent

nocent, and if full second-order logic can be obtained simply by admitting second-order parameters in schematic definitions, then the naturalness of that extension might be taken to suggest that second-order logic too is epistemologically innocent. But, of course, my opponent will see things differently: If it is only because we have disallowed second-order parameters that Arché logic is any weaker than full second-order logic, and if allowing them is indeed so natural, then the resources needed for an understanding of Arché logic might seem to be sufficient for an understanding of full second-order logic, and there is no epistemologically relevant difference between them to be found.

But this response overlooks an important difference between Arché logic and this extension of it, namely: Before we can consider whether schematic definitions containing such parameters should be permitted, we must first enrich the *language*. The langauge of Arché logic is the language of first-order logic, and the only sorts of relations that are definable in it are first-order relations, that is, relations between objects. The language of the extension just mentioned is not even the language of second-order logic: Except for the second-order quantifiers, the primitive expressions of second-order logic are just those of first-order logic. What we need in the language if we are to permit parameters in schematic definitions are primitives like $\Phi_y(x, Fy)$ that express relations not just between objects but between objects and concepts. Such expressions therefore sit at roughly the same level as the second-order quantifiers themselves: The expression Φ in $\Phi_y(x, Fy)$ is a free *third*-order variable. It would not be at all surprising, then, if the resources needed to understand expressions of this kind were similar to the resources needed to understand the second-order quantifiers. But let me not pursue that question now.[47] I am, once again, just trying to establish that there are issues here that are worth discussing.

A similar concern is that that our understanding of the introduction rule for the ancestral involves a conception of the full power-set of the domain. How else, it might be asked, are we to understand

$$\forall x(\phi ax \to Fx) \wedge \forall x \forall y(Fx \wedge \phi xy \to Fy) \to Fb$$

as it occurs in the premise of the rule (*+), except as involving a tacit initial second-order quantifer? Does it not say, explicit quantifier or no, that all concepts F that are thus-and-so are so-and-thus? Doesn't understanding that claim therefore require the disputed conception of the full

---

to it. We then schematically define $\Psi(x_1, \ldots, x_n, F_2, \ldots, F_m)$ in the obvious way:

$$\Phi(x_1, \ldots, x_n, F_1, F_2, \ldots, F_m) \vdash \Psi(x_1, \ldots, x_n, F_2, \ldots, F_m)$$

$$\Psi(x_1, \ldots, x_n, F_2, \ldots, F_m) \vdash \Phi(x_1, \ldots, x_n, F_1, F_2, \ldots, F_m)$$

where the former is, of course, subject to the usual restrictions. By earlier remarks, $\Psi(x_1, \ldots, x_n, F_2, \ldots, F_m)$ is then equivalent to $\forall F_1 \phi(x_1, \ldots, x_n, F_1, F_2, \ldots, F_m)$.

[47] If second-order logic is interpreted *via* plural quantification, as suggested by Boolos (1998l), then expressions of this kind are 'plural predicates', as discussed by Rayo (2002).

power-set? No, it does not. A better reading would be: A concept that is thus-and-so is so-and-thus. What understanding this claim requires is not a capacity to conceive of *all* concepts but simply the capacity to conceive of *a* concept: to conceive of an arbitrary concept, if you like. The contrast here is parallel to that between arithmetical claims like $x + y = y + x$, involving only free variables, and claims involving explicit quantification over all natural numbers. Hilbert famously argued that our understanding of claims of the former sort involves no conception of the totality of all natural numbers, whereas claims of the latter sort do, and that there is therefore a significant conceptual and epistemological difference between these cases. I am making a similar point about claims involving only free second-order variables as opposed to claims that quantify over concepts (or whatever).

But one might still want to insist that, if we do not have a definite conception of the full power-set of the domain—if, in particular, there is nothing in our understanding of free second-order variables that guarantees that they range over the full power-set of the domain—then the meanings of the predicates we introduce by schematic definition will be radically underdetermined, at the very least. It was stipulated earlier that $\mathcal{A}_\phi(a)$ is true if, and only if, $\phi_x(Fx, a)$ is true for every assignment of a subset of the first-order domain to F. But why? If we have no conception of the full power-set, why not take the domain of the second-order variables to be smaller? Why not restrict it to the definable subsets of the domain? Surely the axioms and rules of Arché logic do not require the second-order domain to contain every subset of the first-order domain.

Obviously, there is a technical point here that is incontrovertible: The existence of non-standard models is a fact of mathematics, and a very useful one at that. But the philosophical significance of this technical point is not so obvious. It seems to me that there *is* something about the axioms and rules of Arché logic that requires the second-order domain to be unrestricted, and "*un*restricted" is the crucial word. The difference between the standard model and the various non-standard models is to be found not in what the standard model *in*cludes but in what non-standard models *ex*clude: Since the second-order domain of a standard model includes all subsets of the first-order domain, a non-standard model must *ex*clude certain of these subsets. (It isn't as if it could include more.) That, however, is incompatible with the nature of the commitments we undertake when we introduce a new predicate by schematic definition. Those commitments are themselves *unrestricted* in the sense that we accept no restriction on what formulae may replace $B(x)$ when we infer $\phi_x(B(x), a)$ from $\mathcal{A}_\phi(a)$.[48] One might be tempted to object that, if so, we must somehow conceive of the totality of all such formulae in advance. But that would simply repeat the same error: No such conception of the totality of

---

[48] And $B(x)$ might be demonstrative: Those ones. Definability considerations are therefore out of place here, too.

all formulae is needed; what is needed is just the ability to conceive of *a* formula—an arbitrary formula, if you like.

A neutral referee might well declare a draw at this point, although both sides have moves remaining. But that is enough for my purpose here. What I am trying to do is not to convince the reader of any particular position. I am trying, rather, to convince the reader of the *interest* of a certain position, namely, the position that full Arché logic is epistemologically innocent in whatever way such positions as the Neo-logicist's need logic to be. It is no part of my position that full second-order logic is not epistemologically innocent in that sense. It is my position that there is *enough of a difference* between Arché logic and second-order logic that it would not be unreasonable to regard them as epistemologically unequal in this same sense. I take myself to have accomplished that much. The resources deployed above in the defense of full Arché logic against the predicativist skeptic are not resources that are obviously sufficient to defend full second-order logic. Perhaps they can be built upon for that purpose. I don't necessarily say otherwise. But perhaps they cannot be.

The critical difference between Arché logic and second-order logic thus turns out to lie not so much in the logical principles that distinguish them but, rather, in the expressive power of the underlying languages. What we have seen is not just that the full deductive strength of second-order logic is not needed for the proof of Frege's Theorem; that has been known for some time. What we have seen is that not all of the *expressive* power of second-order languages is needed, either. Quine's view was that second-order quantification is not even a logical notion: If second-order variables range over sets, then second-order quantification is quantification over sets, and second-order logic is set-theory in sheep's clothing, quite independently of its proof-theoretic strength (Quine, 1986, ch. 5).[49] Even if one interprets second-order quantifiers in terms of plurals, however, one might have other reasons to suppose that plural quantifiers are non-logical constants (Resnik, 1988), perhaps reasons connected with the expressive power of plural quantifiers. Again, it is no part of my view that plural quantifiers are not logical constants. What I am claiming is that the question whether axioms for arithmetic can be derived, purely logically, from HP does not depend upon how such issues are resolved. The language of Arché arithmetic has a significantly stronger claim to be a logical language than the language of second-order logic does.

---

[49]That is not to say that Quine didn't make a great deal of the proof-theoretic strength of second-order logic. He seems, in particular, to have regarded completeness as an essential property of anything rightly regarded as a 'logic'. But, so far as I know, Quine never says what is supposed to justify this restriction.

Appendix: Proof of *Begriffsschrift*, Proposition 124

As above, I am not including all the steps that would be necessary to make the argument completely formal. My purpose is simply to illustrate how schematic logics work.

The proof is in full schematic ancestral logic. It could also, of course, be carried out in full Arché logic.

[1]	(1)	$\phi^* ab$	Premise
[2]	(2)	ϕac	Premise
[3]	(3)	$\forall x \forall y \forall z (\phi xy \wedge \phi xz \to y = z)$	Premise
[1]	(4)	$\forall x(\phi ax \to Fx) \wedge \forall x \forall y (Fx \wedge \phi xy \to Fy)$ $\to Fb$	$(1, *-)$
[1]	(5)	$\forall x(\phi ax \to \phi^{*=} cx) \wedge$ $\forall x \forall y(\phi^{*=} cx \wedge \phi xy \to \phi^{*=} cy) \to \phi^{*=} cb$	$(4, \text{subst})$
[6]	(6)	ϕax	Premise
$[2,3,6]$	(7)	$x = c$	$(2,3,6)$
$[2,3,6]$	(8)	$\phi^{*=} cx$	def $\phi^{*=}$
$[2,3]$	(9)	$\phi ax \to \phi^{*=} cx$	$(6,8) \to +$
$[2,3]$	(10)	$\forall x(\phi ax \to \phi^{*=} cx)$	$(9)\forall +$
[]	(11)	$\forall x \forall y(\phi^{*=} cx \wedge \phi xy \to \phi^{*=} cy)$	transitivity
$[1,2,3]$	(12)	$\phi^{*=} cb$	$(5,9,11)$

The substitution rule, applied at line (5), is essential to this proof, which therefore collapses in minimal schematic ancestral logic. Since the substituted formula contains *, it cannot be replicated in predicative schematic ancestral logic either.

Bibliography

Anscombe, G. E. M. and Geach, P. T. (1961). *Three Philosophers*. Ithaca NY, Cornell University Press.

Avron, A. (2003). 'Transitive closure and the mechanization of mathematics', in F. Kamareddine (ed.), *Thirty Five Years of Automating Mathematics*. New York, Kluwer Academic Publishers, 149–71.

Barwise, J. and Cooper, R. (1981). 'Generalized quantifiers and natural language', *Linguistics and Philosophy* 4: 159–219.

Benacerraf, P. (1973). 'Mathematical truth', *Journal of Philosophy* 70: 661–80.

—— (1995). 'Frege: The last logicist', in Demopoulos 1995, 41–67.

Berkeley, G. (1930). *A Treatise Concerning the Principles of Human Knowledge*. London, Open Court.

Bolzano, B. (1950). *Paradoxes of the Infinite*, trans. F. Prihonsky. London, Routledge and Kegan Paul.

Boolos, G. (1972). 'Logicism and second-order logic'. Manuscript.

—— (1998a). 'The advantages of honest toil over theft', in Boolos 1998e, 255–74.

—— (1998b). 'The consistency of Frege's *Foundations of Arithmetic*', in Boolos 1998e, 183–202.

—— (1998c). 'Frege's theorem and the Peano postulates', in Boolos 1998e, 291–300.

—— (1998d). 'Is Hume's principle analytic?', in Boolos 1998e, 301–14.

—— (1998e). *Logic, Logic, and Logic*, ed. R. Jeffrey. Cambridge MA, Harvard University Press.

—— (1998f). 'Must we believe in set-theory?', in Boolos 1998e, 120–32.

—— (1998g). 'On second-order logic', in Boolos 1998e, 37–53.

—— (1998h). 'On the proof of Frege's theorem', in Boolos 1998e, 275–91.

—— (1998i). 'Reading the *Begriffsschrift*', in Boolos 1998e, 155–70.

—— (1998j). 'Saving Frege from contradiction', in Boolos 1998e, 171–82.

—— (1998k). 'The standard of equality of numbers', in Boolos 1998e, 202–19.

—— (1998l). 'To be is to be a value of a variable (or some values of some variables)', in Boolos 1998e, 54–72.

Boolos, G. and Heck, R. G. (1998). '*Die Grundlagen der Arithmetik* §§82–83', in Boolos 1998e, 315–38.

Bromberger, S. (1992a). *On What We Know We Don't Know: Explanation, Theory, Linguistics, and How Questions Shape Them*. Chicago, University of Chicago Press.

—— (1992b). 'Types and tokens in linguistics', in Bromberger 1992a, 170–208.

Bromberger, S. and Halle, M. (1992). 'The ontology of phonology', in Bromberger 1992a, 209–28.

Burge, T. (2005a). 'Frege on apriority', in Burge 2005d, 356–87.

—— (2005b). 'Frege on knowing the foundation', in Burge 2005d, 317–55.

—— (2005c). 'Frege on truth', in Burge 2005d, 83–132.

—— (2005d). *Truth, Thought, Reason: Essays on Frege*. New York, Oxford University Press.

Burgess, J. P. (1984). 'Review of *Frege's Conception of Numbers as Objects*', *Philosophical Review* 93: 638–40.

—— (2005). *Fixing Frege*. Princeton NJ, Princeton University Press.

Burgess, J. P. and Rosen, G. (1997). *A Subject with No Object: Strategies for Nominalistic Interpretation of Mathematics*. Oxford, Oxford University Press.

Carey, S. (1995). 'Continuity and discontinuity in cognitive development', in G. Smith and D. Osherson (eds.), *An Invitation to Cognitive Science: Thinking*, 2d edition. Cambridge MA, MIT Press, 101–29.

Carnap, R. (1950). 'Empiricism, semantics, and ontology', *Revue Internationale de Philosophie* 4: 20–40.

Cook, R. T., ed. (2007). *The Arché Papers on the Mathematics of Abstraction*. Dordrecht, Springer.

Dedekind, R. (1963). 'The nature and meaning of numbers', trans. W. W. Beman, in *Essays on the theory of numbers*. New York, Dover.

Demopoulos, W. (1998). 'The philosophical basis of our knowledge of number', *Noûs* 32: 481–503.

—— (2000). 'On the origin and status of the concept of number', *Notre Dame Journal of Formal Logic* 41: 210–26.

Demopoulos, W., ed. (1995). *Frege's Philosophy of Mathematics*. Cambridge MA, Harvard University Press.

Dummett, M. (1978a). 'Nominalism', in Dummett 1978b, 38–49.

—— (1978b). *Truth and Other Enigmas*. London, Duckworth.

—— (1981a). *Frege: Philosophy of Language*, 2d edition. Cambridge MA, Harvard University Press.

—— (1981b). *The Interpretation of Frege's Philosophy*. Cambridge MA, Harvard University Press.

—— (1991a). 'Frege and the paradox of analysis', in *Frege and Other Philosophers*. Oxford, Clarendon Press, 17–52.

—— (1991b). *Frege: Philosophy of Mathematics*. Cambridge MA, Harvard University Press.

—— (1998). 'Appendix on the term 'Hume's Principle'', in Schirn 1998, 386–7.

Eklund, M. (2006). 'Neo-Fregean ontology', *Philosophical Perspectives* 20: 95–121.

Etchemendy, J. (1990). *The Concept of Logical Consequence*. Cambridge MA, Harvard University Press.

Evans, G. (1982). *The Varieties of Reference*, ed. J. McDowell. Oxford, Clarendon Press.

—— (1985a). *Collected Papers*. Oxford, Clarendon Press.

—— (1985b). 'Things without the mind: A commentary upon chapter two of Strawson's *Individuals*', in Evans 1985a, 249–90.

—— (1985c). 'Understanding demonstratives', in Evans 1985a, 291–321.

Feferman, S. (1964). 'Systems of predicative analysis', *Journal of Symbolic Logic* 29: 1–30.

—— (1991). 'Reflecting on incompleteness', *Journal of Symbolic Logic* 1991: 1–49.

Ferreira, F. (2005). 'Amending Frege's *Grundgesetze der Arithmetik*', *Synthese* 147: 3–19.

Field, H. (1980). *Science Without Numbers: A Defense of Nominalism*. Princeton NJ, Princeton University Press.

Fine, K. (1998). 'The limits of abstraction', in Schirn 1998, 503–629.

—— (2002). *The Limits of Abstraction*. Oxford, Oxford University Press.

Frege, G. (1964). *The Basic Laws of Arithmetic: Exposition of the System*, ed. and trans. M. Furth. Berkeley CA, University of California Press.

—— (1970). *Translations from the Philosophical Writings of Gottlob Frege*, ed. P. Geach and M. Black. Oxford, Blackwell.

—— (1984). *Collected Papers on Mathematics, Logic, and Philosophy*, ed. B. McGuiness. Oxford, Basil Blackwell.

—— (1997). *The Frege Reader*, ed. M. Beaney. Oxford, Blackwell.

—— (Bg). 'Begriffsschrift: A formula language modeled upon that of arithmetic, for pure thought', trans. J. van Heijenoort, in van Heijenoort 1967, 5–82.

—— (CO). 'On concept and object', trans. P. Geach, in Frege 1984, 182–194. Also in Frege 1997, 181–93.

—— (FTA). 'Formal theories of arithmetic', trans. E.-H. W. Kluge, in Frege 1984, 112–21.

Frege, G. (Geo1). 'On the foundations of geometry: First series', trans. E.-H. W. Kluge, in Frege 1984, 273–84.

—— (Geo2). 'On the foundations of geometry: Second series', trans. E.-H. W. Kluge, in Frege 1984, 293–340.

—— (Gg). *Grundgesetze der Arithmetik*. Hildesheim, Georg Olms Verlagsbuchhandlung. Translations of Part I are in Frege 1964, and of portions of Part III, in Frege 1970. Extracts appear in Frege 1997, 194–223.

—— (Gl). *The Foundations of Arithmetic*, 2nd revised edition, trans. J. L. Austin. Evanston IL, Northwestern University Press.

—— (PCN). 'On Mr. Peano's conceptual notation and my own', trans. V. H. Dudman, in Frege 1984, 234–48.

—— (PMC). *Philosophical and Mathematical Correspondence*, ed. G. Gabriel *et al.* Chicago, University of Chicago Press.

—— (RevCan). 'Review of Georg Cantor, *Zum Lehre vom Transfiniten*', trans. H. Kaal, in Frege 1984, 178–81.

—— (RevHus). 'Review of E. G. Husserl, *Philosophie der Arithmetik*', trans. H. Kaal, in Frege 1984, 195–209. Extracts in Frege, 1997, 224–26.

—— (SM). 'On sense and meaning', trans. M. Black, in Frege 1984, 157–77. Also in Frege 1997, 151–71.

—— (WMR). 'What may I regard as the result of my work?', trans. P. Long and R. White, in H. Hermes *et al.* (eds.), *Posthumous Writings*. Chicago, University of Chicago Press, 184.

Geach, P. T. (1951). 'Frege's *Grundlagen*', *Philosophical Review* 60: 535–44.

—— (1955). 'Class and concept', *Philosophical Review* 64: 561–70.

—— (1976). 'Critical notice of Michael Dummett, *Frege: Philosophy of Language*', *Mind* 85: 436–49.

Gelman, R. and Galistel, C. R. (1978). *The Child's Understanding of Number*. Cambridge MA, Harvard University Press.

Gödel, K. (1990). 'What is Cantor's continuum problem?', in S. Feferman *et al.* (eds.), *Collected Works*, vol. 2, 3d edition. Oxford, Oxford University Press, 254–70.

Goodman, N. (1972). 'A world of individuals', in *Problems and Projects*. Indianapolis, Bobbs-Merrill, 155–72.

—— (1977). *The Structure of Appearance*, 4th edition. Boston MA, D. Reidel Publishing.

Goodman, N. and Quine, W. V. O. (1947). 'Steps toward a constructive nominalism', *Journal of Symbolic Logic* 12: 105–22.

Hale, B. (1988). *Abstract Objects*. Oxford, Blackwell.

—— (2001). '*Grundlagen* section 64', in Hale and Wright 2001b, 91–116.

Hale, B. and Wright, C. (2001a). 'Postscript: Eighteen problems', in Hale and Wright 2001b, 421–36.

—— (2001b). *The Reason's Proper Study*. Oxford, Clarendon Press.

—— (2001c). 'To bury Caesar...', in Hale and Wright 2001b, 335–96.

—— (2008). 'Abstraction and additional nature', *Philosophia Mathematica* 16: 182–208.

—— (2009a). 'Focus restored: Comments on John MacFarlane', *Synthese* 170: 457–82.

—— (2009b). 'The metaontology of abstraction', in D. Chalmers *et al.* (eds.), *Metametaphysics*. Oxford, Oxford University Press, 178–212.

Hazen, A. (1985). 'Review of Crispin Wright, *Frege's Conception of Numbers as Objects*', *Australasian Journal of Philosophy* 63: 251–4.

Heck, R. G. (1993a). 'Critical notice of Michael Dummett, *Frege: Philosophy of Mathematics*', *Philosophical Quarterly* 43: 223–31.

—— (1993b). 'The development of arithmetic in Frege's *Grundgesetze der Arithmetik*', *Journal of Symbolic Logic* 58: 579–601.

—— (1995a). 'Definition by induction in Frege's *Grundgesetze der Arithmetik*', in Demopoulos 1995, 295–333.

—— (1995b). 'The development of arithmetic in Frege's *Grundgesetze der Arithmetik*', in Demopoulos 1995, 257–94.

—— (1996). 'The consistency of predicative fragments of Frege's *Grundgesetze der Artithmetik*', *History and Philosophy of Logic* 17: 209–20.

—— (1997). 'The Julius Caesar objection', in R. Heck (ed.), *Language, Thought, and Logic: Essays in Honour of Michael Dummett*. Oxford, Clarendon Press, 273–308.

—— (1998a). '*Grundgesetze der Arithmetik* I §§29–32', *Notre Dame Journal of Formal Logic* 38: 437–74.

—— (1998b). 'The finite and the infinite in Frege's *Grundgesetze der Arithmetik*', in Schirn 1998, 429–66.

—— (1998c). 'That there might be vague objects (so far as concerns logic)', *The Monist* 81: 277–99.

—— (1999). '*Grundgesetze der Arithmetik* I §10', *Philosophia Mathematica* 7: 258–92.

—— (2000a). 'Cardinality, counting, and equinumerosity', *Notre Dame Journal of Formal Logic* 41: 187–209.

—— (2000b). 'Syntactic reductionism', *Philosophia Mathematica* 8: 124–49.

Heck, R. G. (2005). 'Formal arithmetic before *Grundgesetze*'. Presented as the George S. Boolos Memorial Lecture, Massachusetts Institute of Technology, January 2005.

—— (2007). 'Meaning and truth-conditions', in D. Greimann and G. Siegwart (eds.), *Truth and Speech Acts: Studies in the Philosophy of Language*. New York, Routledge.

—— (2010). 'Frege and semantics', in M. Potter and T. Ricketts (eds.), *The Cambridge Companion to Frege*. Cambridge, Cambridge University Press, 342–78.

—— (2011). 'Ramified Frege arithmetic', *Journal of Philosophical Logic* 40: 715–35.

Heck, R. G. and May, R. (2012). 'The function is unsaturated', in M. Beaney (ed.), *The Oxford Handbook of the History of Analytic Philosophy*. Oxford, Oxford University Press. Forthcoming.

Hodes, H. (1981). 'Logicism and the ontological commitments of arithmetic', *Journal of Philosophy* 1984: 123–49.

—— (1990). 'Ontological commitment: Thick and thin', in G. Boolos (ed.), *Meaning and Method*. Cambridge, Cambridge University Press, 235–60.

Kaplan, D. (1990). 'Words', *Proceedings of the Aristotelian Society,* sup. vol. 64: 93–119.

Koellner, P. (2010). 'Strong logics of first and second order', *Bulletin of Symbolic Logic* 16: 1–36.

Korcz, K. A. (2010). 'The epistemic basing relation', *Stanford Encyclopedia of Philosophy*, http://plato.stanford.edu/entries/basing-epistemic/

Kripke, S. (1976). 'Is there a problem about substitutional quantification?', in G. Evans and J. McDowell (eds.), *Truth and Meaning: Essays in Semantics*. Oxford, Oxford University Press, 325–419.

—— (1980). *Naming and Necessity*. Cambridge MA, Harvard University Press.

—— (1992). 'Logicism, Wittgenstein, and de re beliefs about numbers'. Transcription of the 1992 Whitehead Lectures, given at Harvard University, Cambridge MA.

Lewis, D. (1986). *On the Plurality of Worlds*. Cambridge MA, Blackwell.

—— (1991). *Parts of Classes*. Oxford, Basil Blackwell.

—— (1999). 'Noneism or allism?', in *Papers in Metaphysics and Epistemology*. New York, Cambridge University Press, 152–63.

Linnebo, Ø. (2004). 'Predicative fragments of Frege arithmetic', *Bulletin of Symbolic Logic* 10: 153–74.

—— (2006). 'Epistemological challenges to mathematical platonism', *Philosophical Studies* 129: 575–4.

Linnebo, Ø. and Uzquiano, G. (2009). 'Which abstraction principles are acceptable?', *British Journal for the Philosophy of Science* 60: 239–53.

MacBride, F. (2000). 'On finite Hume', *Philosophia Mathematica* 8: 150–9.

MacFarlane, J. (2009). 'Double vision: Two questions about the neo-Fregean program', *Synthese* 170: 443–56.

Mancosu, P. (2009). 'Measuring the size of infinite collections of natural numbers: Was Cantor's theory inevitable?', *Review of Symbolic Logic* 2: 612–46.

McGee, V. (1997). 'How we learn mathematical language', *Philosophical Review* 106: 35–68.

Parsons, C. (1980). 'Mathematical intuition', *Proceedings of the Aristotelian Society* N.S. 80: 145–68.

—— (1981). 'Sets and classes', in *Mathematics in Philosophy*. Ithaca NY, Cornell University Press, 209–20.

—— (1994). 'Intuition and number', in A. George (ed.), *Mathematics and Mind*. New York, Oxford University Press, 141–57.

—— (1995a). 'Frege's theory of number', in Demopoulos 1995, 182–210.

Parsons, T. (1995b). 'On the consistency of the first-order portion of Frege's logical system', in Demopoulos 1995, 422–31.

Peacocke, C. (1992). *A Study of Concepts*. Cambridge MA, MIT Press.

Prior, A. (1960). 'The runabout inference-ticket', *Analysis* 21: 38–9.

Putnam, H. (1979). 'Philosophy of logic', in *Mathematics, Matter, and Method: Philosphical Papers, vol. 1*, 2d edition. New York, Cambridge University Press, 323–57.

Quine, W. V. O. (1953). 'On what there is', in *From a Logical Point of View*. Cambridge MA, Harvard University Press, pp. 1–18.

—— (1960). *Word and Object*. Cambridge MA, MIT Press.

—— (1969). 'Epistemology naturalized', in *Ontological Relativity and Other Essays*. New York, Columbia Unversity Press, 69–90.

—— (1986). *Philosophy of Logic*, 2nd edition. Cambridge MA, Harvard University Press.

Rayo, A. (2002). 'Words and objects', *Noûs* 36: 436–64.

Resnik, M. (1988). 'Second-order logic still wild', *Journal of Philosophy* 85: 75–87.

Rumfitt, I. (2001). 'Hume's principle and the number of all objects', *Noûs* 515–41.

Russell, B. (1903). *The Principles of Mathematics*. Cambridge, Cambridge University Press.

Russell, B. (1905). 'On denoting', *Mind* 14: 479–93.

Schirn, M., ed. (1998). *Philosophy of Mathematics Today*. Oxford, Oxford University Press.

Schönfinkel, M. (1967). 'On the building blocks of mathematical logic', in van Heijenoort 1967, 255–66.

Schroeder-Heister, P. (1987). 'A model-theoretic reconstruction of Frege's permutation argument', *Notre Dame Journal of Formal Logic* 28: 69–79.

Segal, G. (2000). *A Slim Book About Narrow Content*. Cambridge MA, MIT Press.

Shapiro, S. (1991). *Foundations without Foundationalism: A Case for Second-order Logic*. Oxford, Oxford University Press.

—— (2001). 'Classical logic II—higher-order logic', in L. Goble (ed.), *The Blackwell Guide to Philosophical Logic*. Malden MA, Blackwell Publishing, 33–54.

Sider, T. (2007). 'Neo-Fregeanism and quantifier variance', *Proceedings of the Aristotelian Society,* sup. vol. 81: 201–32.

Smiley, T. J. (1988). 'Frege's "series of natural numbers"', *Mind* 97: 388–9.

Sullivan, P. and Potter, M. (1997). 'Hale on Caesar', *Philosophia Mathematica* 3: 135–53.

Sundholm, G. (2001). 'Frege, August Bebel, and the return of Alsace-Lorraine: The dating of the distinction between *Sinn* and *Bedeutung*', *History and Philosophy of Logic* 22: 57–73.

Tait, W. W. (1996). 'Frege versus Cantor and Dedekind: On the concept of number', in M. Schirn (ed.), *Frege: Importance and Legacy*. New York, de Gruyter, 70–113.

Tappenden, J. (1995). 'Geometry and generality in Frege', *Synthese* 102: 319–61.

—— (2005). 'The Caesar problem in its historical context: Mathematical background', *Dialectica* 59: 237–64.

Tarski, A., Mostowski, A., and Robinson, A. (1953). *Undecidable Theories*. Amsterdam, North-Holland Publishing.

Tennant, N. (1997). 'On the necessary existence of numbers', *Noûs* 31: 307–36.

Uzquiano, G. (2005). 'Semantic nominalism', *Dialectica* 59: 265–82.

van Heijenoort, J., ed. (1967). *From Frege to Gödel: A Sourcebook in Mathematical Logic*. Cambridge MA, Harvard University Press.

Visser, A. (2009). 'Why the theory R is special', *University of Utrecht Logic Group Preprint Series*, 279, http://www.phil.uu.nl/preprints/preprints/PREPRINTS/preprint279.pdf

—— (2011). 'Hume's principle, beginnings', *Review of Symbolic Logic* 4: 114–29.

Weir, A. (2003). 'Neo-Fregeanism: An embarrassment of riches', *Notre Dame Journal of Formal Logic* 44: 13–48.

Whitehead, A. N. and Russell, B. (1925). *Principia Mathematica*, vol. 1. Cambridge, Cambridge University Press.

Wiggins, D. (1980). *Sameness and Substance*. Cambridge MA, Harvard University Press.

—— (2001). *Sameness and Substance Renewed*. New York, Cambridge University Press.

Wilson, M. (1995). 'Frege: The royal road from geometry', in Demopoulos 1995, 108–59.

Wright, C. (1969). 'Numbers as objects'. Unpublished B. Phil. thesis, Oxford University.

—— (1983). *Frege's Conception of Numbers as Objects*. Aberdeen, Aberdeen University Press.

—— (2001a). 'Field and Fregean platonism', in Hale and Wright 2001b, 153–68.

—— (2001b). 'Is Hume's principle analytic?', in Hale and Wright 2001b, 307–32.

—— (2001c). 'On the harmless impredicativity of N= (Hume's Principle)', in Hale and Wright 2001b, 229–55.

—— (2001d). 'On the philosophical significance of Frege's theorem', in Hale and Wright 2001b, 272–306.

—— (2001e). 'Response to Dummett', in Hale and Wright 2001b, 256–71.

—— (2004). 'Warrant for nothing (and foundations for free)?', *Proceedings of the Aristotelian Society,* sup. vol. 78: 167–212.

Index

Lightning Source UK Ltd.
Milton Keynes UK
UKOW04f0337031014

239521UK00002B/3/P